UNDERSTANDING EMBRYOLOGY

UNDERSTANDING EMBRYOLOGY

By

Dr. Amita Sarkar

Dept. of Zoology
Agra College
Agra (U.P.)
(India)

DISCOVERY PUBLISHING HOUSE PVT. LTD.
NEW DELHI-110 002

Reprinted - 2019
First Published-2010

ISBN 978-81-8356-512-7

Published by:
DISCOVERY PUBLISHING HOUSE PVT. LTD.
4831/24, Ansari Road, Prahlad Street,
Darya Ganj, New Delhi-110002 (India)
Phone: 23279245 • Fax: 91-11-23253475
E-mail: parul.wasan@gmail.com
info@discoverypublishinggroup.com
Website: www.discoverypublishinggroup.com

Printed at:
Sachin Printers
Delhi

Preface

The present title "Understanding Embryology" has been written for those students interested in careers in diverse fields of biological sciences. It provides a structured approach to learning by covering all the important topics in a uniform, systematic format. The book has been comprehensively designed incorporating recent advances in this fast moving field. It also provides accessible information on embryology in compact form for undergraduate students in biology and related life sciences. It is intelligible to the educated layman, though it deals with some complex ideas. It is an adequate text for all the requirements of students in this area. In addition, busy lecturers who require a quick reference compendium will find it useful, particularly for tutional planning. Simple, yet hopefully clear figures and tables are provided throughout the book.

The over-riding goal of this book, and indeed of the whole *Understanding series*, is to present the essential information concering embryology in a compact, readily accessible form which leads itself to student learning and revision. The convergence of various approaches has generated a rich panorama of detail, the significance of which we are still attempting to unraval. The present text has been written as an introduction to this rapidly growing field.

To make the work more comprehensive and informative, the author has consulted many authoritative books, research journals, abstracts, monographs etc., so there can be no claim to originality except in the manner of treatment.

The author expresses his thanks to his friends and colleagues whose continue inspirations have initiated him to bring out this book.

The author expresses his gratitude to Mr. Wasan and staff of M/s Discovery Publishing House Pvt. Ltd. for their whole hearted co-operation in the publication of this book.

In the mean time, the author will remain sincerely responsible for any shortcomings of the book and be grateful to the readers for their suggestions and constructive criticism for the continuous betterment of the book. He takes this opportunity to appeal to the readers to send their suggestions straightaway to his Publisher.

Author

Contents

1

REPRODUCTION

THE MALE REPRODUCTIVE ORGANS

The primary male organs consist of paired glands called the testes, and these produce the male gametes or spermatozoa. The testes develop in the abdominal cavity adjacent to the adrenal glands but in most mammals they migrate downwards through the body cavity into a special fold of skin called the scrotum, or scrotal

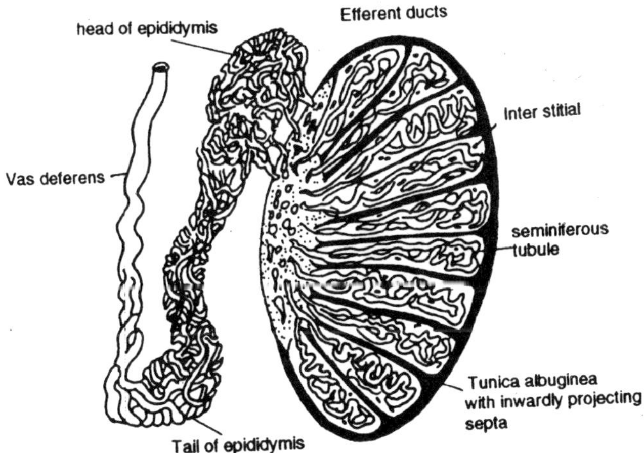

Figure 1.1 : Diagram of a vertical section of the testis.

sacs. The reason for this is that spermatogenesis will not take place at body temperature, and within the thin walled scrotum the temperature is several degrees below that of the body cavity. If for some reason the descent does not occur then the animal is infertile. In amphibia, reptiles and in some mammals e.g. the elephant, the testes remain within the abdominal cavity. The testis is guided in its descent by a long cord, the gubernaculum, which extends from the lower pole of the testis to the scrotum, and the gubernaculum anchors the mature testis in the scrotum.

The secondary sexual organs include a variety of ducts and glands which convey the spermataozoa in special secretions to the exterior of the body, so that they can deposited within the female; the secondary sex organs also, include those characters of the male which distinguish it from the female e.g. the long mane of the male' lion. These secondary sex organs are developed and maintained by,means of sex hormones produced within the testis itself. Male sex hormones are called androgens.

Structure of the Testis

The testis is a tubular gland surrounded by a fibrous capsule the tunica albuginea. It is divided into several hundred compa-rtments by means of fibrous tissue septa and each compartment contains several tubules, called seminiferous tubules. Each tubule is about 50 cms. long in man and is coiled upon itself, hence the name convoluted seminiferous tubules.

All the tubules drain into one border of the testis, into larger collecting tubules which are coiled together in a mass called the epididymis which is applied to the surface of the testis. Between the seminiferous tubules inside the testis, there is connective tissue containing blood vessels and the glandular cells which are called interstitial cells or Leydig's cells and are responsible for the production of the male sex hormone.

Structure of the Tubule

In the adult the seminiferous tubule consists of a basement membrane lined by the seminiferous epithelium; this seminiferous epithelium consists of two types of cell. First the germ cells themselves, secondly the cells of Sertoli which support and nourish the germ cells.

The youngest germ cells are those lying close to the wall of the tubule and these divide to produce cells which pass nearer to the lumen of the tubule where the mature spermatozoa occur. The details of this transformation are described in more detail.

The Sertoli cells are slender pillar like cells attached at their base to the basement membrane. At from the lumen of the seminiferous tubules into the larger collecting tubules of the epididymis, whose walls bear a ciliated epithelium which moves the spermatozoa into the vas deferens.

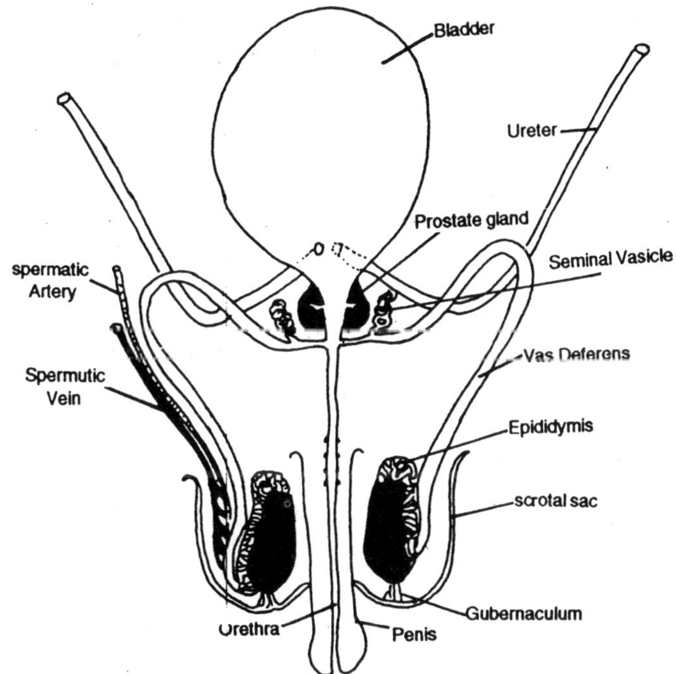

Figure 1.2 : Diagram of the male reproductive organs.

Secondary Sex Organs

The spermatozoa are conveyed by the vasa deferentia to the base of the bladder where they can pass into the urethra. In structure the vas deferens is a hollow muscular tube, which, by *means of muscular.* contraction, rapidly transports the spermatozoa before they are deposited in the female.

Opening into the vas deferens before it joins the bladder neck is the duct of the seminal vesicle; the seminal, vesicles were once thought to store sperm but their function is to produce a thick secretion to provide the bulk of the fluid in which the sperms are transported.

Surrounding the base of the bladder where the vasa deferentia open into the urethra is another gland, the prostate gland, whose secretions are discharged into the urethra together with the spermatozoa and the secretions from the seminal vesicles.

The prostate gland produces a thin alkaline secretion; this helps to neutralize any acid urine remaining in the urethra and also to neutralize some of the acid secretions of the female vagina after the sperms have been placed into the female.

These various secretions, spermatozoa and the products of the seminal vesicles and prostate gland are together called semen; the semen is discharged through the urethra which passes through the penis. The penis, through which both urine and semen pass, contains in its walls, sponge-like systems of blood spaces which can become filled with blood so making the penis a more rigid organ so that the semen can be deposited within the female.

THE FEMALE REPRODUCTIVE ORGANS

The primary sex organs in the female consist of paired ovaries situated within the abdominal cavity. The germ cells are liberated from the surface of the ovary into the peritoneal, cavity from whence they pass into the secondary sex organs- the Fallopian tubes, the uterus and vagina, leading to the exterior of the body. Paired secretory organs, the mammary glands, are included in the secondary sexual organs of the female, and serve to nourish the newly-born mammal.

Structure of the Ovary

The ovary is attached to the wall of the body cavity by a fold of peritoneum. The free surface of the ovary bulges into the peritoneal cavity into which the germ cells are liberated. The ovary is studded with follicles in various stages of development, containing the germ cells. When the follicles are ripe they come to the surface of the ovary where they rupture.

The free surface of the ovary is covered by a thin layer of germinal epithelium from which the germ cells arise in the embryonic period. In some species of mammal it seems that even in the adult the germinal epithelium can give rise to successive crops of new germ cells which pass inwards to mature in the tissues of the ovary.

The timing of the successive crops of new germ cells coinci-des with the rupture of mature Graafian follicles in which follicular fluid rich in the hormone oestradiol pours over the surface of the ovary. This hormone has been called a 'mitogenic' hormone because of its effect on the germinal epithelium in stimulating cell division and the formation of new crops of germ cells.

Beneath the germinal epithelium is a layer of dense connective tissue, the tunica abhiginea. Beneath the tunics albuginea the thicker outer part of the ovary, or cortex, contains the follicles in various stages of development. The central part of the ovary or medulla contains a loose connective tissue containing masses of blood vessels.

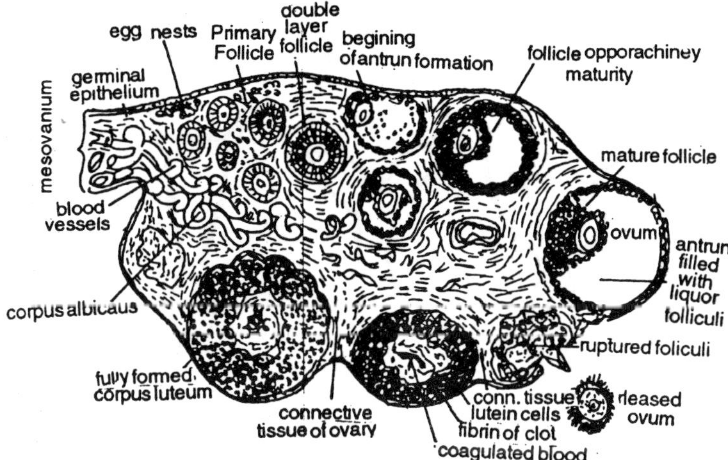

Figure 1.3 : Mammalian Ovary

The interstitial connective tissue of the ovarian cortex consists of connective tissue fibres and various types of cells. Some of these cells, large polyhedral 'epithelioid' cells are given the name interstitial cells. The number of these cells varies throughout the life of the female mammal. In some mammals with large litters e.g. rodents, there may be enormous numbers of these cells. They may arise from the walls of degenerating Graafian follicles.

Not all of the follicles in the ovary undergo the course of the development into mature Graafian follicles as described. Very many of them undergo a degenerative change called atresia in which there is a hypertrophy of the cells forming the wall of the follicle together with a degeneration of the ovum. The interstitial cells and the cells of the atretic follicles are considered to have an endocrine function.

The oviduct or Fallopian Tube

The oviducts are muscular tubes which serve to convey the germ cells from the ovaries to the uterus. The outer end of the tube, nearest to the ovary is expanded, and its edge is split up into fringes, the fibriae, which are closely applied to the surface of the ovary.

The lumen of the oviduct is lined by a secretory mucous membrane, in which there are many ciliated epithelial cells. The germ cells are conveyed down the tube to the uterus by means of peristaltic movements of the tube itself and by the effect of the ciliated epithelium.

The Uterus

The uterus is a thick-walled muscular structure within which the embryo develops. Its wall has three layers, an outer serous coat, a thick middle coat consisting of interlaced smooth muscle fibres (the myometrium) and an inner vascular mucous layer, the endometrium.

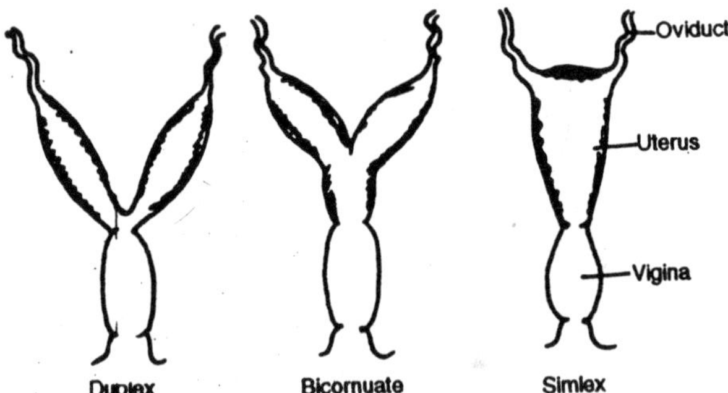

Figure 1.4 : Types of mammalian uteri.

In primitive mammals there are two uteri, each opening into the vagina and this is called the duplex condition and is found in marsupials, many rodents (e.g. rats, mice, rabbit) and bats. In most mammals the distal end of the two uteri is fused to give a biconuate uterus. In higher primates, including man, the two uteri are completely fused together to give a single organ, the uterus simplex.

The vagina is a distensible tube lined by squamous epithelial cells which connects the uterus to the outside world.

GAMETOGENESIS

The process of formation of gametes is called gametogenesis; the formation of eggs is called oogenesis and the formation of sperms is called spermatogenesis. Gametogenesis may be conveniently divided into three stages. The first stage is one in which the cells of the germinal epithelium divide and is called the stage of multiplication. The second stage is one of growth when each of the tiny cells produced in the multiplication stage grows to a larger size.

The cell at the end of this stage is called the primary oocyte in the case of the female, and the primary spermatocyte in the case of the male. The third stage in gametogenesis is a period of maturation; during this period very important changes occur in the nucleus with the result that the chromosome number is reduced from the diploid to the haploid number.

This reduction in chromosome number takes place in the so-called reduction division of a specialkind of cell division called meiosis. When the primary oocyte divides by the reduction division it does so unequally;

the nucleus divides into two equal parts, but almost all of the cytoplasm goes with one half of the nucleus, whereas the remaining half of the nucleus has very little cytoplasm.

The latter is called the first polar body. The nucleus with most of the cytoplasm is now called the secondary oocyte and it contains only the haploid-number of chromosomes. In mammals it is at this stage that the female gamete is released from the ovary; and before this gamete can be considered as fully mature another division of the nucleus has to take place, and this again is an unequal division resulting in the production of a second polar body.

The production of this second polar body takes place, in a mammal, when the egg is fertilized by the sperm. We have seen that the development of the female gamete involves three stages, the first of multiplication, the second of growth ending in the formation of the primary oocyte, and the third of maturation involving the production of the secondary oocyte with polar bodies.

The first stage of gametogenesis is complete in the female embryo by the end of intra-uterine life, and she is born with all' the oogonia already formed within the ovary. The second stage of growth of the oogonia continues throughout the life of the mammal and we will now look at this growth phase in more detail.

Development of the Graafian Follicle

In the cortex of the ovary of the mature mammal are many small clusters of cells called the primary follicles, which have been formed during the embryonic period from invaginations of the germinal epithelium, called sex cords. The primary follicles are very small and there are about 400,000 in the human female at maturity.

At birth the first phase of oogenesis, the phase of multiplication has already started producing small collections of germinal cells, called primary follicles. One of the cells in the primary follicle is larger than the rest and is the oogonium, whilst the smaller surrounding cells are called follicular cells.

In the second phase of oogenesis, the growth phase, the primary follicle develops and changes occur, in the oogonium as it becomes the primary oocyte, in the follicular cells and also in the connective tissue which surrounds the follicles. The oogonium enlarges, its nucleus gets bigger, and a few yolk granules begin to appear in its cytoplasm.

At this stage a well defined shining layer appears around the surface of the oogonium called the zona pellucida. In the primary follicle the oogonium was surrounded by a simple columnar epithelium but as the follicle grows the follicular cells multiply to produce an epithelium which is several layers in thickness. The cells of this epithelium secrete a follicular fluid which accumulates in spaces which begin to appear between the cells.

The follicle by this stage in the human female is about 2 m. in diameter and is now called the Graafian follicle. The follicle increases in size with the accumulation of more fluid within it and the oogonium is pushed to one side of the follicle where it is attached to the wall of the follicle in a group of columnar cells called the discus proligerus.

The cavity of the follicle is lined by a few layers of columnar cells called the membrana granulosa. The connective tissue surrounding the Graafian follicle has become organized into a membrane called the theca (consisting of two layers, the *theca* interna and externa).

Eventually the Graafian follicle may reach a size of 10 mm. in diameter in the human female and bulges from the surfaces of the ovary; by this time the oogonium has grown to its full extent and is called the primary oocyte. The fluid within the follicle is formed at a faster rate than the follicle wall grows and the follicle eventually ruptures.

By the time the follicle has ruptured the primary oocyte has undergone the meiotic division producing the first polar body and is now called the secondary oocyte. When the secondary oocyte is released it is surrounded by a few columnar cells which form the corona radiata, which may have a nutritive function similar to that of Sertoli cells in the male.

When the Graafian follicle has ruptured and liberated the oocyte it collapses and the hole left by the departing oocyte becomes plugged with a blood dot. There is now a multiplication of the remaining cells of the follicle, the granulosa and theca cells. The cells enlarge and develop deposits of a yellow pigment called lutein.

The whole structure so produced is called the corpus luteum, a solid ball of-yellow pigment cells, which produces hormones which prepare the uterus to receive the fertilized oocyte.

Spermatogenesis, the formation of spermatozoa. Unlike the female, where the germinal epithelium forms the outermost layer of the ovary, in the male the germinal epithelium lines the walls of the seminiferous tubules. The cells nearest to the wall of the tubule, the spermatogonia, are the most primitive, undifferentiated cells of the tubule and they give rise to the other cells by mitotic division. As in oogenesis there are three stages in spermatogenesis.

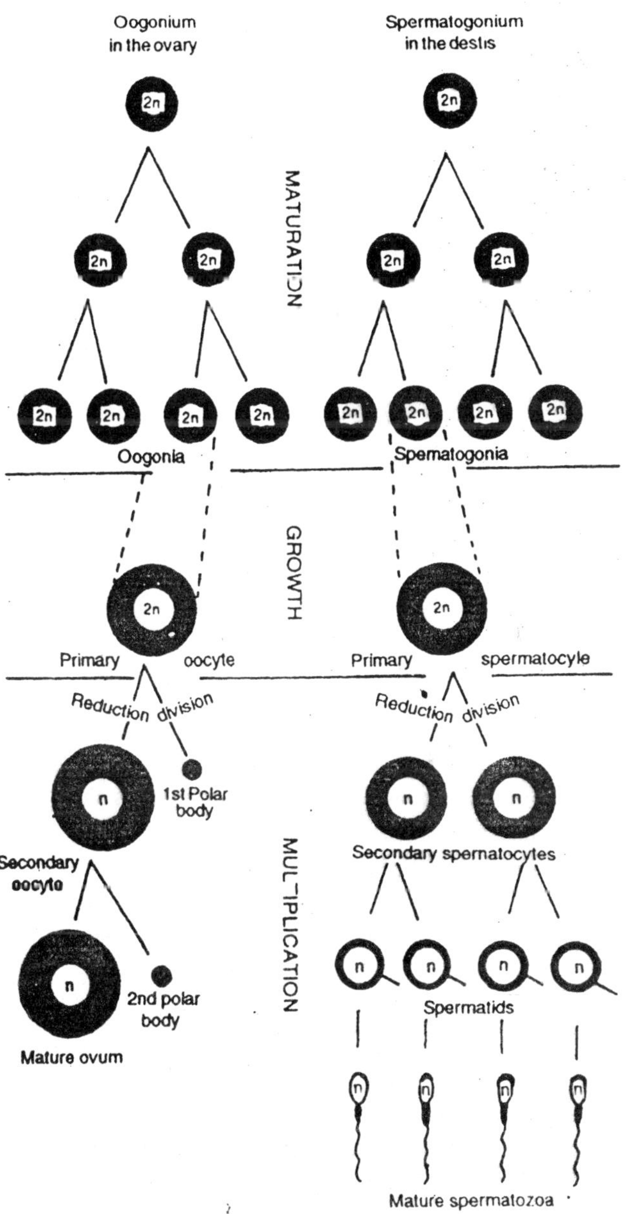

Figure 1.5 : Structure of a mature sperm.

The first stage of spermatogenesis is that in which the spermatogonia multiply by mitotic division. Some of the products of these divisions pass inwards nearer the lumen of the tubule where they enter the second phase of spermatogenesis, the growth phase.

During this phase the spermatogonia become larger, producing the the primary spermatocytes. In the third phase each primary spermatocyte undergoes a reduction division producing two secondary spermatocytes containing the haploid number of chromosomes.

Each secondary spermatocyte then divides to produce two spermatids. The spermatids do not undergo divis on but a series of changes occur which transform the spermatid into the mature sperm. During this maturation of the spermatids they are attached to the cells of Sertoli.

The Mature Sperm (human). The mature sperm consists of a head, middle piece and tail. The head consists of the condensed nucleus of the spermatid and is a flattened ovoid structure about 5 microns long. The head is capped by a sheath of material called the head cap. In the middle piece of the sperm there is a centriole from which arises a long axial filament which passes through the middle piece and the tail. Surrounding the axial filament in the middle piece is wound a sheath of mitochondrial material, mitochondrial sheath, which is probably concerned in the respiration of the sperm. In the tail the axial filament is covered by a sheath.

The spermatozoa remain inactive until they pass from the testis. During their passage from the testis to the penis they are activated by the secretions of the accessory glands. The sperm is capable then of active swimming during which S-shaped waves pass along the tail. The energy for this is derived from the anaerobic breakdown of fructose which is present in the prostate secretions.

THE PHYSIOLOGY OF REPRODUCTION

Hormones and Reproduction in the Male

In the sexually immature mammal the primary and secondary sex organs are small and undeveloped. The growth and develop-ment of the sex organs is dependents upon the activity of the pituitary gland. The pituitary exerts its effect by the production of hormones called gonadotrophic hormones because of their growth effects upon the gonads.

The gonadotrophic hormones are complex protein substances which have been isolated in a relatively pure state; their chemical structure is not yet elucidated. There are almost certainly two gonadotrophic hormone. One of these hormo-nes is called the follicle stimulating hormone (or F.S.H.) because of its effect in the female in stimulating the growth of the follicles in the ovary. F.S.H. stimulates the growth of the seminiferous tubules of the testis and stimulates the activity of the germinal epithelium.

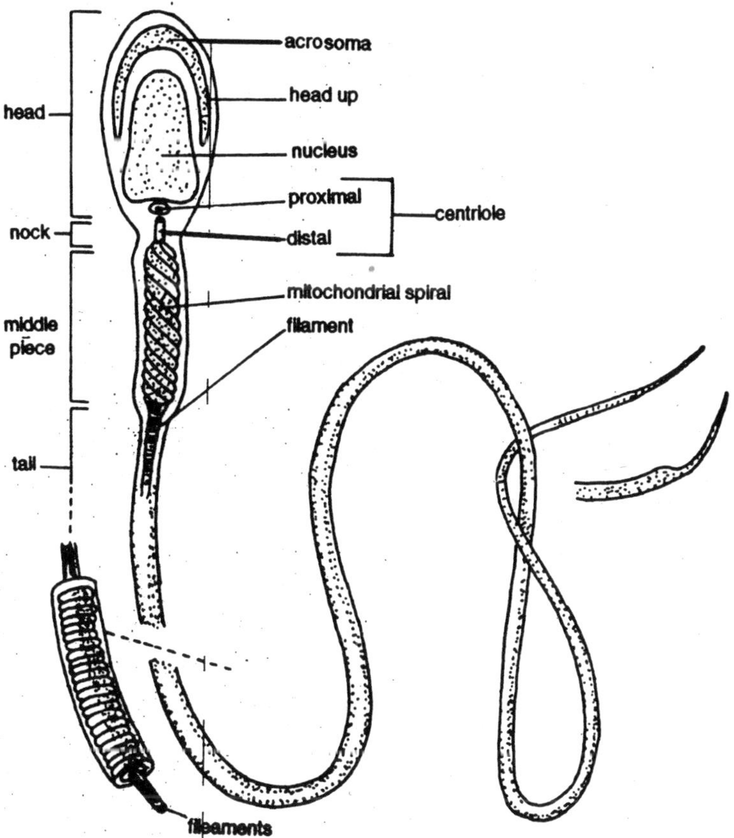

Figure 1.6 : Formula of testosterone.

The other pituitary hormone is called the luteinizing hormone (L.H.) or interstitial cell stimulating hormone. L.H. stimulates the growth and secretory activity of the interstitial or Leydig cells of the testis. These cells produce certain steroid hormones called sex hormones, because of their effect on the sex organs.

The most important sex hormone in the male is called testost-erone. However, there are also female sex hormones or oestrogens pro-duced in the male. The effect of testosterone is to promote the growth of the various sex organs the vas deferens, seminal vesicles, prostate gland, penis etc. It also promotes the development of those other secondary sex characters which vary from one mammalian

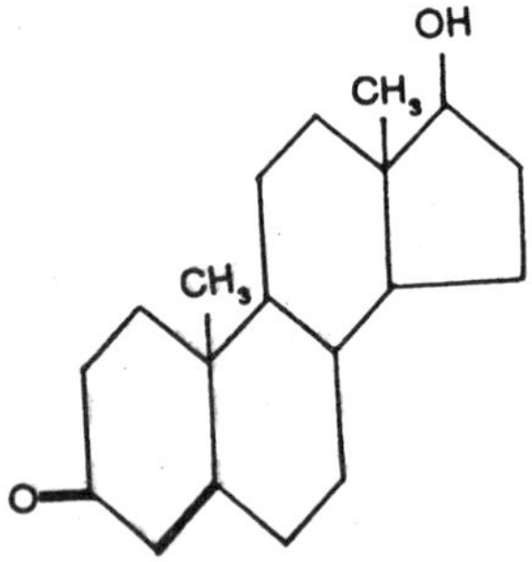

Figure 1.7 : Formula of testosterone.

species to another; in man these include the growth of the beard, enlargement of the larynx with the development of the deeper voice, the male distribution of hair on the body, and the greater development of muscle. The effect on muscle growth is due to the fact that testosterone is what is called a protein anabolic horm-one, that is it promotes the retention and incorporation of protein in the tissues.

In the immature animal testosterone will promote the precocious development of the sex organs, and in those species of mammals in which the testes do riot descend into the scrotum until sexual maturity it stimulates the descent of the testis. In animals which have been castrated the sex organs gradually atrophy and the adm-inistration of testosterone can reverse these charges.

Testosterone has also a direct effect on the pituitary gland and a certain level of testosterone in the blood will inhibit the pituitary from producing gonadotrophic hormones; when the pituitary production of leuteinizing hormone falls then the Leydig cells of the testis stop producing testosterone and the blood level of testosterone falls.

With the falling producing of testosterone by the Leydig cells the pituitary gland is now released from the inhibitory effect of testosterone and it begins to produce the gonadotrophic hormones again. By this feedback mechanism. the secretion of testosterone is controlled.

This effect of testosterone upon the pituitary gland also explains the varying results that experimenters have had in the administration of testosterone to animals. Some workers have found that in some animals testosterone will cause a stimulation of the germinal epithelium of the testis and the production of spermatozoa; because of the appearance of large numbers of dividing cells in the germinal epithelium the hormone has been called 'mitogenic', that is one which stimulates mitotic division within the cells. In large doses however testosterone can depress the growth of the testis, presu-mably because it inhibits the anterior pituitary gland from producing the gonadotrophic hormones, by the feed-back mechanism.

Testosterone also appears to be responsible for behaviour chan-ges in animals, and administration of the hormone to sexually immature males results in the appearance of male breeding behaviour. In some species even the female will show a masculine pattern of breeding or sexual behaviour when given testosterone.

The aggressiveness of many male mammals can be promoted by giving testosterone. In higher primates, including man, sexual behaviour cannot be controlled so simply, and cultural factors play a much more important role in the determination of the direction of sexual impulses.

HORMONES AND REPRODUCTION IN THE FEMALE

In the mature female mammal sexual activity tends to be an intermittent phenomenon during the breeding season, which is the period when mating can occur. The breeding season may consist of several weeks, or months of the year, and there may be more than one breeding season in the year. During the breeding season itself sexual activity is a cyclic phenomenon, with periods of sexual activity or oestrus ('heat') alternating with periods of sexual inactivity.

These cycles of activity stop of course as soon as a pregnancy is started. In some animals, including primates and rodents, these oestrous cycles are continuous throughout the sexual life of the animal and are not restricted to breeding seasons; these animals are said to experience poly-oestrus.

In man, because of the unusual feature of menstruation of the oestrous cycles are known as menstrual cycles; but the menstrual cycle is fundam-entally similar to the oestrous, cycle of other mammals. The variat-ions in breeding activity are best illustrated by referring to particular examples:

Horse

The mating season of the horse is between March and August, although some breeds will mate in the autumn and winter in England. During the mating **season,** if pregnancy does not occur then there is a regular occurrence of oestrus, each period of oestrus lasting about 20 days. The horse is said to be seasonally poly-oestrous.

Dog

The domestic dog has two breeding seasons in the year, in late Winter and early spring and in the Autumn. During each season there is only one period of oestrous, and" the dog is said to be monoestrous.

Golden Hamster

This rodent is polyoestrous,-coming -into heat at all times of the year, the oestrous cycles recurring every ten days, each cycle lasting about four days.

Roe Deer

Like the dog the Roe Deer is monoestrous. The breeding season is in July and August during which there is only one period of oestrous.

The oestrous cycle

During the oestrous cycle there are wide-spread changes in the structure and behaviour of the female; all the changes that occur are under the control of hormones, produced mainly by the anterior pituitary gland and the ovary. The aim of these changes is to mature an ovum and prepare the uterus to receive and nurture the ovum if it is fertilized by a sperm.

In the early phase of an oestrous cycle the ovary is activated by the secretion of follicle-stimulating hormone (F.S.H.) from the anterior pituitary gland. The ovary responds by the progressive growth of one or more Graffian follicles, depending upon the species of mammal. Small amounts of luteinizing hormone (L.H. is said to have a synergistic effect.

The ripening Graafian follicles secrete a steroid sex hormone called oestradiol which has wides-pread effects upon the secondary sex organs, and also has an effect on the secretory activity of the pituitary gland. Oestradiol is only one of a number of substances isolated from the ovary, blood and urine of the female mammal, all of which have some effect on the sex organs; the name oestrogen have beenused to describe this class of substances, although oestradiol is the most potent naturally occurring oestrogen.

Oestradiol stimulates the growth of the uterus; the myometriuni increases in, thickness because of growth of its individual cells, its vascularity increases and the endometrium thickens, becoming more vascular as its secretory glands grow in length.

The Fallopian tubes and vagina are also stimulated. In the sexually inactive phase or anoestrus, the epithelial lining of the vagina is thin, only one or two cells thick, but after stimulation by oestrogen the epithelium thickens and cornifies, and flattened eornified squames appear in the vaginal secretions. The mammary glands also increase in size under the influence of oestrogen, which stimulates the growth of he duct system.

The rising level of oestradiol in the blood has important effects upon the pituitary gland. It inhabits the formation of F.S.H. and at the same time stimulates the production of further amounts of L.H. which brings about ovulation and the formation of the corpus luteum.

This first phase of the oestrous cycle, as described above, is called thc *follicular phasc,* bccausc of thc growth of the Graafian follicles, or the oestrogen phase, because of the importance of this hormonc in this part of the cycle. In the uterus the follicular phase is associated with growth, both of the myometrium and endometrium, and this phase of uterine change is called the proliferative phase.

The follicular phase finishes at ovulation when the egg escapes from the ovary and the remains of the Graffian follicles grow to produce special glandular structures under the influence of the pituitary luteinizing hormone (L.H.), called *corpora lutea.*

We now enter the *luteal phase* of the oestrous cycle. This is often called the progesterone phase, because of the importance of this hormone at this time. The corpora lutea are large yellow pigmented bodies studded in the ovarian cortex. They are formed by the proliferation and growth of cells in the wall of the ruptured Graafian follicle.

Under the influence of another pituitary hormone called luteotrophin or lactogenic hormone the corpora lutea secrete a hormone called progesterone, in addition to small amounts of oestradioL Pugesterone produces further changes in the endometrium of the uterus the increase in thickness and vascularity of the-uterus progresses as the glands of the endometrium become tortuous and begin to persecretions into the cavity of the uterus.

Because of these glandular changes this phase of uterine activity is called the *secretary phase.* Progesterone also has effects on the mammary glands . which haw already been primed by the effects of oestradiol; now, glandular element begin to appear around the ends of the duct systems of the mamma glands.

We have seen that the time of maximal oestradiol activity is at the time of ovulation and after this time the level of oestrogen production gradually falls, as progesterone comes to play a more important par in the cycle. It is at the time of maximal oestradiol production that the female mammal is most willing to receive the male and this is the true period of 'heat' Mating and fertilization usually occur about this time. If fertilization of an ovum does not occur then corpora lutea gradually disintegrate and retrogressive changes occur in the secondary sex organs.

The falling level of blood oestrogen in the luteal phase the cycle, together with some inhibitory effect of progesterone on the anterior pituitary gland, are responsible for a gradual decline in the production of L.H. and therefore the corpora lutea degenerate.

Menstrual Cycle

In the human female when the corpus luteum disintegrates at the end of the luteal phase of the cycle, there is a complete breakdown of the hypertrophied endometrium; blood and broken-down tissues are discharged from the vagina and this constitutes menstruation. In the human the menstrual cycle lasts about 28 days.

In the first half of the cycle there is the follicular phase, culminating in ovulation at about the foul teenth day. In the second half of the cycle, the luteal phase, the corpus luteum is formed and the endometrium enters the secretory phase, and in the absence of fertilization the luteal phase is ended by the appearance of the menstrual flow.

Following menstruation only fragments of endometrium are remaining and these lie in crypts in the myometrium. From these fragments the entire endometrium is reformed during the proliferative phase of the next cycle. The primate endometrium undergoes this almost complete. breakdown because of a peculiarity in its blood supply.

When the supply of progesterone is waning as the corpus luteum degenerates at the end of the luteal phase, certain spiral. arteries of the endometrium go into such intense spasm that the tissues supplied by them die and undergo degenerative changes. The whole of the dead endometrium is then sloughed off from the uterine wall together with blood.

The Oestrous Cycle and Pregnancy

If during the luteal phase of the oestrous cycle an ovum is fertilized and settles in the uterine cavity then the retrogressive changes in the secondary sex organs do not occur, nor does the corpus luteum degenerate. We have seen that in the absence of pregnancy the corpus luteum degenerates; it does this because of the decline in the production of pituitary gonadotrophic hormones.

As soon as there is a union established between the fertilized ovum and the uterine wall increasing amounts of gonadotrophins appear in the maternal blood and this maintains the structure and function of the corpus luteum. The gonadotrophic hormone is produced by the placenta, the organ which unites the mother and foetus, and through which is receives its nourishment.

The placenta also produces large amounts of oestrogen and progesterone and gradually replaces large amounts of oestrogen and progesterone and gradually replaces the corpus luteum as a source of these hormones, so that later in the pregnancy one can remove both ovaries from the female mammal without disturbing the pregnancy: The function of the large amounts of sex hormones produced by the placenta is to promote further growth of the uterus, vagina and mammary glands.

During pregnancy, growth of further Graafian follicles and ovulation is prevented because the large amounts of oestrogen produced by the placenta inhibit the anterior pituitary gland from producing follicle-stimulating hormone.

Reproduction and the Environment

The reasons why animals tend to breed at relatively restricted times of the year have been studied and classified under two headings, internal physiological mechanisms on the one hand, and external environmental factors on the other.

An internal physi-ological 'clock' cannot be the sole factor in determining periodic breeding; seasonal breeding has an adaptive significance in that young are produced at favourable times of the year, and obviously a rigid internal mechanism would, through the course of time, fail to adapt the animal to changing climatic conditions.

It appears that animals have become adapted to respond to certain environmental factors which herald the oncoming favourable season. In temperate latitudes many animals have as it were, harnessed their breeding behaviour to the length of day and respond to an increase in day length by development of the sex organs, so that the young will develop in a favourable season with warmth and an adequate food supply.

The first study of the effect of light on sexual cycles was made on the Canadian bunting, a bird which normally breeds in Spring. By giving these birds extra periods of light in the Autumn it was possible to cause development of the testes, even at a time when the temperature animals, reptiles, fish, birds and mammals it has been possible to cause the develo-pment of the sex organs out of the breeding season.

The light exerts its effect by way of the eyes and connections with the hypothal-amus and anterior pituitary gland. Not only is the increase in day length of importance but in some autumn breeding animals the shortening days may also act as the stimulus to sexual development. When some animals with fixed breeding seasons in temperate latitudes are transferred from the Southern to the Northern Hemisphere, after a However, animals imported from tropical countries tend to continue differences in daylight to which the species inhabiting temperate zones respond.

Thus may be due to the fact that owing to the comparative uniformity of conditions in their own countries they have never acquired the capacity to respond to variations in light intensity or duration, characteristic of many animals living under seasonally changing conditions. Thus Java deer when imported to England continue to produce young in late Autumn, as they are believed to do in Java, a condition which is abnormal for any deer inhabiting temperate countries.

But even in tropical climates animals may have special breeding seasons; if reproduction were continued throughout the year the competition for food for the young might be too intense for survival of the species. A further advantage of seasonal breeding may be the sync-hronization of the male and female sexual cycles by the fact that they are both adapted to the same environmental factor.

What these environmental factors are in tropical climates in uncertain. That breeding cycles in some tropical animals are harnessed to some environmental factor and not dependent upon an internal rhythm seems certain; a tropical insectivorous bat lives throughout the period of daylight until about ten minutes before sunset in dark and almost thermo-static caves and yet it was found that in one year of observation no pregnancies occurred until a few days at the beginning of September.

It seems impossible that such synchron-ization of the sexual cycles of such a group of bats could be achieved by an internal physiological mechanism. The homiothermic vertebrates (warm blooded) have become almost independent of the temperature of their surroundings and are able to carry out breeding at any time of the year.

They do, however, tend to breed in Spring in temperate latitudes to ensure a favourable environment with adequate food for the young. But even in homiothermic animals temperature may play some role in the timing of breeding behaviour.

If Autumn weather is warm and food supply is good, sexual behaviour in the robin and other birds may be pronounced and may even lead to reproduction in a few birds, although they normally only breed in Spring. In this Autumn breeding phase sexual activity is developing hen daylight is decreasing.

We have now seen something of the way in which sexual' cycles are controlled by the endocrine organs, and the way in which the endocrine organs, by way of the hypothalamus of the brain, are synchronized with external environmental factors so that the young are cared for under favourable conditions.

THE PLACENTA

The uterus has been preparing to receive a fertilized ovum throughout the oestrous cycle. In the follicular phase of the cycle there is growth both of the myometrium and endometrium and an increase in the blood supply of the uterus. The endometrium is thickened, new blood vessels grow and its glands increase in length. In the luteal phase of the cycle these changes continue and the endometrium takes up secretary character. The glands become tortuous and their epithelium becomes active, secretions passing into the uterine cavity.

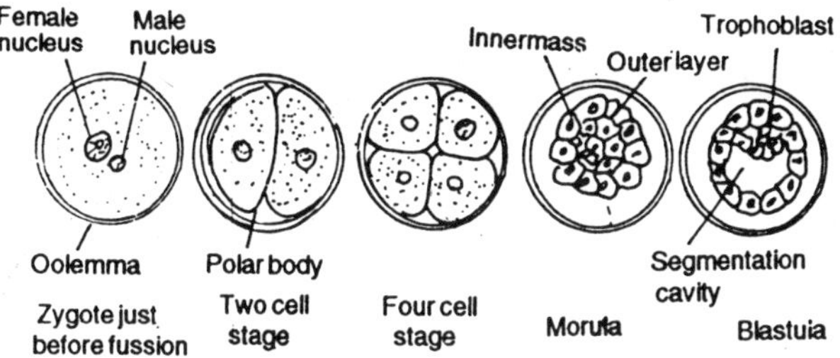

Figure 1.8 : Diagrams showing the early development of the fertilized egg of the mammal up to the blastula stage.

The early development of the fertilized egg occurs during its passage through the Fallopian tube, since fertilization is usually achieved high up in the Fallopian tube. The fertilized egg or zygote undergoes divisions to produce a ball of cells called the morula.

Then morula than differentiates into an outer layer and an inner cell mass producing the blastula. The cells of the outer layer form the trophoblast or trophoblastic ectoderm. This enters into the formation of the chorion, the outermost, covering of the developing zygote.

This outer layer is called *trophoblast* because it enters into the formation of placenta, the organ concerned in nourishing the foetus. The inner cell mass of the blastula contains the cells from which the embryo will develop.

During its passage down the Fallopian tube the nutrition of the zygote is dependent on the small amounts of food stored in the protoplasm. When the morula reaches the uterine cavity it is bathed in the secretions of the uterine glands, and these secretions probably have some nutritive function. In some species the blastocyst lies in the cavity of the uterus, and in contact, by means of its trophoblast, with the endometrium all over its surface. In some other species the blastocyst becomes attached to the endometrium on one surface, and projects freely into the uterine cavity or its other serface.

In other types the blastocyst sinks into the endom-etrium and becomes surrounded completely by maternal tissues; this type of implantation of the blastocyst is found in man and is called the interstitial *type*. In order to understand the varied types of placenta found in mammals it is necessary to examine in some detail the formation of the various foetal membrane which take part in the formation of the placenta.

There are four foetal membranes concerned in the adaptation of the foetus to life in the uterus, the amnion and chorion, formed from the original embryonic body wall, and the yolk sac and allantois, parts of the original gut of the embryo.

We had left the development of the embryo at the blastula stage, consisting of an *inner* cell mass and an outer layer, the trophoblast; a two layered vesicle is produced by a growth of endoderm cells around the inner layer of the trophoblast (enclosing the yolk of the egg, when this is present) Only a part of the wallof the blastula is destined to form the embryo and this is called the embryonic area.

This embryonic area gradually sinks into the centre of the blastula as it becomes covered over by the amniotic folds. In man the embryo is not covered by the amniotic folds and there is merely a hollowing out of the cells of the embryonic area to produce a cavity, the amnion. The amniotic folds meet and fuse above the embryo and because of the double walled nature of the folds the embryo becomes to be surrounded by two membranes, an outer chorion and an inner amnion.

The chorion thus becomes the membrane which is in contact with the endometrium of the uterus, and projections called chorionic villi grow out from its surface to make a more intimate contact with the maternal tissues. The amniotic membrane surrounds a fluid filled cavity, the amniotic cavity, in which the embryo floats. We have seen above how the endoderm grows round the inner surface of the trophoblast to produce a two layered vesicle.

In the heavily yolked eggs of birds and reptiles this layer encloses the yolk of the egg and is called the yolk sac. The primitive gut of the embryo, a groove on the under surface of the embryonic area, is open into this yolk sac but later as the gut is folded off it is separated from

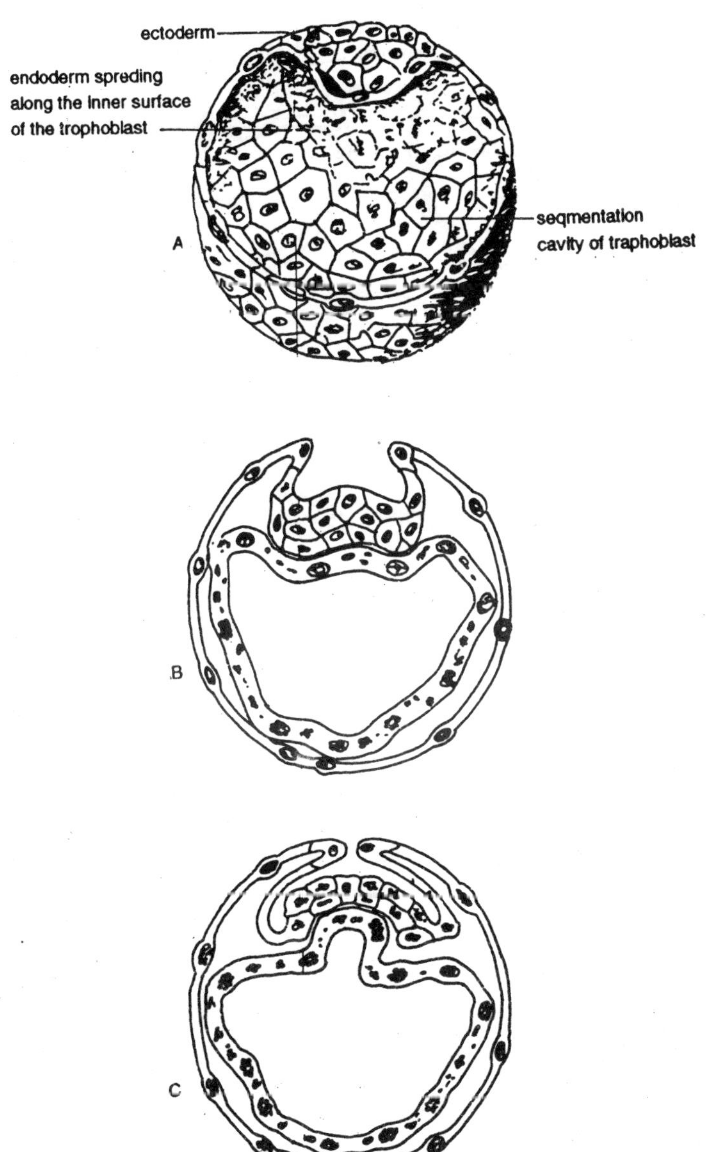

Figure 1.9 : Development of the late blastula. Note the embryonic area gradually sinking into the blastula in A, B and C, gradually becoming covered by the amniotic folds. The endoderm is shown progressively spreading over the inner surface of the trophoblast.

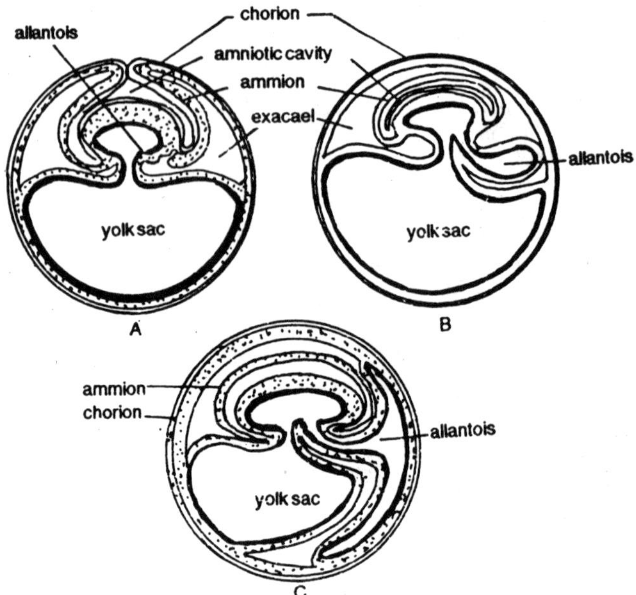

Figure 1.10. Showing the development of extra embryonic structure.

the yolk sac except for a narrow passage, the yolk sac stalk. Later mesoderm grows out from the embryo between ectoderm and endoderm; this process provides a double layer of mesoderm in the amniotic folds and a layer of mesoderm between the ectoderm and endoderm of the yolk sac.

It is in this layer of mesoderm that the embryonic blood vessels develop. In the heavily yolked eggs of birds and reptiles the inner layer of the yolk sac becomes vascular in order to absorb the nourishment from within the yolk sac. But in mammals there is very little nourishment within the yolk sac and it is from outside that it has to look for nourishment.

Thus the outer layer of the trophoblast becomes highly developed and supplied with blood vessels from the mesoderm. In primate embryos, including man, the yolk sac is a rudimentary structure only. In Marsupial mammals which have a less developed placenta the yolk sac is a more important organ, and is the major organ for nourishment of the embryo. In rodents the yolk sac becomes partly vascularized; the non vascularized area becomes eroded away so that the cavity of the yolk sac opens into the uterine cavity. The inner wall of the yolk sac is applied directly to the endometrial lining of the uterus, producing what is known as an inverted yolk sac placenta.

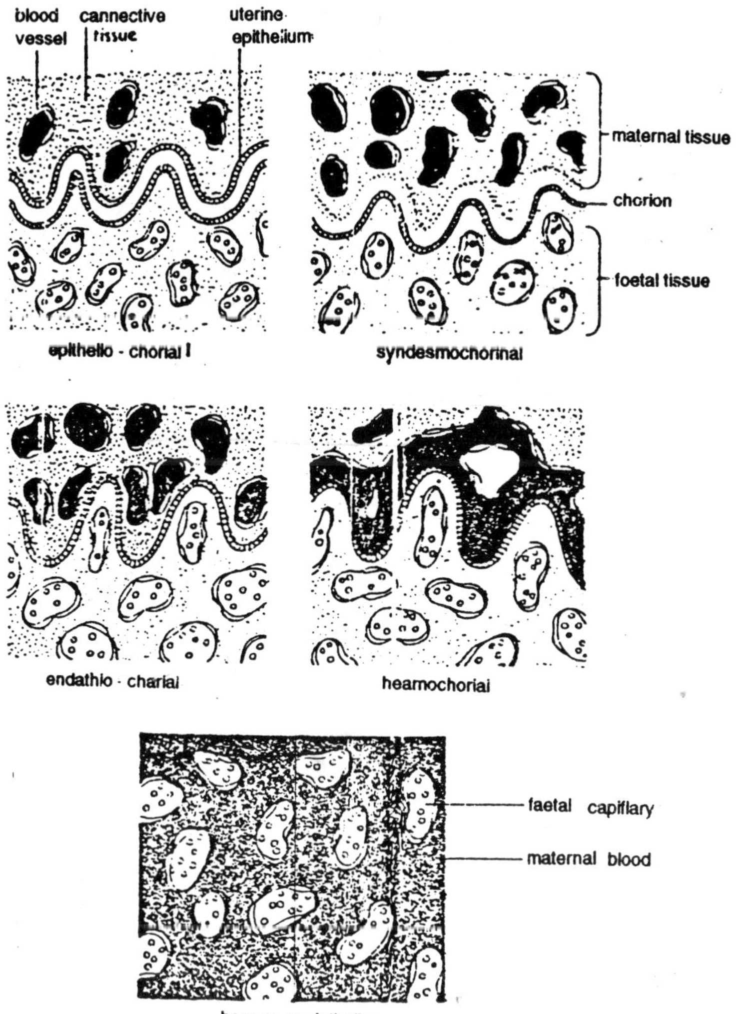

Figure 1.11 : Diagrammatic representation of different placental types.

In most mammals the chief absorbing surface occurs on a structure called the allantois. This is a blind ended outgrowth of the hind gut carrying mesoderm on its surface as it grows into the extra embryonic coelom. In the eggs of reptiles and birds the allantois functions as a bladder for the storage of excretory products, but it also serves to transport oxygen which diffuses through the shell, through the blood vessels of the allantois to the developing embryo. In mammals the allantois has become progressively more important in the nourishment of the foetus, and blood vessels of the allantois pass into the villi of the chorion which make connection with the lining of the uterus.

The processes which grow out from the trophoblast (i.e. extra-embryonic ectoderm of the chorion) are called trophoblastic or chorionic villi. The villi are connected to the embryonic blood vessels through the yolk sack or allantois, depending upon the species of mammal. Not all of the surface of the trophoblast is covered by villi and the arrangement of the villi varies from species to species.

In the horse and the pig the villi are diffusely arranged over the surface of the trophoblast producing a diffuse placenta. In the sheep and cow the villi are localized in patches called cotyledons. producing the cotyledonary placenta. In carnivores the villi are restricted to a band encircling the embryo producing a zonary placenta. In man (also rodents and insectivores) the villi are at first scattered over the whole trophoblast and later become limited to a disc shaped area producing a discoidal placenta.

Types of Placental Union with the Uterine Wall

There is a great variation in the intimacy of contact between the foetal and maternal tissues in the placenta of mammalian species. On the foetal side of the placenta there are three layers of tissue, the chorion, the mesenchymal tissues and the endothelium of the 'capillary vessels. On the maternal side there are uterine secretions, the endometrial epithelium, connective tissues and the endothelium of the blood vessels.

There are thus at least six possible layers of tissue separating the foetal and maternal blood. In the diffuse placenta found in the pig and horse these six layers which separate the maternal and foetal blood streams persists, producing what is called an epithelio-chorial placenta.

This description of the placental structure of the pig and the horse give a rather exaggerated picture of the separation of maternal and foetal blood. The foetal cells of the chorion are low cuboidal in shape and foetal capillaries ramify close to the chorion and may actually penetrate between the chorionic cells to form intra-epithelial plexuses. The capillaries at places displace the cuboidal cells and compress their cytoplasm into thinned out plates.

Thus over a large part of the placental surface only a very thin plate of epithelial cytoplasm separates, the foetal and maternal blood. In other mammals the barrier between maternal and foetal blood is reduced by the erosion of the layers of the uterine wall by the trophoblastic villi. In the syndesmochorial placenta of cattle and sheep the endometrial epithelium is eroded and the epithelium of the trph oblastic villi is in contact with the uterine connective tissue and in the endotheliochorial placenta of the cat and dog the uterine tissues are further eroded and the epithelium of the trophoblastic villi is separated from the maternal blood only by the endothelium of the maternal capillaries.

In insectivores, rodents and man the maternal capillaries are eroded so that the trophoblastic villi are bathed in lakes of maternal blood. The epithelial cells of the chorion are provided with many minute projections called microvilli which interdigitate with maternal tissue and provide a very intimate form of contact between the two.

The long processes of foetal tissue, called trophoblastic villi, run vertically from the surface of the placenta into maternal tissues. Between these trophoblastic plates are the maternal capillaries with a thick endothelium. These capillary endothelial cells show, in electron micrographs, an interesting structure which may well be related to the way in which some substances are transferred by the placenta between maternal and foetal blood.

The cytoplasm of the maternal capillary endothelium is spongy, containing many vacuoles. In the cytoplasm between these vacuoles are mitochondria and Golgi material. The vacuoles are sometimes seen to connect directly with the lumen of the capillary and it seems probable that the vacuoles are formed by invagination of the surface membrane of the cell. Thus substances included in the vacuoles may be transferred by these means from maternal blood to foetal tissues without having to pass through the 'cytoplasm' of the endothelial cells.

The basement membrane of these endothelial cells varies a great deal in thickness, being absent in some places so that the plasma membrane of the maternal endothelial cells may be in direct contact with the cells of the foetal trophoblastic villi, The irregular basement membrane may represent erosion by the trophoblastic villi.

In the human placenta the trophoblast consists of an inner layer of cells (Langhans cells) lying next to the foetal connective tissue and blood vessels and an outer sycitial layer which is bathed by the maternal' blood. The Langhans cells become flattened and many disappear as pregnancy advances. The cytoplasm of the syncytium is foamy and many ovoid nuclei are scattered through it. The foaminess of -the cytoplasm is due to the presence of many vacuoles, each with a granular outline, the granules probably being of R.N.A. Synthesis of foetal protein may occur at these sites until the foetal liver takes over this role.

In addition to these smaller vesicles there larger vacuoles, usually near to the outer surface of the syncytium under the microvilli which occur here. These vacuoles may rise by invaginations of the surface membrane of the cell and may be concerned with the 'absorption' of maternal plasma into the cell by the process of pinocytosis. The large number of microvilli increases the surface area of the cell for absorption or secretion.

In the rabbit, guinea pig and rat there is even more intimate contact since the epithelium and mesenchymal tissues of the chori-onic villi disappear leaving only the endothelium of the trophoblastic capillaries separating maternal and foetal blood, producing the haemoendothelial placenta.

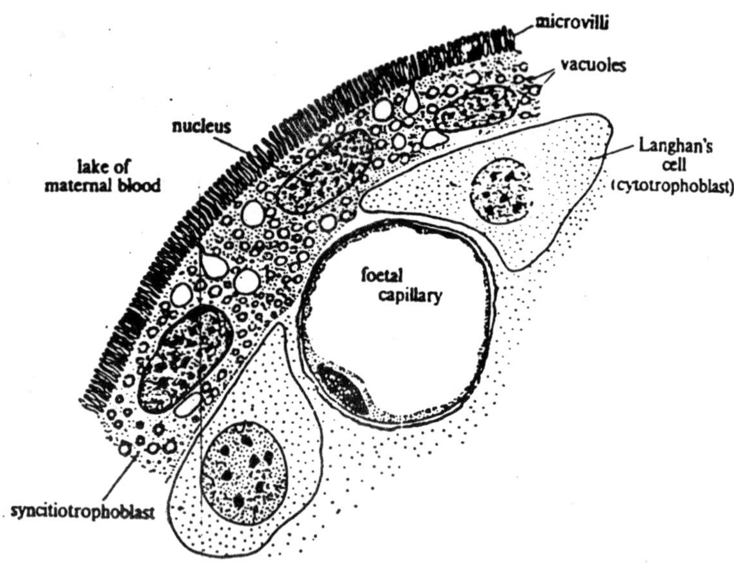

Figure 1.12 : Diagram of a small area of human placenta made from electron micrographs. Note the microvilli of the syncytiotrophoblast projecting into a pool of maternal blood. Some of the vacuoles of the syncitio-trophoblast are shown opening between the bases of the microvilli.

Type of placenta, efficiency and length of gestation period. In the ungulates with epithelio-chorial or syndesmochorial placentae the barrier between maternal and foetal circulations is much greater than in the other placenta: types.

Nutrition of the foetus in ungulates is assisted by secretions of the uterine glands called uterine milk It is difficult to correlate the efficiency of the placenta with the type of placenta; even in one species the placental structure is not constant throughout development, and in some species e.g. rat, there are additional structures to the chorio-allantoic placenta in the form of the yolk sac placenta.

However there is a general tendency for animals possessing a more highly developed placenta (i.e. fewer layers) to have a shorter gestation period, although the young are born in a helpless condition. This feature is illustrated in the following list of gestation periods :

Mouse	gestation period 19 days. Haemoendothelial plac-enta. Young very immature at birth.
Rat	gestation period 21 days. Haemoendothelial placenta.
Dog	gestation period 63 days. Endothelio-chorial placenta. Young helpless and blind.
Cat	gestation period 65 days. Endothelio-chorial placenta. Young are helpless and blind.

In the above examples the young are in a very similar condition at birth, although the gestation period of the cat and dog is three times as long as that of the mouse and rat. This difference is explained on the grounds of a more efficient placenta is rats and mice.

Summary of Types of Placenta

Placental type	*Epithelia chorial*	*Syndesmo- chorial*	*Endothelio- chorial*	*Haemo- chorial*	*Haemo Endothelial*
Maternal tissues					
Endothelium	+	+	+	–	–
Connective tissue	+	+	–	–	–
Epithelium	+	–	–	–	–
Foetal tissues					
Chorionic epithelium	+	+	+	+	–
Mesenchyme	+	+	+	+	–
Endothelium		+	+	+	+
	+				
Examples				Insectivores	
	Horse	Cattle	Dog,	Man,	Rat
	Pig	Sheep	Cat	Lower	Rabbit
				Rodents	

+ *indicates presence*

– *indicates absence*

Guinea Pig. gestation period approx. 68 days. The young are born active with eyes open. There is an accessory yolk sac placenta throughout the gestation period. The additional period of gestation compared to the mouse and rat is used to further the development of the young.

Pig. gestation period 119 days. Epitheliochorial placenta. This is a primitive type of placenta and a long gestation period is necessary.

Thus it is seen that the more primitive type of placenta is associated with a longer gestation period, the guinea pig being an exception to this general rule perhaps because the young are so well developed by the time they are born.

PHYSIOLOGY OF THE FOETUS

The placenta and metabolism of the foetus

Through the placenta the foetus obtains its nourishment; water, carbohydrates, proteins, fats, mineral salts, vitamins and oxygen, and through the placenta pass the waste products of the foetus, including carbon-dioxide and urea, to be excreted eventually by the lungs and kidneys of the mother.

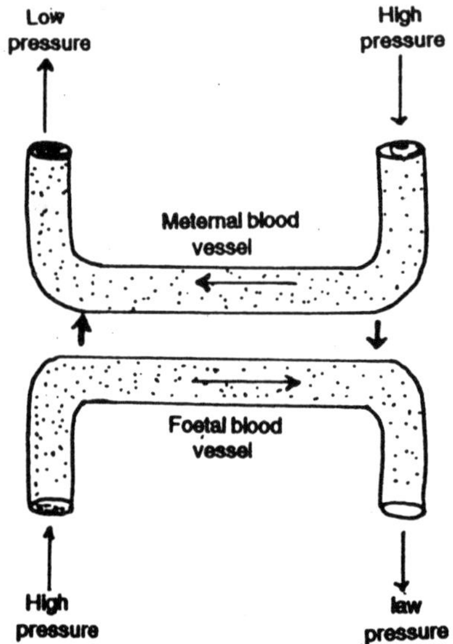

Figure 2.13 : Diagram to illustrate the principle of the arrangement of blood vessels in the placenta. The two short thick arrows indicate the direction of the exchange of materials.

The blood vessels of the foetus and mother do not join one another and all these substances must diffuse across the barrier between the two sets of blood vessels. There are three devices in the placenta which promote the passage of materials across the barrier between the two sets of vessels. We have already seen two of these devices: first the large surface area of the placenta provided by the branching villi and secondly the reduction in the thickness of the placental barrier with varies from species to species.

The third device consists in the arrangement of maternal and foetal blood vessels so that the maternal blood vessels containing blood at relatively high pressure are first adjacent to foetal vessels containing blood at low pressures, whilst foetal blood vessels containing blood at relatively high pressure lie adjacent to maternal vessels containing blood at low pressure: this arrangement encourages an efficient transfer of substances between the two series of blood vessel.

Respiration of the Foetus

The need for an increasing supply of oxygen by the foetus is somewhat anticipated by the maternal part of the placenta in that the maternal vascular bed in the uterus grows more rapidly than the foetal contribution to the blood vessels of the placenta. Thus, at an early stage in gestation the maternal blood leaving the placenta still contains large amounts of oxygen.

As the gestation period progresses the maternal blood leaving the placenta becomes progressively more de-oxygenated; the placenta reaches its maximum size at a time when the foetus is still growing rapidly, and at the end of the gestation period the supply of oxygen is critical.

The foetus is adapted in two main ways to obtain oxygen efficiently from the placenta. First there is the special arrangement of foetal blood vessel which has already been described above. Secondly, in many species foetal haemoglobin has somewhat different properties from adult haemoglobin, in that it takes up oxygen with a greater avidity and gives it up less readily. foetal haemoglobin can take up oxygen at partial pressures at which maternal haemoglobin would give up oxygen.

The foetus is a relatively inactive creature and can exist with haemoglobin which retains its oxygen more avidly than does adult haemoglobin. But once the animal is born it may need to use oxygen as rapidly as does an active adult, particularly in those species such as the horse where the young are very active at birth.

It has been found that in some animals, e.g. the goat, there is a gradual alteration of the properties of foetal haemoglobin produced towards the end of the gestation period to approximately those of the adult, so that the animal will be better adapted to an active life breathing air.

Carbohydrate metabolism

As in the adult, the foetus uses glucose as an important source of energy. Glucose can pass across the placenta. from mother to foetus in all animals. There may be persistent differences between the blood sugar level of the foetus and mother which may imply that the transfer of glucose across the placental membranes is not merely one of diffusion. In most mammals the foetal blood sugar tends to be lower than that of the mother but in some epitheliochorial placentae, e.g. pig, the blood sugar is higher on the foetal side.

In early foetal life the maternal side of the placenta acts as a temporary liver for the foetus in that it holds stores of glycogen at a time when the foetal live contains little or no glycogen. Later in the gestation period when the foetal liver has developed stores of glycogen, less in stored in the placenta. From experiments it has been found that the foetus exercises a strict glycogen economy, drawing upon glycogen stores only in emergencies.

Lipid metabolism

The foetus usually has good stores of fat. This may come from several sources. It may be able to synthesize fat from carboh-ydrates and amino acids. Some lipids pass across the placenta and become available for the synthesis of fat, but the mechanism of transfer across the placental membranes is not understood. Most of the fat fed to the mother and stored in her tissues does not pass unchanged across the placental barrier.

If the food fat is stained with the dye Sudan III than the mother's stores of fat are intensely stained red, although none appears in the foetus. Further evidence which indicates that fat is not transferred directly across the placental membranes is the different chemical constitution of foetal and maternal fat.

Protein metabolism

Foetal protein may be derived directly by transfer of intact protein molecules from the mother or by transfer of amino acids. Antibody proteins (gamma globulins) are known to be transferred intact from mother to foetus in some species (e.g. rabbit) and this confers on the foetus a passive type of immunity.

In other species much or all of the antibody proteins are received by way of the colostrum, the first pale secretion of the mammary glands of the mother. Amino acids probably do not pass across the placenta by simple diffusion and active transport mechanism may be involved.

Endocrine Function of the Placenta

In addition to serving the nutrition of the foetus the placenta is an endocrine organ producing several hormones in large quantities. Gonadotrophin is produced from the chorion and is called chorionic gonadotrophin.

In the human this hormone is produced in such large quantities by the end of the second month of pregnancy that a specimen of urine, in which the hormone is excreted, will cause the growth of the ovaries and ovulation when injected into a sexually immature mouse. Modifications of this effect are used in the early diagnosis of pregnancy.

When the urine is injected into a female toad called Xenopus, eggs are shed into the surrounding water within 24 hours. Even mature male amphibians have been used in pregnancy tests and injections of chorionic gonadotrophin are followed by a release of sperms, which have to be identified in the urine microscopically.

In addition to gonadotrophin the placenta also produces large amounts of oestrogen and progesterone. The exact significance of these various hormones is not well understood but they undoubtedly play their part in the development and maintenance of the sex organs during the pregnancy and prepare the mammary glands for their function after the birth of the young.

The high local concentration of progesterone at the placental site, by increasing the membrane potential of the uterine smooth muscle cells renders them less excitable. Thus the uterus is mechanically inactive and the foetus is safeguarded.

The circulation of the Foetus and Modifications at Birth

The circulation is adapted to life in utero, in which the placenta is the organ of nutrition and respiration; the foetal lungs and gastro-intestinal tract do not function as the respiratory and nutritive organs in utero and their blood supply is correspondingly small. But immediately at birth the placenta ceases to have these functions and there must be a rapid adjustment of the organism to an independent life.

The lungs are rapidly converted from semi-solid organs with a small blood supply into air filled organs with a.large blood supply and are responsible for the gaseous exchange of the whole organism. The gastro-intestinal tract takes over the nutritive functions of the placenta as the young animal begins to feed.

In order to make these transformations possible there must be special devices within the cardio-vascular system of the foetus so that blood which once went to the placenta can now be diverted to the lungs and gastro-intestinal tract. Foetal oxygenated blood is returned from the placenta by way of the umbilical vein which passes to the liver where it joins the hepatic portal vein.

Some of the oxygenated blood goes to the liver but most of it is shunted away through a connection of the umbilical vein with the inferior vena cava called the ductus venosus, a feature only of the foetal circulation. In the inferior vena cava the oxygenated blood mixes with venous blood draining from the lower part of the body.

When this blood reaches the heart most of it is shunted from the right auricle through an opening in the septum between the two auricles, into the left auricle. The opening in the septum, called the foramen ovale, is a special feature of the foetal heart and it censures that the oxygenated blood returning from the placenta is diverted away from the right ventricle and lungs to supply the head (i.e. central nervous system) and upper limbs of the foetus by way of the left ventricle and aorta.

This flow of blood has been confirmed using X-ray studies of the foetus after radio-opaque material has been injected into the circulation. The blood is diverted from the right auricle into the left auricle by a valvular arrangement in the wall of the right auricle.

The venous blood which drains into the right auricle from the head and neck of the foetus by way of the superior vena cava passes through the right auricle into the right ventricle, from which it passes out into the pulmonary artery.

But since the foetal lungs are not functioning as organs of gaseous exchange, only a small amount of blood is needed, sufficient to meet the metabolic needs of the tissues in the lung and much of the blood in the pulmonary artery passes directly into the aorta by way of a connection called the ductus arteriosus. This blood supplies the lower part of the body, and much of the it passes into the umbilical arteries (branches of the internal iliac arteries) supplying the placenta with deoxyge-nated blood.

There are thus four special features of the foetal circulation:

1. The placental circulation.
2. The ductus venosus which shunts blood from the umbilical vein away from the liver into the inferior vena cava.
3. The foramen ovale through which oxygenated blood returning to to the heart via the inferior vena cava passes into the left auricle and so to supply the upper part of the body.
4. The ductus arteriosus which shunts blood from the pulmonary arch into the aorta, so by-passing the lungs.

It will be seen that the right ventricle of the foetus, pumping blood to the lungs and to the lower part of th body and placenta, performs more work than the left ventricle. This position is reversed after birth.

At birth there are dramatic changes in this circulation. The umbilical vessels become increasingly irritable towards the end of the gestation period and with the physical stimuli of birth these vessels go into spasm, thus excluding the placental circulation from the foetus. This means that when the mother bits through the umb-ilical cord to free the young animal it does not bleed to death.

The ductus venosus also goes into spasm so that blood in the hepatic portal vein now has to pass through the tissues of the liver and cannot be directly diverted into the inferior vena cava. With the expansion of the lungs at birth larger amounts of blood pass into the lungs from the plumonar arch, and the wall of the ductus arteriosus contracts so that blood in the pulmonary arch can no longer be diverted into the aorta.

With the increased supply of blood to the lungs there is an increasing amount of blood returning to the left auricle and the pressure of blood in the left auricle rises. This rise in pressure pushes a loose flap of tissue against the foramen ovale, closing the connection between the left and right auricles. The adult type of circulation is now achieved.

The ductus arteriosus, ductus venosus and umbilical vessels, initially closed by muscular spasm are gradually permanently obliterated as fibrous tissue grows in the lumen of the vessel, and the flap of tissue which is closing the foramen ovale gradually fuses with the septum.

Parturition

At the end of the gestation period parturition occurs when the young are expelled from the uterus into the outside world by means of powerful intermittent contractions of the myometrium

What initiates the process of parturition is not known but it has been suggested that the declining production of the hormone proge-sterone, which has an inhibitory effect on uterine motility, sensitizes the uterus to the posterior pituitary hormone oxytocin, which stimulates intermittent powerful contractions of the uterus.

In cases of slow labour associated with weak uterine contractions, injections of oxytocin are used to produce more powerful uterine contractions.

In those species in which there is an intimate mingling of foetal and maternal tissucs in thc placenta the uterine contractions not only propel the young through the birth canal but they serve to separate placenta from the uterine wall, and after birth of the young the persistent contraction of the uterine muscles serve to close the blood vessels which have been torn upon during the separation of the placenta.

Lactation

The mammary glands have been prepared for their function during the pregnancy by the effect of the sex hormone produced by the placenta. As mentioned previously oestrogen causes a growth of the duct system of the glands and progesterone initiates the development of the glandular elements around the ducts.

Btu the glands are unable to produce milk during pregnancy because the large amounts of oestrogen inhibit the pituitary gland from producing a hormone vital for lactation, the lactogenic hormone. Towards the end of pregnancy the oestrogen production by the placenta gradually falls, and after parturition, when the placenta is expelled, the level of oestrogen in the body falls still further Lactogenic hormone is now produced by the anterior pituitary gland and lactation starts soon after parturition.

The first pale secretions called colostrum are rich in proteins, and in some species they contain antibodies, which serve to protect the young for some months until they have been broken down by the body.

During the lactation period milk is secreted by the glandular cells of the mammary gland continuously.

Milk, however, only passes along the larger ducts that open onto the nipple during suckling. When the infant mammal grasps the nipple in its mouth stimulation of sense organs in the nipple cause a reflex liberation of the hormone oxytocin from the posterior pituitary gland.

This hormone reaches the mammary gland by way of the blood stream. On reaching the mammary gland the hormone stimulates certain contractile cells surrounding the glandular cells causing an emptying of milk into the larger ducts within open onto the nipple. Milk now appears from pores on the nipple and the young, by compressing the dilated ducts which underly the nipple increase the flow of milk.

2

ORIGIN OF GERM CELLS

Developmental biology is a multidisciplinary science concerned with analyzing the progressive acquisition of specialized structure and function by organisms and their various components. The multidisciplinary approach to the study of development first emerged before the turn of the century as an integration of embryology with cytology and later with the new science of genetics.

Literally, *embryology* means "the study of embryos," but the word cannot be rigidly defined, because it implies a point of view and is frequently used to connote the study (descriptive or experimental) of changes in the form or shape (*morphogenesis*) of animals during their embryonic phase.

The cytologists of the late 1800s saw the development of the embryo as a manifestation of the changes occurring in the individual cells that compose the embryo. They believed that the cell is the key to all ultimate biological problems and that the fundamental principles underlying development would emerge by studying the properties and behavior of cells.

Every multicellular organism begins life as a single cell, the fertilized egg, which is essentially similar to all other cells but differs by its potential to divide and produce all the cells of the body. The cells diverge from one another structurally and functionally and become organized into an adult organism.

Accordingly, these basic processes of development must first be understood at the cellular level before an overall understanding of development is possible.

Since the fertilized egg (or zygote) is derived from fusion of the male and female gametes, the cytologists assumed that the factors that control development could be identified by tracing the elements of the gametes and following their behavior in the cells of the developing organism.

The foremost proponent of the cytological approach was E.B. Wilson, and the most eloquent statements of this philosophy were contained in the three editions (1896, 1900, and 1925) of his classic, *The Cell in Development and Inheritance*. He wrote in the first edition:

Every discussion of inheritance and development must take as its point of departure the fact that the germ is a single cell similar in its essential nature to any one of the tissue-cells of which the body is composed. That a cell can carry with it the . . . heritage of the species, that it can in the course of a few days or weeks give rise to a mollusk or a man, is the greatest marvel of biological science.

In attempting to analyze the problems that it involves, we must from the onset hold fast to the fact . . . that the wonderful formative energy of the germ is not impressed upon it from without, but is inherent in the egg as a heritage from the parental life of which it was originally a part. The development of the embryo is nothing new.

It involves no breach of continuity, and is but a continuation of the vital processes going on in the parental body. What gives development its marvelous character is the rapidity with which it proceeds and the diversity of the results attained in a span so brief.

Thus, Wilson understood that the characteristics of the organism emerge during development by utilization of inherited information, and the only way to understand development fully is to comprehend the nature of that information and the ways in which it is utilized. These concepts have had a rocky history, both before and after Wilson's time.

GERM CELLS: BRIDGING THE GENERATION GAP

The idea that the form of an embryo gradually emerges during development originated with Aristotle. However, nearly 2000 years later, in the seventeenth and eighteenth centuries, many embryologists rejected this concept and proposed that the egg contains a miniature, fully formed embryo. Development to them was analogous to the unfolding of a flower bud. Bonnet formalized this concept in the theory of *emboitement,* or encasement. He stated that since the egg contains the complete embryo, it must also contain similar preformed eggs for all future generations, like an infinite series of boxes encased one inside the other.

Others of his contemporaries believed that a preformed embryo exists inside the sperm, and many microscopists of the time claimed to have seen a tiny creature (homunculus) curled up in the sperm head.

These theories of *preformation* were not universally held, however. Caspar Wolff championed the theory of an alternative mechanism of development-epigenesis. Epigenesis means that the adult gradually develops from a rather formless egg as originally proposed by Aristotle.

Wolff made careful observations on the development of the chick and showed that the early embryo is entirely different from the adult and that development is progressive, with new parts being continually formed. Wolff's conclusions were rejected by most of his contemporaries, and the concept of epigenesis was not accepted by a majority of biologists until the early part of the nineteenth century.

Even after epigenesis was determined to be the mechanism for the formation of the external structure of the organism during development, the nature and the significance of the germ cells remained unclear for nearly a century.

The fact that the egg itself is a cell was recognized by Schwann in 1839. Likewise, the cellular nature of sperm was determined in 1865 by Schweigger-Seidel and St. George. Another decade elapsed before Oscar Hertwig (1876) established that fertilization results from the union of the egg and sperm. Thus, each sex contributes a single cell from its own body.

These cells, which carry the complete set of instructions for production of another generation similar to the preceding one, fuse to form a single cell, the zygote, which undergoes division or *cleavage*, to produce the cells that form the embryo.

Within the fertilized egg, two nuclei, one derived from the egg and the other derived from the sperm, combine to form a single zygote nucleus, which gives rise by division to all nuclei of the body. In contrast to the equal nuclear contribution to the zygote made by both gametes, the cytoplasmic contributions are decidedly uneven; the egg contributes virtually all of the cytoplasm.

These observations led Hertwig to the conclusion that the germ cell nuclei, and not the cytoplasm, are the vehicles of inheritance.

Hertwig's discoveries marked the beginning of a new era of

Figure 2.1 : Homunculus in human sperm.

investigation into the role of the nucleus in fertilization and development. Much of the attention of cytologists during this era was directed to the primary nuclear constituents, the *chromosomes*.

The first detailed description of the behavior of the chromosomes in fertilized eggs was by van Beneden (1883), who used the nematode, *Ascaris megalocephala*. This species was a splendid choice for chromosome study, since it has only four chromosomes, which are large and stain intensely.

Van Beneden observed that the egg nucleus and the sperm nucleus each provides two of the four chromosomes that align on the metaphase plate at the first cleavage division. Each chromosome splits lengthwise, and the daughter chromosomes are transported to the opposite poles of the spindle where they are incorporated into the nuclei of the two-celled stage. Therefore, each of these nuclei receives an equal number of maternal and paternal chromosomes.

Soon after publication of van Beneden's work on *Ascaris,* reports based upon studies using several species of animals and plants established the rule that each gamete nucleus contributes one half the number of chromosomes that is characteristic of somatic cells.

The behavior of chromosomes at fertilization led Hertwig and three other German scientists-Strasburger, Kolliker, and Weismann-to conclude independently that chromosomes are the means for transmission of inherited information. As Wilson stated in the 1896 edition of *The Cell,* these observations on chromosomes during fertilization could not logically lead to any other conclusion:

> These remarkable facts demonstrate the two germ-nuclei to be in a morphological sense precisely equivalent, and they not only lend very strong support to Hertwig's identification of the nucleus as the bearer of hereditary qualities, but indicate further that these qualities must be carried by the chromosomes; for their precise equivalence in number, shape, and size is the physical correlative of the fact that the two sexes play, on the whole, equal parts in hereditary transmission. And we aTe finally led to the view that chromatin is the physical basis of inheritance.

THE ROUX-WEISMANN THEORY

The genetic material was beginning to take on a definite physical form in the theories of biologists. However, the absence of hard information about the nature of genetic material and its function in development led to speculation that was based on logic that Wilson characterized as bordering on the metaphysical.

Wilson reserved most of his contempt for a theory of development that originated with Wilhelm Roux in 1883. Roux believed that the hereditary material represents different characteristics of the organism. He assumed that the fertilized egg receives all of these substances, which, as cell division ensues, become linearly aligned on the chromosomes.

The substances are then distributed unequally to daughter cells, Roux proposed. This "qualitative division" fixes the fate of the cells and their descendants, since a portion of determinants is lost to a cell at each division.

August Weismann later elaborated on Roux's ideas. Weismann developed a highly complex scheme for the hereditary material, or *germ plasm*. The primary hereditary units were biophores, which aggregated to form determinants, the determinants to form ids, the ids to form the larger idants, which were equivalent to chromosomes.

He believed that there are two kinds of division: qualitative and quantitative. Weismann proposed that the id gradually disintegrates during development, splitting into smaller and smaller groups of determinants, which are isolated into different daughter cells during cell division. Finally, only one kind of determinant remains in a particular cell or group of cells.

The determinant then breaks up into its constituent biophores and imparts specific characteristics on the cell. To account for the formation of germ cells that must contain the entire set of heritable information, Weismann proposed that the germ cell line is set aside early in

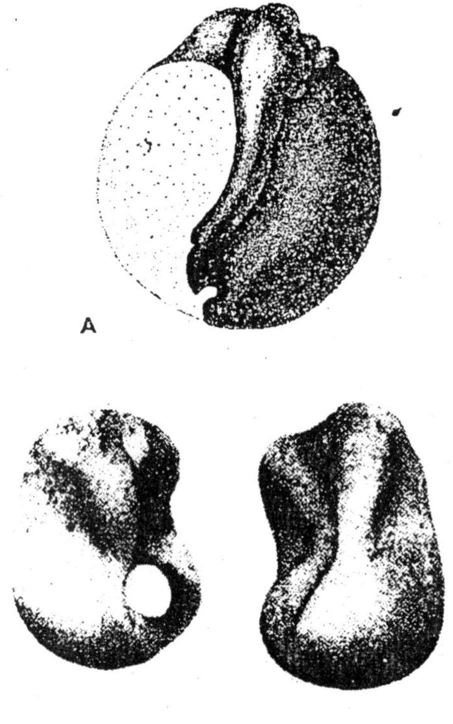

Figure 2.2 : Developmental potential of half embryos of the frog. A, Half embryo produced by Roux following destruction of one blastomere with a hot needle. B, *Two whole embryos produced by Schmidt following ligation at the two-cell stage.*

development by quantitative, rather than qualitative division. By equal distribution of nuclear constituents, the germ plasm remains intact.

This separate developmental pathway for the germ cells was an important concept. Weismann astutely concluded that although the body is derived from the germ plasm, the germ plasm itself is passed on without modification (the concept of mutation was unknown at that time) from one generation to another.

Hence, he concluded that inheritance of acquired characteristics, as proposed by Lamarck, was impossible. The usefulness of this concept far outlived that of the Roux-Weismann theory of qualitative division, which was soon disproven.

For his part, Roux appeared to have confirmed his theories by an experiment he conducted on frog eggs. He used a hot needle to destroy one of the two cells that result from the first division after fertilization. In some cases the uninjured half continued developing to form a half-embryo that lacked structures corresponding to the side that was damaged. Roux concluded that this result supported his theories, since the undamaged half appeared to lack the information that was necessary to produce the other half of the embryo.

Hans Driesch approached the problem differently with sea urchin embryos, dissociating them by mechanical shaking at the twocell stage. These half-embryos developed into normally formed dwarf larvae. Driesch subsequently modified his technique, separating the cells in calcium-free seawater, and found that isolated cells at the four-cell stage also develop normally. Thus, Driesch concluded that each cell retains all the developmental potential of the zygote.

The conflict between these two opposing views of development has been settled in favor of Driesch's interpretation by numerous cell separation experiments on several animal species. These have included experiments on the frog embryo, the same kind of embryo that Roux used. When the cells of the two-cell frog embryo are separated with a ligature, both halves proceed to develop normally.

The experiment conducted by Roux illustrates the importance of proper experimental design. Roux had introduced an artifact into his experiment by allowing the damaged half of the embryo to remain attached to the uninjured half, interfering with its development.

This fact was demonstrated by Hertwig, who repeated Roux's experiment, but instead of leaving the damaged cell attached to the normal half, he removed it. Development of the remainder of the embryo resulted in a complete, half-sized embryo. The interference caused by the death of one cell in Roux's original experiments was only temporary, for Roux himself admitted that the missing half of the damaged embryo was restored during later development, indicating that the frog embryo has the ability to regulate its development by formation of missing parts.

Roux should have realized that this observation contradicted his hypothesis. He instead attempted to rationalize the result. Perhaps this was his greatest failure. Wilson dismissed Roux's rationalization as "an artificial explanation."

Roux's work did have one major lasting effect on developmental biology, however: the *experimental* approach to development. For the first time, an embryologist had manipulated embryos and observed the effects of these manipulations on them. For this reason, many embryologists consider Roux to be the "Father of Experimental Embryology."

THE MENDELIAN ERA

It was now clear that in cell division the inherited information is equally distributed to all cells of the embryo. But the central question still remained: How does the inherited information participate in development? Frustration began to replace the optimism generated by the discoveries of the significance of the nucleus and chromosomes in inheritance. The year 1900 was the turning point.

During that year, the paper that Mendel had presented to an unreceptive audience in 1865 was discovered. But now the scientific world was more receptive, for scientists needed new concepts. Mendel's major new insight was that characteristics of organisms are determined by factors that retain their identity through generations of breeding. Each individual inherits a single factor for a trait from each parent.

If the parents have "antagonistic" characters, the offspring (hybrid) displays the unaltered trait of one parent, while the trait of the other parent is missing. However, if the hybrid generation is bred with itself, the previously missing secondary character could reappear unchanged in the next generation.

Mendelism did not immediately take the scientific world by storm, since existing ideas about heredity were often held tenaciously. William Bateson, a Cambridge biologist, was one of the most active proselytizers of the new faith, arguing convincingly and passionately in favor of Mendelism.

Bateson coined the term "genetics" and worked hard to establish this new biological discipline. Gradually, nebulous terms such as "biophores" and "idants" were to be replaced with genes and linkage groups. Mendelism made possible a new precision in thought, and soon the modern concepts of genetics were formulated.

One other experimental observation was also important in setting the stage for this new era. This was a paper written by Theodor Boveri and published in 1902. Although published two years after the rediscovery of Mendelian inheritance, Boveri was apparently unaware of Mendelism. Boveri's work established more precisely the role of chromosomes in development.

His experiments utilized sea urchin embryos with variable numbers of chromosomes (a condition that today is called *aneup-loidy*). This abnormal situation can be produced when sea urchin eggs are fertilized by two sperm. The frequency of the *dispermy* is increased by adding an excess number of sperm when fertilizing sea urchin eggs *in vitro.*

The two sperm nuclei and the egg nucleus form an abnormal division figure, usually with four centrioles, so that the first mitotic division produces four cells that rarely receive the normal diploid number of chromosomes. Since the distribution of chromosomes to these four cells is random, cells with variable numbers of chromosomes are produced from a single zygote.

Boveri separated the four cells from one another by placing the dispermic egg in calcium-free seawater and observed the development of each. Subsequent cleavage divisions were normal, thus fixing the abnormal chromosome number in the embryos developing from the separated cells. Never did these embryos develop normally.

But *most importantly,* each typically developed in a different way, arresting at a different stage. Boveri concluded that the different developmental patterns were due to the different chromosomal combinations produced in the abnormal first division. Thus, normal development is dependent upon the *normal combination of chromosomes.* Clearly, each chromosome must have qualitatively unique effects on development.

That conclusion seems self-evident to us today-each chromo-some contains a linear sequence of genes, and the gene sequences on the chromosomes are unique. But only a minority of biologists of the early 1900s (Wilson among them) accepted the conclusion that the hereditary factors are on the chromosomes. In 1914, Wilson presented a lecture in London in which he gently chided those who did not accept the chromosome theory:

I am well aware that some eminent students of genetics are still reluctant to accept this theory, at least in its more detailed applications. I am not disposed to reproach them for such scepti-cism. The cytologist suffers under the disadvantage of working in so unfamiliar a field that some of his conclusions, even among those most certainly established and most readily verifiable, are apt to give a certain expression of unreality, even to his fellow naturalists.

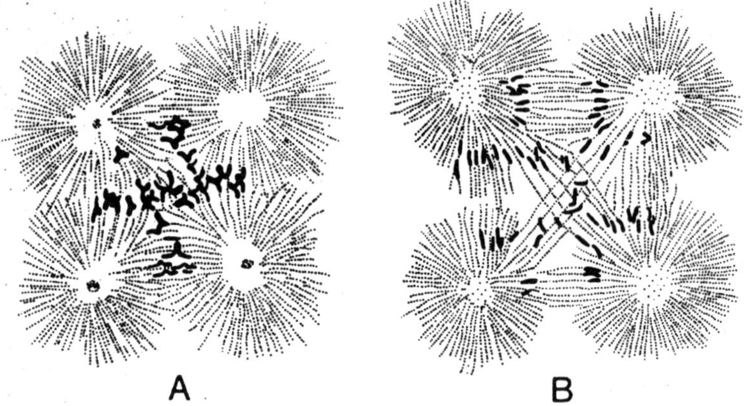

Figure 2.3 : Mitosis in the first cleavage of dispermic sea urchin eggs. A, metaphase; B, *anaphase. Each sperm contributed two centrioles, each of which attracts chromosomes at random.*

It is undeniable, too, that in this subject, for better or for worse, hypothesis and speculation have continually run far in advance of observation and experiment. It is quite possible that some of my hearers may consider some of the views I have touched upon as a fresh illustration of that fact.

If so, I beg them to bear in mind that no conclusion which I have considered has been reached as a merely logical or imaginative construction. I have endeavoured to limit myself to matters of observed fact, and to conclusions that are either demonstrated by facts or directly and naturally suggested by them.

To those who have had opportunity to come into intimate touch with both cytological and genetic research the conclusion has become irresistible that the chromosomes are the bearers of the "factors" or "gens" with the investigation of which genetics is now so largely occupied.

The evidence for the chromosome theory was based on both cytological and genetic data. Although Wilson enthusiastically promoted Boveri's conclusions, they were dependent upon indirect evidence; more convincing proof was needed. Walter Sutton, a graduate student working with Wilson, was the first to make a clear correlation between the behavior of Mendel's genetic factors and the behavior of chromosomes. Sutton pointed out that the parallel behavior of hereditary factors and chromosomes must mean that the factors are localized on the chromosomes.

Thus, in a haploid gamete, genes and chromosomes are singly represented. Following fertilization the cells of the diploid zygote contain a set of chromosomes that are derived from each parent; the chromosomes are present in homologous pairs, and the members of each chromosome pair possess the alleles, which specify the maternal and paternal traits, respectively.

During the formation of gametes in the mature diploid organism, the homologous chromosomes pair, and as a result of meiosis, every gamete receives one chromosome of each homologous pair (Mendel's law of segregation). Furthe-rmore, since the behavior of each homologous pair of chromosomes during meiosis is independent of every other pair, the chromosomes are distributed randomly to the gametes, resulting in several combinations of maternal and paternal chromosomes (Mendel's law of independent assortment).

Sutton also noted that if two genes were located on the same chromosome, they would be inherited together and would not behave according to Mendelian principles. He had, therefore, foreshadowed the concept of linkage.

The final proof of the chromosome theory of inheritance resulted from experiments conducted in the laboratory of Thomas Hunt Morgan, utilizing the fruit fly, *Drosophila melanogaster.* Mostly as a result of work by Morgan and his associates A. H. Sturtevant, C. B. Bridges, and H. J. Muller, the basic concepts of transmission genetics were rapidly discovered. These concepts have been summarized by Moore. They are:

1. Inheritance is the transmission of genes from parents to offspring.
2. Genes are located on chromosomes.
3. Each gene occupies a specific site (locus) on a chromosome.
4. Each chromosome has many genes that are arranged in linear order.
5. Somatic cells of diploid organisms contain two of each kind of chromosome (homologues), and thus each gene locus will be represented twice.

6. During each mitotic cycle, each gene is replicated.
7. Genes can exist in several alternative states (alleles). The change from one state to another is a mutation.
8. Genes can be transferred from one homologous chromosome to the other by crossing-over during meiosis.
9. Every gamete receives one chromosome of each homologous pair; the distribution of chromosomes to the gametes is random.
10. The distribution of chromosomes of one homologous pair to the gametes has no effect on the distribution of chromosomes of other pairs.
11. At fertilization, gametes unite randomly; the zygote receives one chromosome of each homologous pair from its father and one from its mother.
12. When the cells of an organism contain two different alleles of the same gene (heterozygous condition), one allele (the dominant) usually has a greater phenotypic effect than the other (the recessive).

The choice of *Drosophila* by Morgan was indeed a fortunate one. These remarkable creatures are tailor-made for genetics research: They are easily bred and reproduce rapidly (a single pair can produce several hundred progeny in a couple of weeks), and large numbers of them can be maintained in small vials or bottles containing simple inexpensive "fly food." Although *Drosophila* are small, their external characteristics are simple to score with no more than a hand lens, which Morgan used in his early work.

Geneticists and cytologists were soon to discover other qualities that made *Drosophila* a nearly ideal organism for genetic research. For example, the haploid chromosome number is four, meaning only four linkage groups. Furthermore, certain larval cells have giant chromosomes called *polytene chromosomes*.

These structures were described in the larval salivary glands by T. S. Painter in 1934. The salivary gland chromosomes are about 100 times longer and thousands of times wider than chromosomes of normal body cells. Actually, Painter had "rediscovered" this; it was known to the cytologists in 1881, and a drawing of polytene chromosomes can be found in the *1896* edition of *The Cell.*

Although one would expect salivary gland nuclei to contain eight chromosomes, Painter found only four. This is due to pairing of homologous chromosomes. The most important aspect of polytene chromosomes is the presence of cross-bands on them, as shown in Painter's first drawing of salivary gland chromosomes.

The cross-banding extends across both paired chromosomes; the pairing of the homologues is so exact that each band appears as a single structure. But for the early *Drosophila* geneticists, the significance of polytene chromosomes was that chromosome markers were now available.

The bands are in various shapes and sizes, and interband distances are variable. However, in different cells of the same larva, and even in different larvae, the band pattern itself is constant. The pattern apparently represents the essential organization of the chromosome, and as was soon discovered, it marks the organ-ization of genes within the chromosomes.

Once again, cytology and genetics joined forces, and a new branch of genetics called *cytogenetics* emerged. For the first time, the concepts of transmi-ssion genetics, such as gene deletions and duplications, transloc-ations, and inversions, could be correlated with chromosome markers. These correlations have resulted in very elaborate chromosome maps showing genes in a linear order and in definite sequence.

Genes have been localized to a portion of the chromosome within a single band, and genes have been detected only in regions containing bands. These observations led to the hypothesis that bands represent single gene loci. Although this hypothesis has been shown to be an oversimplification, it is still thought to be generally valid.

EMBRYOLOGY: LOSING THE FAITH

One of the most important concepts to emerge from Wilson's writing concerns the nature of the relationship among genes, cells, and the developing embryo. Gene function determines cellular characteristics in a process called *cell differentiation*.

The characteristics of the whole organism are, in turn, the net result of the characteristics of its individual cells. Thus, genes function at the cellular level to cause development of the organism, or as stated by Wilson, development is "the appearance of hereditary traits in regular order of space and time."

What is the specific role of the genetic material in this process? How is temporal and spatial coordination of cell differentiation regulated by the nucleus? Wilson pointed out that the nucleus itself does not cause development. Although the nucleus alone suffices to transmit inherited information, it must work in concert with the cytoplasm. Wilson proposed

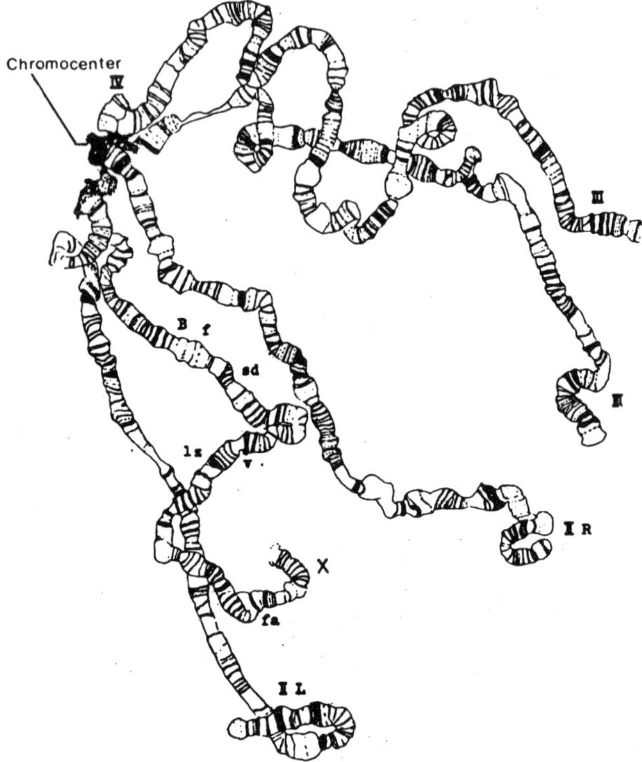

Figure 2.4 : Painter's first drawing of the salivary gland chromosomes of Drosophila melanogaster. The chromosomes radiate out from the chromocenter. The X is attached to the chromocenter by one end so it appears as a single long structure. Both the II and III chromosomes are attached by their middle portions. Consequently, both of these chromosomes have two arms extending from the chromocenter. The tiny IV chromosome is attached by its end to the chromocenter. The approximate location of several X chromosome genes (B, f, sd, etc.) is shown.

that the specific role of the nucleus is the regulation of "constructive metabolism" (i.e., synthetic metabolism) of the cell, which is fundamentally a problem of biochemistry and was well beyond the level of understanding of scientists of that time.

Wilson and his contemporaries had expected that genetics and cytology would together lead the way to a solution of the "how" and "why" questions of development. However, Wilson admitted that they had underestimated the magnitude of the problem of development.

Just as the questions of heredity required the Mendelian principles and the chromosome theory, explanation of the mechanisms of development would require the understanding of new, and as yet undiscovered, principles. Studies on the role of genes in development could not progress significantly until the nature of the gene was discovered and advances were made in understanding how genes function to influence cell biochemistry. An era had ended.

The embryologists' disillusionment with genetics was complete. Many embryologists considered genetics to be peripheral to the fundamental mechanism of development. They believed that the basic structure of an embryo is produced by *embryological* mechanisms and that the only function of genes is to add the nonessential finishing touches, such as eye and hair color, number of bristles on a leg segment, and color of flower petals.

Emphasis during the 1920s, 1930s, and 1940s was on the descriptive and comparative aspects of embryonic development. Experimentation was concentrated on the interactions and rearrangements of the cells and tissues that form embryos. Chemical analyses of embryos were largely devoted to descriptions of the chemical components of embryos, comparisons of chemical components among embryos of different taxonomic groups, and attempts to establish the chemical basis for embryonic induction.

The importance of this era of classical embryology should not be underestimated. It has provided us with a vast amount of information about embryonic development, without which cellular differentiation is an isolated phenomenon that loses much of its significance. The student of development should remain aware that cell differentiation is a process that occurs during a phenomenon of much greater consequence-the formation of a living, functioning adult organism.

GENETICS: KEEPING THE FAITH

Since very few embryologists concerned themselves with gene function in development, it became incumbent upon geneticists to keep alive the concept that development occurs under genetic control. That this was truly a "rear guard action" can be deduced from this quote by Richard Goldschmidt (1958): "In spite of such isolationism or possessiveness [by embryologists], the geneticists will continue to worry about the problem of genetic action and take the risk of climbing over the fence erected by some jealous embryologists, who, while claiming the kingdom for themselves, do not set out to till its soil."

It is not surprising that geneticists became strong advocates of the idea that genes play an important role in basic developmental processes, because the new mutations they were discovering provided strong evidence

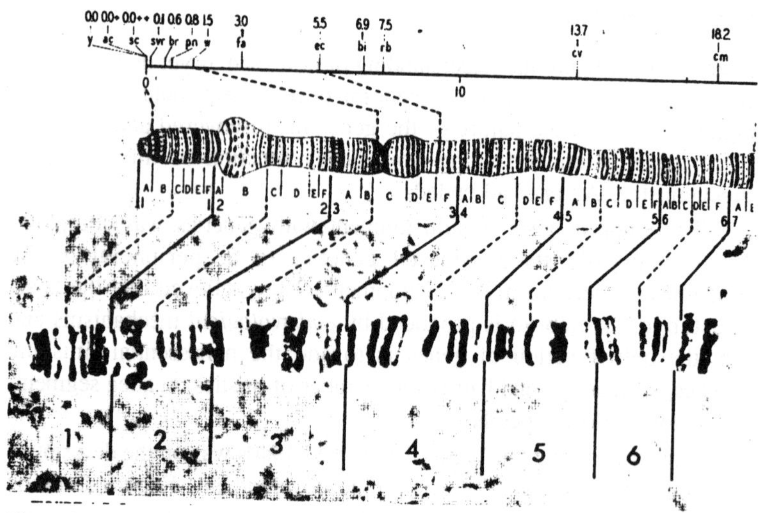

Figure 2.5 : Photographic map of a portion of the Drosophila X *chromosome, correlated with Bridges' (1935) drawings. The cytological positions of several of the common sex-linked genes are indicated.*

for this fundamental principle. In *Drosophila,* for example, there are mutants that affect virtually every aspect of development from the most general down to the smallest detail. In fact, mutants can even alter the basic body plan that characterizes *Drosophila* as a fly belonging to the insect class Diptera. The body of an adult dipteran consists of a head, thorax, and abdomen.

The thorax is made up of three segments: the *prothorax*, the *mesothorax*, and the *metathorax*. Each of these segments bears a pair of legs; in addition, the mesothorax bears the wings, and the metathorax has a pair of rudimentary appendages called *halteres*. The halteres are balancing organs that maintain the fly in an upright position while flying.

The combination of two wings and two halteres is unique to the dipterans; most insects have four wings-one pair on the mesothorax and a second pair on the metathorax. A mutation called *bithorax (bx)* converts *Dros-ophila* from a typical dipteran to a four-winged creature by produ-cing a pair of wings on the metathorax in place of the halteres. Combination of *bithorax* with certain other mutant genes (e.g., *postbithorax* and *ultrabithorax)* can change the metathorax so that it resembles a second mesothorax.

These genes can apparently bring about a major switch when the body plan is being laid out, causing the group of cells that normally form a metathorax to form mesothoracic structures instead. The completion of any organ from the initial structure formed during organization of the basic body plan to the definitive organ is under genetic control at virtually every step.

Literally hundreds of genes have been discovered that affect the detailed development of *Drosophila* body parts. The detection of a mutant gene affecting a developmental process implies that the normal allele of the gene is involved in control over that developmental step. The nature of the defect often indicates which developmental process is affected.

The sequence of gene-controlled steps in development of a body part can be reconstructed by careful analysis of a series of mutants that affects its development. Analysis of development in this manner is called *genetic dissection.*

PHYSIOLOGICAL GENETICS

Many geneticists who were interested in development during the 1930s and 1940s were asking how genes produce their effects. The most prominent among these scientists was Richard Goldschmidt. Although most of Goldschmidt's specific ideas about the role of genes in development have been disproved, he served an important role by drawing attention once again to this basic problem. A new field emerged-physiological *genetics.*

As we have already noted, biologists had long suspected that genes are involved in regulating cellular metabolism. Therefore, the discovery of the specific function of genes could be found only by utilizing a biochemical approach.

New insight into the nature of cellular metabolism began emerging from biochemical studies at approximately the same time that the new science of genetics was making its rapid discoveries about the mechanism of inher-itance.

These critical biochemical discoveries involved the role of enzymes in metabolism and the concept of intermediary metabolism. Biochemical reactions in cells were found to require specific enzymes that temporarily combine with the reactants and accelerate their conversion.

The synthesis or degradation of complex molecules was found to occur by elaborate metabolic pathways in which a series of consecutive reactions (each contro-lled by a specific enzyme) are linked in sequence. The study of these complex pathways (intermediary metabolism) required the development of techniques to study the rapid intermediate reactions. Poisons were found that would destroy or inhibit specific enzymes.

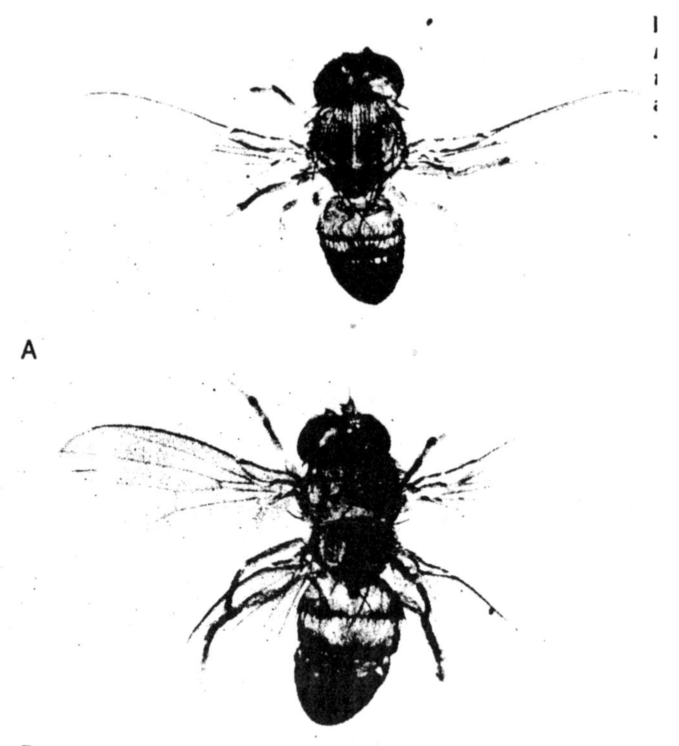

Figure 2.6 : Photographs of wild-type (A) and bithorax (B) Drosophila. The fly in B has the genotype bx^3pbx/Ubx^{105}. Code: bx, bithorax; pbx, postbithorax; Ubx, Ultrabithorax.

Consider the following sequence of reactions:

$$A \xrightarrow{\text{Enzyme 1}} B \xrightarrow{\text{Enzyme 2}} C$$

If the action of enzyme 1 is prevented, B and C will not be formed, but A may accumulate so that it can be detected. Similarly, if the action of enzyme 2 is prevented, C will not be formed and B will accumulate. In this way the intermediates formed during metabolism could be detected and the complex pathways clarified.

How could genes be involved in regulating these pathways? Could the regulation be mediated by the enzymes? This possibility was first suggested by an English physician, Sir Archibald Garrod. He was interested in inherited metabolic diseases of humans, which he called "inborn errors of metabolism."

In 1909, Garrod published a book in which he discussed the rare disease alkaptonuria. The disease is detected early in life, since the urine of alkaptonuric babies stains their diapers black. The staining results from the presence in the urine of homogentisic acid (or alkapton), which blackens upon exposure to the air. Homogentisic acid is normally converted to acetoacetic acid.

Garrod proposed that this conversion is accomplished by an enzyme, and this enzyme is missing in the affected individual. The effect is similar to that of a metabolic poison-the metabolic conversion is blocked, resulting in the buildup of homogentisic acid, which is excreted.

Garrod also observed that parents of affected children were often first cousins, and he speculated that the disease is inherited. He consulted the geneticist Bateson, who concluded that alkaptonuria is inherited as a simple Mendelian recessive gene.

Thus, Garrod correlated the metabolic defect with the mutant gene. He did not categorically state that the mutant gene resulted in the absence of the enzyme, but he certainly implied it. Unfortunately, Garrod's work lay dormant in the medical literature and was ignored by a generation of geneticists.

Serious experimental analysis of the role of genes in metabolism began in the 1930s. The initial efforts utilized mutant genes that had an obvious effect on metabolism: eye color mutants of *Drosophila.* The modification of color by mutation suggested that the genes were involved in some way in the synthesis of the pigments.

The task was to identify the pigments, clarify the synthetic pathways, pinpoint the metabolic block in the mutant, and thus establish the relationship between the block and the mutant gene. Several mutants were known to affect the production of brown eye pigment. George Beadle and Boris Ephrussi reasoned that some of these mutants must be involved in regulating the various steps in the synthesis of brown pigment from its precursor.

Through a series of ingenious experiments, Beadle and Ephrussi identified the genes affecting the synthetic pathway and established the sequence in which the mutant genes blocked the synthesis of the brown pigment. As we have seen, the blockage of a synthetic step implies that the enzyme for that conversion is missing.

If the mutation removes the enzyme, it stands to reason that the function of the normal allele of that gene is to produce the enzyme. Therefore, they suggested the hypothesis: *one gene-one enzyme*, that is, the primary function of a gene is to produce a specific enzyme. They could not, however, test their hypothesis on *Droso-phila,* since they were unable to unravel the pathway for brown pigment synthesis. It was

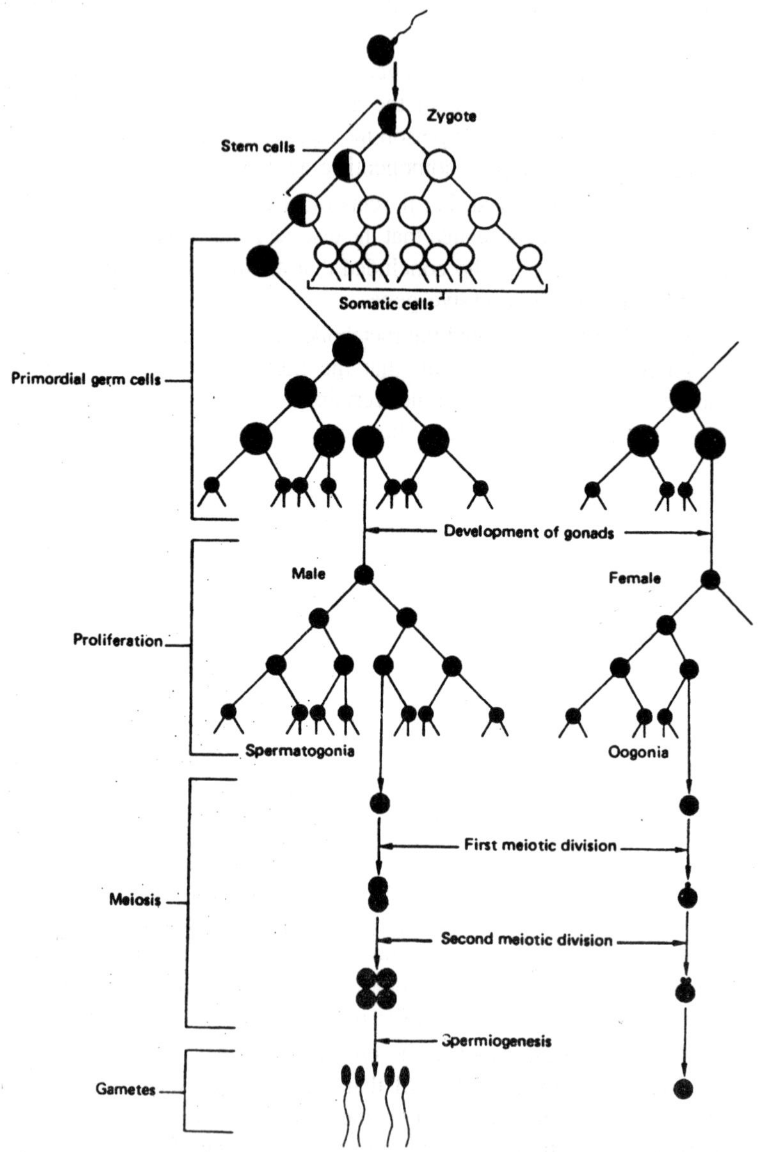

Figure 2.7 : Origin and fate of the male gametes (left) and the female gametes (right). The number of mitotic divisions is larger than depicted.

impossible to say with certainty that a gene specifies the enzyme that catalyzes a reaction, because they did not know what the reaction was, let alone which enzyme was involved.

A new approach to gene function was needed. Beadle, working with Edward Tatum, decided to reverse the usual procedure. Instead of beginning with known mutants and attempting to identify the biochemical reactions they affect, they would start with known reactions, induce mutations, and find the genes that control the reactions. The organism they selected for this work was the red bread mold, *Neurospora crassa.*

Just as *Drosophila is* an ideal organism for transmission genetics, *Neurospora is* tailor-made for biochemical genetics. Two aspects of *Neurospora* particularly contribute to this desirability: its life cycle and its ease of maintenance in the laboratory.

Neurospora is haploid throughout most of its life. Hence, every mutant gene will be expressed; none will be suppressed by a dominant allele. The haploid organism exists as a colony that profusely reproduces asexually by fragmentation or by forming asexual spores.

In addition to the asexual mode of reproduction, *Neurospora* also reproduces sexually. There are two *Neurospora* mating types-A and *a.* If A and *a* colonies are grown together, parts of the colonies will fuse with one another, and A nuclei will fuse with *a* nuclei to form diploid zygote nuclei. The diploid stage is abbreviated, and each zygote nucleus immediately undergoes meiosis to produce four haploid nuclei.

These nuclei then undergo mitosis, and a cell wall forms around each nucleus to form eight haploid ascospores. Each ascospore will then form a separate colony.

Beadle and Tatum soon found that the nutritional requirements of *Neurospora* are modest; the mold can be maintained *in vitro* on a *minimal medium* of water, inorganic salts, sucrose, and the vitamin biotin. The mold is capable of synthesizing all of its chemical constituents from this minimal medium (i.e., it is autotrophic).

Since the pathways for the synthesis of basic metabolic compounds such as amino acids were well known, Beadle and Tatum decided to search for mutations in genes that regulate these metabolic reactions. Since such basic compounds are essential to life, mutants could survive only if they had an external source of the missing compound.

Thus, these mutants would be *conditional lethals*; that is, they would be lethal unless compensation was made for the defect. The experimental procedure Beadle and Tatum adopted has become the standard way to select for conditional lethals. First, the *Neurospora* were irradiated with x-ray to induce mutation. Next, the irradiated colony was mated with a colony of opposite mating type.

Ascospores from this cross were individually placed in separate vials containing a complete medium, which includes all 20 essential amino acids plus certain other biologically important organic compounds. The ascospores formed colonies in these vials. To determine whether a colony carried a metabolic defect, bits of it were removed to a vial containing the minimal medium.

If the colony could survive on the complete medium but not on the minimal medium, it meant that it was unable to produce some component of the complete medium, presumably owing to a mutation induced by the radiation. These nutritional mutants are called *auxotrophs*. To determine whether the deficiency was in amino acid metabolism, a bit of the colony was placed on minimal medium supplemented with a mixture of essential amino acids.

If the colony grew, it meant that it was unable to synthesize one of the 20 amino acids. The pathway for the synthesis of the amino acid arginine had been discovered shortly before the *Neurospora* work was begun. Arginine is formed from citrulline, citrulline from ornithine, and ornithine from an unknown precursor. Each step requires a specific enzyme. The reaction is written as follows:

$$\text{precursor} \xrightarrow{E_1} \text{ornithine} \xrightarrow{E_2} \text{citrulline} \xrightarrow{E_3} \text{arginine}$$

Amino acid auxotrophs were tested systematically to determine whether any of them required arginine for growth. Several arginine-requiring mutants were found, and all of them fell into one of three categories: One class could grow only if arginine was added to minimal medium. Another would grow with either arginine or citrulline. The third would grow with arginine, citrulline, or ornithine.

The three classes of mutant genes affected different steps in the arginine synthetic pathway. The mutants that would survive only with the addition of arginine were unable to convert citrulline to arginine. The mutants that would survive with citrulline were able to make arginine from citrulline but could not convert ornithine to citrulline.

Likewise, the mutants that would survive with ornithine lacked the ability to convert a precursor into ornithine. Thus, the wild-type genes specify the enzymes (E_1 - E_3) for the reactions, and the mutant genes lack this ability. The one gene-one enzyme hypothesis had been verified.

Subsequently, the rule was generalized to "one gene-one protein," since all proteins, not only enzymes, are gene-dependent. However, proteins are frequently composed of unlike subunits (polypeptides), which are specified by different genes. For example, the hemoglobin molecule is composed of four subunits representing two different polypeptide chains, each specified by a different gene. Hence, the rule can be restated as "one gene-one polypeptide."

THE MODERN ERA

Now that the function of genes in the cell had been established, one of the key stumbling blocks to a clear understanding of the role of genes in development was removed. It now remained to determine the composition of genes and the way that the genetic information is utilized.

During the late 1940s, 1950s, and early 1960s, several important breakthroughs emerged that laid the groundwork for a renewed onslaught of investigations into the role of genes in development and the mechanisms involved in cell differentiation. These contributions resulted mainly

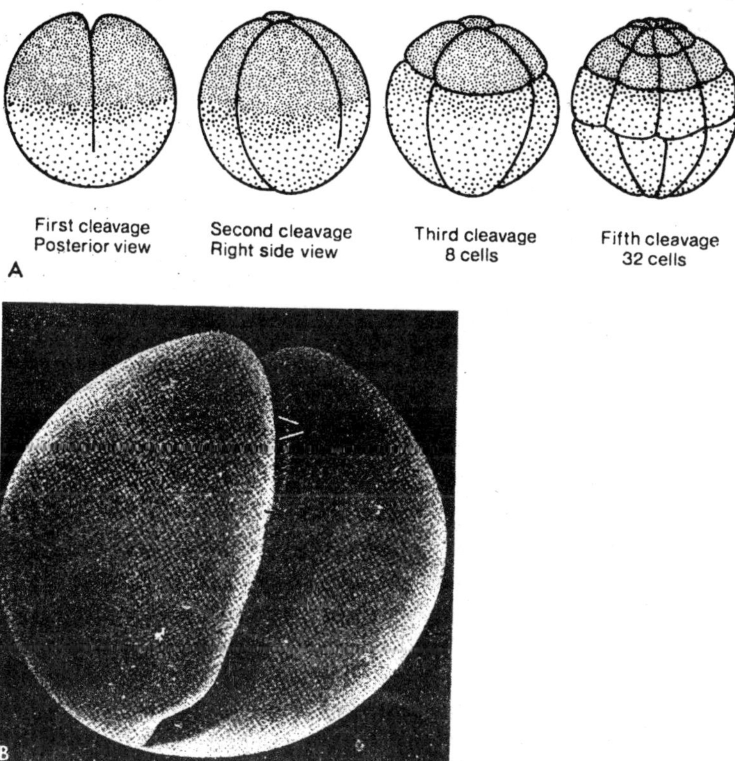

Figure 2.8 : Cleavage in the frog embryo. A, Pattern of first five divisions.

from work in biochemistry and two new disciplines-cell biology and molecular biology.

Biochemists gathered valuable data on the relationships between genes and proteins and the role of enzymes in cellular metabolism. Cell biologists described the structure and function of cellular components, aided by a powerful new tool, the electron microscope. But most importantly, molecular biologists determined the nature and structure of the genetic material, the genetic code was broken, and the protein synthetic machinery was unraveled. Nearly all of the basic principles of molecular biology were obtained from work on bacteria and viruses.

But these principles also apply to multicellular plants and animals, with certain variations. The rapid developments in molecular biology are too numerous to discuss here in detail, but the contemporary student of development should be intimately familiar with the discoveries that form the basic foundations of molecular biology. Some of the more important ones are the following:

1. Genetic information is coded in deoxyribonucleic acid (DNA) as two antiparallel polynucleotide strands in a double-helical structure (the Watson-Crick double helix).
2. Genetic information is stored in a linear sequence of purine and pyrimidine bases. The genetic alphabet consists of four letters (bases): Adenine, Thymine, Guanine, Cytosine.
3. During replication of DNA, each strand serves as a template for the formation of the complementary strand. Each base of the template strand specifies a nucleotide bearing the complementary base: A is complementary to T, G is complementary to C, and vice versa.
4. Genetic information in the chromosomes is expressed by transcription of the sequence of bases in DNA into a complementary sequence of bases in ribonucleic acid (RNA). During transcription, each base of the DNA template strand specifies a ribonucleotide bearing the complementary base: A is complementary to Uracil, T is complementary to A, G is complementary to C, and C is complementary to G.
5. The genetic information for synthesis of a specific protein is a *structural gene*. Structural genes are transcribed into messenger RNA, which is transported into the cytoplasm, where it is translated into protein.
6. Proteins are composed of a linear sequence of amino acids. The placement of an amino acid in protein is designated by a

triplet of bases in mRNA called a *codon*.

7. The genetic code is universal (mitochondria are exceptions), non-overlapping (except in some viruses), and degenerate (i.e., redundant), and it contains codons that punctuate protein synthesis.
8. Protein synthesis occurs on ribosomes, which are composed of protein and ribosomal RNA. The genetic code is read by transfer RNA molecules. These molecules recognize a specific codon and bear the corresponding amino acid, which is added to the sequence of amino acids by formation of a peptide bond. Like messenger RNA, ribosomal and transfer RNA are transcribed from genes. Thus, not all genes code for polypeptides.
9. Recent investigations have revealed the presence of nucleotide sequences whose function is to regulate the transcription of structural genes. These regulatory sequences may not produce transcripts themselves.

By the late 1950s and early 1960s, enough of the critical pieces were in place. Molecular biologists began asking how gene function can be controlled to produce the wide variety of cell types in adult organisms. Biochemists began analyzing the changes in cellular biochemistry that occur during development. Cell biologists began monitoring the structural and functional changes that accompany cell differentiation. Genes that modify development were recognized as valuable tools in understanding normal developmental events. Developmental biology emerged as a vital, exciting, broadly based science.

Interdisciplinary barriers fell as investigators came to realize that plant and animal development have much in common and that simple organisms such as algae and slime molds are excellent model systems for studying cell differentiation. Horizons were broadened with the realization that developmental events occur during all phases in the life span of an organism, not only during embryogenesis.

In such a broadly based science the student is expected to understand concepts that are diverse in nature but have a common goal. That goal is to interpret the processes that produce a tree, a frog, or a human from a fertilized egg.

Although the analysis is multileveled, each approach analyzes the same phenomenon from its own particular perspective. The formation of an eye is viewed differently by the biochemist, the electron microscopist, and the molecular biologist. Yet the eye develops nevertheless, and it is up to the developmental biologist to integrate the information provided by these diverse analyses and describe how and why the eye develops.

Developmental biology has made remarkable strides in its short life. It has been a robust science, benefiting immensely from periodic transfusions of new technological advances originating in its contributory disciplines. Among these technological advances are:

1. *In vitro* analyses of development.
2. Improvements in ultrastructural analysis.
3. Radiotracer technology.
4. Refinements in separation sciences, allowing for resolution and identification of small amounts of cell-specific molecules.
5. Ingenious procedures to trace the fates of specific cells during development.
6. Nucleic acid hybridization technology.
7. Isolation of messenger RNA molecules and their use as templates for synthesis of DNA.
8. Recombinant DNA technology, allowing for isolation and amplification of DNA nucleotide sequences of developmental significance.
9. Nucleic acid sequencing techniques, which allow the evaluation of possible control sequences in DNA.
10. Application of immunochemistry for the identification and quantification of cell-specific molecules.
11. Cell-cell hybridization techniques, allowing for introduction of chromosomes into foreign nuclei.
12. Microinjection, enabling analyses of the function of exogenous genes.
13. Acquisition of a new understanding of the cell surface, facilitating the study of the surface in cellular interactions.
14. Sophisticated monitoring devices to trace the movements of minute amounts of ions during development.

The rapid development of technology in recent years and the judicious application of this technology to the study of development have helped to make developmental biology one of the most exciting and dynamic sciences in recent years.

In their ready acceptance of the ideas and technology of allied sciences, however, contemporary students of development owe a debt of

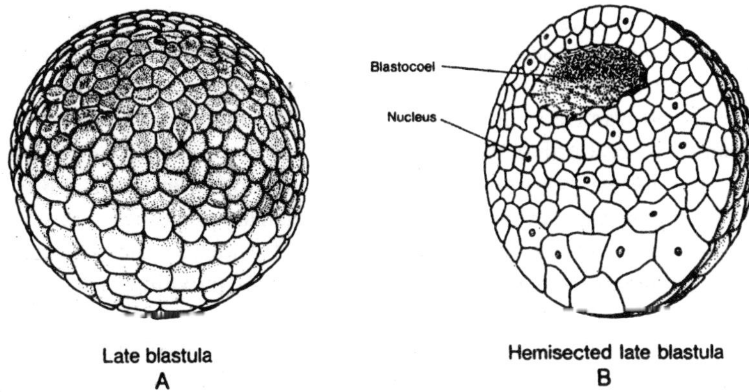

Figure 2.9 : The frog blastula. A, Surface view of late blastula. B, Hemisected blastula of comparable stage.

gratitude to E. B. Wilson, who championed the idea that development is best understood from a multidisciplinary approach.

OVERVIEW OF ANIMAL DEVELOPMENT

Development is reflected by continual change in functional and structural properties, not only in the organism itself but also in its individual cells. To study development, we devise means of monitoring these changes, either quantitatively or qualitatively.

When a number of changes are occurring simultaneously to form a pattern of events that is unique to that period of development, we call it a *developmental stage*. Although stages are not real entities, they are useful aids for discussing development because they allow us to refer to embryos at particular times without describing all of the separate events or morphological features that characterize the embryo at those times.

Thus, it is worthwhile to describe briefly the stages of animal development at this point. We shall discuss later in this book the detailed structural changes that occur in developing animal embryos and the mechanisms that produce those changes.

1. *Gametogenesis* is arbitrarily designated the first stage of animal development. The question of "Which came first, the chicken or the egg?" is one that has intrigued man for centuries. In embryology, the gametes (eggs and sperm) are usually discussed first, since they provide both the blueprint and the raw material from which the embryo is formed. Gamete formation in the two sexes is tailored to the roles of their gametes in reproduction.

The male gametes are usually small and mobile. They are dispensed from the male reproductive organoften into a hostile environment-and they must locate the female gamete, make contact, and fuse with it. The female gamete is usually less mobile than the sperm and larger, often by several orders of magnitude.

The female gamete must be "competent" to be fertilized, which means that it must develop a number of specialized properties to enable it to interact with the sperm. Both classes of gametes make an equal contribution to the nucleus of the zygote, each providing a haploid genome.

However, the male gamete makes a minimal contribution to the cytoplasm; the female gamete provides the zygote with virtually all of the cytoplasm, which contains the constituents from which the embryo is fashioned.

Reproduction is a prime concern for any species, since it ensures the species' survival. The *germ cell line* is, therefore, a precious commodity, and its formation is an important developmental event, which is often one of the first orders of business for the embryo after fertilization.

The germ cell line may derive its specificity from a specialized cytoplasmic constituent, the germ plasm, which may preexist in the egg prior to fertilization and become segregated into the germ cell line during cleavage. Determination of the germ cell line results in two distinct categories of cells in the embryo. The nongerm cells are called the *somatic* cells. This distinction is retained throughout the life of the organism.

The initial cells in the germ line are called *primordial germ cells*. The primordial germ cells of both sexes are indistinguishable from one another. The acquisition by germ cells of sex-specific characteristics occurs at a later stage of development and is culminated by the formation of mature sex cells with distinctly different shapes and organelles.

The primordial germ cells may arise at some distance from the presumptive gonads to which they migrate, become established, and increase in number by mitosis. The establishment of germ cells in the gonads often involves a close association between the germ cells and the somatic cells of the gonad. These somatic cells may serve to support and protect the germ cells and to provide them with nutritive material.

In the female, the somatic cells surrounding the germ cell are called *follicle cells*. In the male, various terms have been used for them. Among the most familiar examples are the *Sertoli cells* in

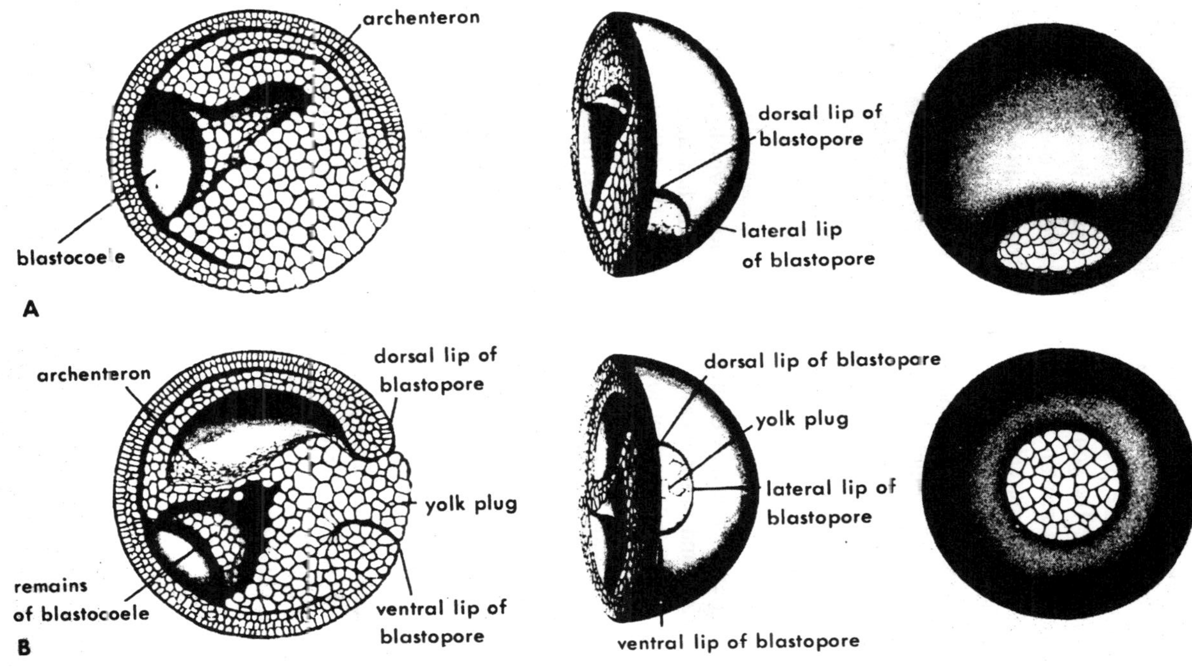

Figure 2.10 : Gastrulation in the frog. Drawings on the left represent embryos cut in the median plane. Drawings in the center represent the same embryos viewed at an angle from the dorsal side (A) or from the posterior end (B).

mammalian testes.

During the proliferative phase the germ cells are called *gonia* (spermatogonia in the testis and oogonia in the ovary) and act as a stem cell population for the production of cells that will differentiate into functional gametes. The gonial cell divisions may be incomplete so that the daughter cells remain in communication with one another via intercellular bridges. Successive incomplete divisions produce very large clones of interconnected cells.

This intercellular communication may serve to synchronize the development of the conjoined cells. The formation of sperm from spermatogonia is *spermatogenesis*, and the formation of ova (or eggs) from oogonia is *oogenesis*. These processes involve the reduction in chromosome number by meiosis and acquisition of the structural and functional characteristics of the distinct sex cells.

In the male, meiosis precedes sex cell differentiation. In the female, however, differentiation may occur early in meiosis, which is completed after ovulation, and in some cases after the sperm has entered the egg at fertilization. It is important to keep in mind the difference in the number of sex cells that result from meiosis in the male and female. In the male, a single spermatogonium enters the first meiotic division as a *primary spermatocyte*.

This division produces two *secondary spermatocytes*, each of which divides to form two haploid *spermatids*. Consequently, *four* haploid cells result from each diploid spermatogonium. Each spermatid differentiates into a *spermatozoon* by the elaboration of structural and functional specializations that enable the sperm to fertilize the egg.

By contrast, in the female, each of the meiotic divisions is uneven, producing only one full-sized cell. During the first meiotic division the *primary oocyte* divides to produce one small polar body and one *secondary oocyte*. The latter enters the second meiotic division to produce the second polar body and the haploid *ovum*, which is the only functional sex cell to result from meiotic reduction of an oogonium.

2. *Fertilization* is the union of male and female gametes, which activates, or initiates, embryonic development and restores the diploid condition.

3. *Embryogenesis* is the phase that encompasses most of the developmental events in animals. Most animal species pass through comparable embryonic stages, although the details vary considerably from one group to another. Since amphibian embryos have been used extensively for research in experimental embryology, the stages of

embryogenesis are illustrated here with amphibian embryos: a. *Cleavage*.

In order for the single-celled zygote to produce a multicellular organism, a number of mitotic divisions must occur in rapid succession. During cleavage the size and shape of the embryo is retained, while the cleavage cells, or *blastomeres*, become smaller at each division. At the completion of each division, the blastomeres are separated from one another by the formation of cleavage furrows.

In the frog embryos, the first cleavage furrow is seen to begin at one pole of the egg and spread to the opposite pole. The egg consists of a dark half and a light half. The dark half of the egg is the *animal hemisphere*, whereas the light half is the *vegetal hemisphere*.

The dark color of the animal hemisphere is due to a layer of pigment granules below the surface of the egg. The nucleus and most of the egg cytoplasm are located in this half of the egg. The unpigmented vegetal hemisphere contains virtually all of the yolk and very little cytoplasm. The first cleavage begins at the *animal pole* of the egg, spreading around both sides to meet at the *vegetal pole*.

The second cleavage furrow also begins at the animal pole, at right angles to the first furrow. The third cleavage is in the *equatorial plane* of the egg, at right angles to both of the first two cleavages. This cleavage is displaced slightly toward the animal pole, causing a size disparity in the cells; animal hemisphere cells are smaller than those in the vegetal hemisphere. Note that the size disparity between animal and vegetal hemisphere cells is retained throughout cleavage.

Small fluid-filled spaces appear between the blastomeres during early cleavage stages. As cleavage proceeds, these spaces coalesce to form a large central cavity, the *blastocoele*, surrounded by a layer of cells, the *blastoderm*. The embryo at this time is called a *blastula*. The frog blastula is represented, both in a surface view and sectioned through the animal-vegetal axis to show internal organization.

b. *Gastrulation*. The cells of the blastoderm undergo extensive rearrangement during this stage to produce three layers of cells known as *germ layers*, from which the various organs of an animal's body are derived.

The three layers are the outer *ectoderm*, which gives rise to the epidermis and the nervous system; the intermediate *mesoderm*, which produces the circulatory system, muscle, skeletal system, and connective tissue; and the inner *endoderm*, which produces the gut and its associated organs.

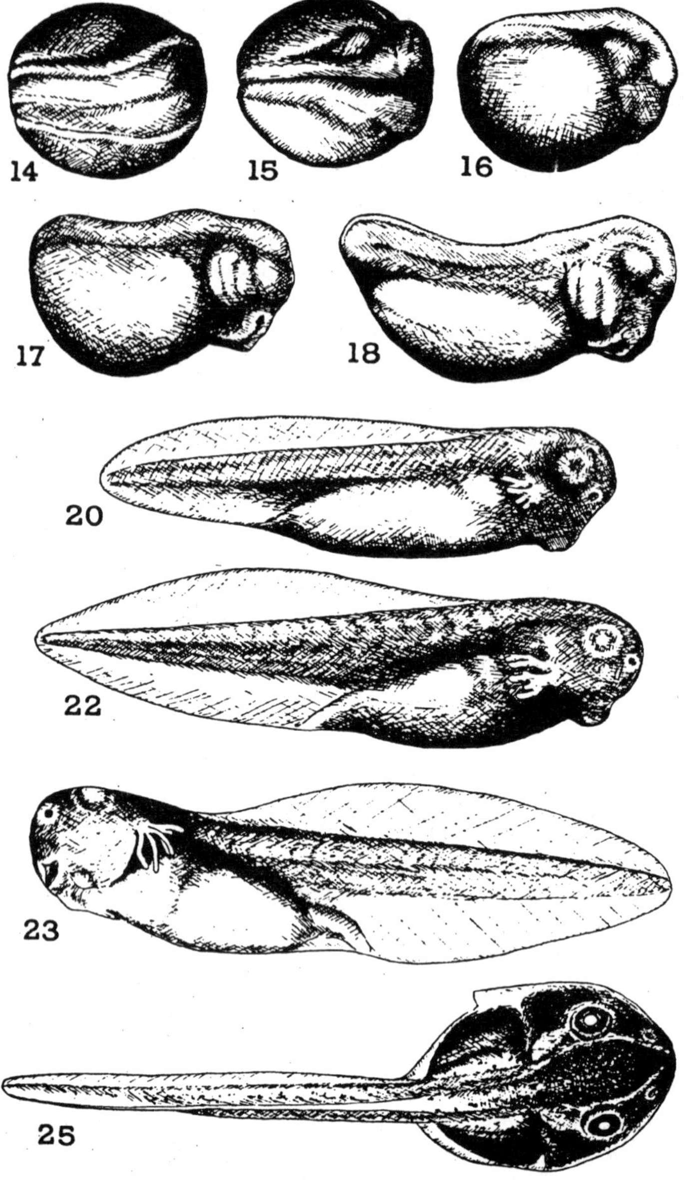

Figure 2.11 : Late stages in the development of the leopard frog, Rana pipiens, *beginning with the neurula stage. Hatching occurs at stage 20, while stage 25 is a feeding tadpole.*

The cell movements of gastrulation involve the inward displacement of endoderm and mesoderm cells and the complete envelopment of the internal cells by the ectoderm. The cell rearrangement mechan-isms employed by animal embryos vary considerably from group to group.

The organization of the embryo during gastrulation often leads to the formation of a new embryonic cavity, the ***archenteron***, which eventually gives rise to the cavity of the alimentary tract. The external opening of the archenteron is the *blastopore*.

c. *Establishment of the basic body plan.* Following gastrulation, the basic body plan of the embryo is laid out along the axis of body symmetry. In a bilaterally symmetrical animal such as an amphibian, the embryo organizes along the anterior-posterior axis, with distinct developmental regions forming along the axis.

Externally, the conspicuous changes in the embryo during this time are caused by the formation of the future nervous system, which extends along the entire dorsal midline of the elongating embryonic axis.

The formation of the nervous system dictates the shape of the embryo. Internally, the archenteron organizes a tubular gut with an anterior and a posterior opening.

The germ layers may dissociate and form clusters of cells that are the precursors, or *rudiments*, from which definitive organs and tissues will be formed. These clusters often consist of cells derived from more than one germ layer, such as endoderm plus mesoderm for internal structures and ectoderm plus mesoderm for peripheral structures.

d. *Organogenesis.* The embryonic rudiments that form after gastrulation acquire the functional and structural characteristics of organs and distinct body parts during the subsequent phase of organogenesis.

4. *Postembryonic development* normally begins when the embryo hatches from its protective coats to become a free-living larva (indirect development). Development of adult features is completed during or near the termination of the larval period. Larvae also serve one or both of two additional functions.

They are often the chief means for geographic distribution of the species. This is particularly true in species whose adults are more or less *sessile*, or stationary. Most larvae, therefore, have well-developed locomotory organs. The larval period may also serve a nutritional function. Larvae may be voracious feeders, usually subsisting on ' a different diet from that of the adult.

Food reserves accumulate and provide raw material and energy for the construction of the adult. Transformation of a larva into an adult is called *metamorphosis*. The amphibian larva is called a *tadpole*, which is a free-swimming aquatic organism.

In animals without a larva (direct development) a juvenile stage may follow embryogenesis. Thus, a miniature organism emerges that closely resembles the adult in appearance (e.g., nematode worms). In other cases the larval life has been replaced by an extended period of development within an egg shell (e.g., birds, reptiles) or a uterus (e.g., mammals).

The adult phase of life is characterized by growth, sexual maturity, and finally, deterioration by a process of senescence, ending in death.

ORIGIN OF GERM CELLS

Two controversial issues with respect to the origin of the PGCs in vertebrates have dominated the literature for many decades: (a) that of the gonadal versus extra-gonadal origin of the PGCs, and (b) that of the one-time versus repeated origin of the PGCs during the life span of the organism. We feel that these controversial issues should be dealt with before the place and mode of origin of the PGCs can be satisfactorily discussed.

GONADAL VERSUS EXTRA-GONADAL ORIGIN OF THE PGCS

The early literature on the origin of the PGCs in vertebrates was dominated by the controversial issue of the gonadal versus extra-gonadal origin of the PGCs. Embryologists were sharply divided into opposing camps, i.e. those supporting Waldeyer's (1870) ideas and those advocating Nussbaum's (1880) concepts.

Waldeyer's theory, which implies an origin of the PGCs from the somatic 'germinal epithelium' of the gonadal anlage, was primarily based on a study of the higher vertebrates, where the PGCs were not clearly distinguishable before they were inside the gonadal anlagen. The concept was later extended to all the vertebrates and even to a number of invertebrate groups.

Nussbaum's notion, which implies an early segregation of the PGCs from the somatic cells of the embryo long before the formation of the gonadal anlagen and topographically separated from them, was chiefly based on studies in the anuran amphibians and teleosts.

The issue remained controversial until the fourth decade of this century, as a result of the rather unspecific staining methods available. During the nineteen-thirties the balance between the two points of view began to shift towards extra-gonadal origin, particularly in the lower vertebrates. When Bounoure's book appeared in 1939 the advocates of Waldeyer's concept had lost nearly all their ground as regards the lower vertebrates, but not as regards the higher vertebrates, especially the mammals, where the PGCs could only be clearly distinguished inside the gonadal anlagen.

Brambell (1960) already stated: 'It is now recognized by a large majority of embryologists that [in the vertebrates] the PGCs arise exceedingly early [in development], long before the rudiments of the gonad are formed and at a distance from the site they will [ultimately] occupy ...Their state of origin is extra-gonadal in all instances and in the amniotes and some of the fishes it is [moreover] extra-embryonic.'

This marked shift in the general consensus was mainly due to the development of new histological and histochemical techniques which allowed identification of the PGCs at much earlier stages of development than was possible until then.

Recent studies have fully confirmed Brambell's conclusions, e.g. in the fishes by Johnston (1951), Gamo (1961b), De Smet (1970), Nedelea & Steopoe (1970) and Pala (1970); in the amphibians by Lacroix & Capuron (1966), Gipouloux (1967) and Sutasurya & Nieuwkoop (1974); in the reptiles by Pasteels (1953, 1957a, 1964) and Hubert (1969, 1976); in the birds by Simon (1960, 1964), Dubois (1965), Rogulska (1968), Fargeix (1969, 1970, 1975), Reynaud (1969), Clawson & Domm (1969), Komar (1969), Dubois & Croisille (1970), Bruel (1973) and Fujimoto *et al.* (1976b); and finally in the mammals by Mintz & Russell (1957), Ozdzenski (1967, 1969), Falin (1969), Semenova-Tian-Shanskaya (1969), Merchant & Zamboni (1973) and Zamboni & Merchant (1973).

ONE-TIME VERSUS REPEATED ORIGIN OF THE GERM CELLS

In the last part of the nineteenth and the first half of the twentieth century the concept of the origin of the PGCs was obscured by the controversial issue of whether there exists only a single source or multiple sources of germ cells during the life span of an individual.

The concept of the continuity of the germ line as postulated in Weismann's *Keimplasma* theory implies a segregation of the germ cells from the somatic cells—preferably during early development—as well as the persistence of the former throughout fertile life. The adversaries of this concept essentially claimed the existence of several phases of asexual and sexual life.

In his 1945 review Everett distinguished four different categories among those working on the origin of the germ cells in vertebrates:

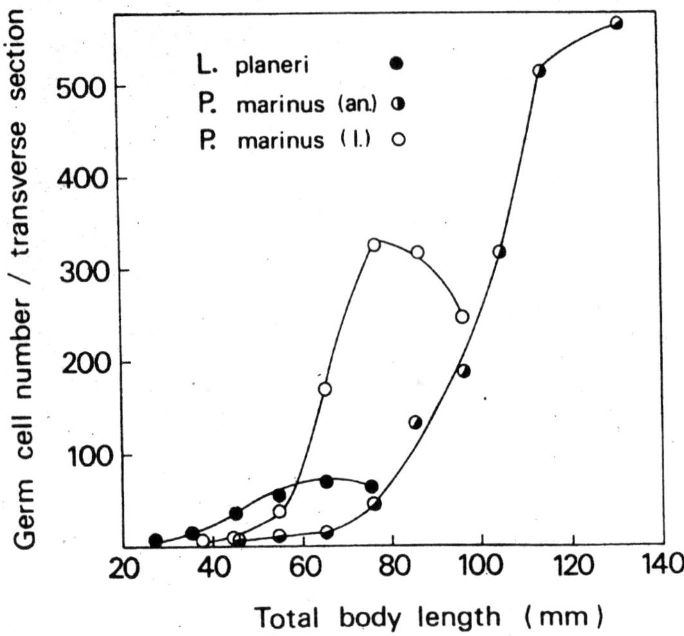

Figure 2.12 : Increase in average germ cell number per transverse body section in relation to total body length, in Lampetra planeri and in Petromyzon marinus of the anadromous (an.) and landlocked (1.) races.

(1) those who deny the early segregation of the PGCs and advocate only a late, somatic origin from the 'germinal epithelium',

(2) those who believe in the early segregation of germ cells but postulate the subsequent degeneration of this first generation and the ultimate formation of a second generation of definitive germ cells from the '*germinal epithelium*',

(3) those who agree to the early segregation of PGCs and their subsequent development into definitive gametes, but in addition postulate a supplementary formation of germ cells from the 'germinal epithelium', and

(4) those who defend a one-time, early segregation of the PGCs, which constitute the only source of the definitive gametes. The issue is further complicated by the assumption either of a periodic or of a continuous formation of germ cells from the 'germinal epithelium' throughout the reproductive period.

We have seen in the preceding section that in the vertebrates the PGCs are formed outside the gonadal anlagen, to which they are subsequently transferred. The problem therefore boils down to the question of whether these PGCs represent the sole source of the definitive gametes or whether they are supplemented with or replaced by one or more generations of germ cells formed *de novo* in the gonadal anlagen during later phases.

This uncertainty was mainly due to the fact that particularly in the higher vertebrates the PGCs lose most of their cytological characteristics during multiplication in the gonadal anlagen. The problem can therefore only be definitely solved by experimental analysis. Nevertheless, the evidence for the existence of a single and uninterrupted germ line in the life cycle of vertebrates is both descriptive and experimental.

The descriptive evidence consists of quantitative data on germ cell history in normal animals and sterile mutants, while the experimental evidence mainly relates to defect and grafting experiments.

Quantitative Data

In the vertebrates *oogenesis follows* a uniform pattern with two main variants. Either germ cell multiplication continues uninterruptedly or cyclically throughout the reproductive period (as in most teleosts, all amphibians, the majority of reptiles and possibly some mammals), or it ceases early in life with the onset of meoisis, so that a definitive, finite stock of germ cells is formed for the entire fertile period (as in cyclostomes, elasmobranchs, a few teleosts, perhaps a few reptiles, all birds, the monotremes, and all eutherian mammals, with a few possible exceptions).

Following Okkelberg's (1921) study on the early history of the germ cells in the brook lamprey, Hardisty & Cosh (1966) followed the germ cell history in several Cyclostomata from the appearance of the germ cells to the establishment of the definitive stock of oocytes in the sexually differentiated gonad.

The final maturation and spawning of the eggs occurs only several years later, i.e, shortly before the end of adult life. When plotting the number of germ cells against the total length of the larva Hardisty (1971) found smooth S-shaped curves.

He concluded from these data and from the frequent observation of mitotic figures that there is actually no need to postulate a .formation of PGCs *de novo* from somatic cells. Their initial increase and later relative decrease in number can easily be accounted for by the observed mitotic proliferation and subsequent degeneration of germ cells.

The observed differences in fecundity of the various species of lamprey must be due to differences in proliferation, since the initial average number of PGCs is nearly the same in the different species. The difference between the actual fecundity of the adult female (number of eggs produced) and the potential fecundity of the ammocoete larva (maximum number of germ cells) can be accounted for by the observed extensive degeneration of germ cells during sexual differentiation.

In the chick Hughes studied germ cell number in the left ovary from day 9 of incubation till the first day after hatching. The formation of oocytes ceases at about the time of hatching.

Oogonial mitoses are responsible for the nearly 25-fold increase in germ cell number between day 9 and day 17, while the subsequent decrease by about 40% between day 17 and the first day after hatching is due to the high incidence of degenerating germ cells.

Hughes concludes that also in the chick ovary there is no need to postulate a second source of germ cells during oogenesis. A similar study, on oogenesis in the rat by Beaumont & Mandl (1962), performed between day 144 *post coitum (p.c.)* and day *2 post partum (p.p.)*, shows that oogonia are mitotically active until day 174 p.c., after which a sharp decline in mitotic activity occurs at the onset of the leptotene stage of meiosis.

The total population increases sixfold between days 144 and 172, reaches a slightly higher maximum value at day 184, and is reduced by about one-third, between day 182 *p.c.* and day *2 p.p.* During this period four different waves of germ cell degeneration were distinguished.

In the rabbit ovary Chretien found two main periods of germ cell multiplication, a first one between days 10 and 12 *p.c.* leading to a fourfold increase, and a second one between days 16 and 18 *p.c.* resulting in an approximately eightfold increase.

The germ cell number reaches a maximum at day 26 p.c., after which a marked decrease takes place by day 28 as a consequence of extensive degeneration. Since in the rabbit PGC migration starts at about day 10 and lasts at least until day 16 to 18, it is evident that PGCs actively multiply during migration. A similar germ cell history was found in the guinea pig by Ioannou, in the monkey by Baker, and in the human foetus by Baker.

The most dramatic increase in germ cell number occurs in the ovary of the human foetus, viz. from 700-1300 during migration to approximately 600000 in the second month of gestation, and further to a maximum of around 7000000 in the fifth month. After several waves of degeneration the number falls to about 2000000 at birth and further to approximately 300000 at the age of seven.

Since only about 500 eggs ovulate during the entire fertile period, there is a tremendous overproduction of female germ cells in human development.

Although in the human foetus oogonial multiplication reaches a maximum in the fifth month of gestation, Gaillard observed extensive germ cell formation in explants of ovarian cortex of 14-week to 36-week human foetuses and concluded that the potency for germ cell multiplication is not restricted to the early foetal period but certainly persists until birth.

His explants first showed complete degeneration of parenchyma and young oocytes, after which new cord-like structures were formed from the 'germinal epithelium' which surrounds the explant. In these sex cords new oogonia appeared in large numbers. He left open the question of whether these oogonia originated from persisting but unrecognisable germ cells or were formed *de novo* from the somatic cells of the 'germinal epithelium'.

In the mouse Rudkin & Griech found that ['H]thymidine is incorporated into oocyte nuclei when administered half-way through the gestation period. The label was still found in the oocytes of female offspring at 6 weeks p.p., suggesting that the PGCs present at the time of labelling had actually formed mature oocytes in the adult female.

Borum found up to 100% labelling in the oocytes of female offspring after ['H]thymidine injections into pregnant mice at days 12 to 15 of gestation, but no incorporation into oocytes upon injection at different times after birth. He concluded that all oocytes present in the adult mouse have persisted from foetal life, so that at birth the mouse is furnished with a definitive population of oocytes from which all mature ova will later be derived.

Peters & Crone showed that DNA synthesis occurs in oogonia in premeiotic interphase, the rate of synthesis falling during the foetal period in the mouse but within the neonatal period in the rabbit. Once synthesized the DNA persists in the growing and maturing oocytes. Kennelly & Foote concluded from [^{3}H]methylthymidine injections into female rabbits on the day of birth and at 4 weeks *p.p.* that most, if not all definitive ova have already been formed at birth and that oocytogenesis *de novo* does not occur in the post-pubertal rabbit.

Unfortunately little is known about PGC numbers in forms which show a continuous or cyclic multiplication of the germ cells during the reproductive period, all the above forms belonging to the second variant of germ cell multiplication.

Spermatogenesis is usually characterised by either continuous or cyclic renewal of the germ cell population. In mammals it has been studied in recent years by Clermont & Leblond, Roosen-Runge, Leblond, Steinberger & Roosen-Runge, Hilscher, Roosen-Runge & Leik and Oakberg, and reviewed by Clermont, Hilscher & Hilscher and Roosen-Runge. In these studies emphasis was placed on renewal of the spermatogonial stem cells and on the phenomenon of germ cell degeneration.

Hilscher & Hilscher state that when female and male gametogenesis in mammals are compared the ' gonia' stage of the female germ cells shows only a single proliferation wave, whereas that of the male germ cells shows a first proliferation wave, which is comparable with that of the oogonia, followed by a second wave after a preparatory interphase.

The second proliferation wave is characterised by both stem cell renewal and differentiation into spermatocytes. According to Clermont the seminiferous epithelium of the mammalian testis is composed of five to six generations of germ cells. The spermatogonial population is renewed by a number of successive mitoses (three in man and up to six in the rat).

Two types of spermatogonial stem cells can be distinguished, the type A 'reserve' and type A 'renewing' stem cells. The former type is not involved in the production of spermatocytes, while the latter type renews itself and simultaneously gives rise to differentiating spermatogonia at each mitotic cycle. However, the 'key' division between stem cell and differentiating spermatogonium does not have the character of a differential division in the morphological sense.

The mechanism which regulates this choice is still unknown. The general conclusion from the quantitative analysis of both oogenesis and spermatogenesis in the vertebrates is that there actually is no need to postulate any formation of germ cells *de novo* from somatic cells during any part of development or fertile life.

It is therefore very likely that all the definitive male and female gametes descend from the initial population of germ cells which has arisen extra-gonadally during early embryonic development.

Experimental Evidence

Elimination of the PGCs has been achieved surgically by removing that portion of the embryo in which the PGCs are located at the time of operation. It is a well-established fact that complete removal of the gonads leads to permanent sterility, but incomplete removal may lead to regeneration of the gonad and recovery of fertility.

In amphibians removal of the extra-gonadal source of the PGCs prior to their migration to the gonadal anlagen leads to the formation of gonads without PGCs and to permanent sterility. In neurulae of *Ambystoma mexicanum* Nieuwkoop removed the presum-ptive lateral plate mesoderm—the source of the PGCs in the urodeles—and obtained larvae which had more or less normal genital ridges but lacked PGCs.

Vivien obtained similar results in *Lebistes* and *Xiphophorus* upon ^{32}P administration. Blackler replaced the fertile region of the endoderm of *Xenopus laevis* neurulae—the source of the PGCs in the anurans - by a similar, more anterior region of the endoderm which does not contain PGCs, and obtained completely normal-looking tadpoles with normal gonadal anlagen but without PGCs. When reared to maturity the animals remained sterile.

In the anurans similar results were obtained by the elimination of the so-called 'germinal plasm', which is assumed to act as a germ cell determinant. Pricking of the vegetal pole or UV-irradiation of the vegetal hemisphere performed at the 1 - to 4-cell stage can lead to sterility of the gonadal anlagen of the larva and to subsequent permanent sterility.

It may therefore be concluded that destruction of the PGCs or their putative 'determinants' leads to permanent sterility. In other words, no formation of germ cells *de novo* occurs in animals which have no PGCs from an early stage of development.

Although this seems a strong argument in favour of the continuity of the germ line, it is essentially negative and therefore inconclusive, for formation of germ cells *de novo* from somatic cells of the 'germinal' epithelium could depend upon a stimulating influence from existing viable germ cells.

This rather unlikely postulate can only be disproved by heteroplastic or xenoplastic recombinations of presumptive germ cells and somatic gonadal tissue or by using specific nuclear markers. Such experiments were performed by Blackler & Fischberg, who exchanged the fertile endoderm region between two strains of *Xenopus laevis* of which one contained the Oxford nuclear marker.

Blackler made a similar exchange between two subspecies of *X. laevis,* while Blackler & Gecking performed the same experiment between *X. laevis* and *X. mulleri,* with subsequent intraspecific and interspecific matings. In all these experiments the germ cells showed the characteristics of the donor type that furnished the fertile endoderm region.

Similar experiments were carried out in urodeles by Smith, who exchanged the ventro-lateral marginal zone (presumptive lateral plate mesoderm, representing the source of PGCs in the urodeles) between gastrulae of the white and the black axolotl. The successful cases in which the exchange was complete showed only progeny of the donor type, as tested by mating as well as histological examination.

Because in Blackler's experiments the overlying ectoderm and mesoderm were also exchanged, while in Smith's experiments the graft not only furnished the source of the PGCs but also presumptive lateral plate mesoderm, it is possible, though rather unlikely, that the gonadal anlage, which is normally formed from the intermediate mesoderm, partly developed from graft tissue.

Disregarding this minor objection, these experiments demonstrate that the presence of PGCs does not lead to a formation of germ cells *de novo* from the somatic components of the gonad. Moreover, no recent public-ations have led to any claim contradictory to the conclusions stated above.

The most elegant proof of the continuity of a single germ line was furnished in birds. First the extra-gonadal and extra-embryonic location of the PGCs in the so-called anterior germinal crescent of the blastoderm at early somite stages was demonstrated by surgical extirpation, cauterisation, UVirradiation, y-irridiation and X-irridiation.

In birds the PGCs are transported by the blood stream from the extra-embryonic germinal crescent to the gonadal anlagen. Reynaud was able to obtain repopulation of the gonadal anlagen after intravenous injection of a PGC suspension made from germinal crescent endoderm into a host embryo which had previously been sterilised by UV-irradiation of its germinal crescent.

When he made xenoplastic recombinates of turkey PGCs with chick hosts or vice versa, he found that the gametes formed in the F1 generation were all of donor type. The colonisation of the host gonadal anlagen by foreign germ cells constitutes a crucial experiment, demonstrating that in birds the PGCs formed during early development are the sole precursors of the definitive gametes.

In mammals the evidence for the existence of a single and uninterr-upted germ line is not as conclusive as in amphibians and birds, but is nevertheless fairly strong. X-irradiation (more than 168 r) of mouse gonadal anlagen after their colonisation with PGCs leads to the development of sterile gonads, which nevertheless show more or less normal proliferation of the gonadal epithelium. Mouse gonadal epithel-ium—prior to colonisation with PGCs—when grafted into the kidney capsule of host embryos remains sterile, whereas similar grafts made after colonisation with PGCs form typical testicular or ovarian tissues containing germ cells. These results plead strongly against any formation of germ cells *de novo.*

Further evidence comes from genetical studies. Mintz & Russell and Mintz described several alleles of a mutation in mice called *dominant white spotting (W),* which in homozygous condition leads to sterility at birth. In the mutants the PGCs are formed at 8 days of gestation in the yolk sac endoderm in the normal number, which ranges from 10 to 100.

The cells behave normally and migrate towards the gonadal anlagen. However, in normal mice the number of PGCs increases exponentially and reaches a value of approximately 5000 at the end of the migration period, whereas in the mutants the number of PGCs does not increase during migration.

They subsequently degenerate, leading to total and permanent sterility of the gonadal anlagen. Twenty-five per cent of the offspring of heterozygous parents of the W series show the defect and are sterile at birth. This germ cell deficiency, which begins to manifest itself on the ninth day of embryonic development and is fully expressed at birth, furnishes a strong argument against any formation of germ cells *de novo* from the gonadal epithelium.

This conclusion is further supported by culture *in vitro* of sterile mutant half-gonads fused with younger normal fertile half-gonads. In these chimaeric explants epithelial cells of the mutant gonad are in close contact with normal PGCs but no proliferation of mutant germ cells was ever observed.

A similar germ cell deficiency seems to exist in the *steel* mutant *(Sl)* described by Bennett. Radiation-induced damage (400 r) likewise results in a depletion of the PGC stock leading to permanent sterility. At lower doses the gonads may become repopulated, but all evidence points towards repopulation by remaining germ cells and not from somatic cells of the gonadal epithelium.

GENERAL CONCLUSIONS

Summarising, it may be concluded that in the vertebrates the PGCs have an extra-gonadal origin during early embryonic development. The concept of a continuous, single germ line originating in early embryonic development has become highly probable for amphibians and birds and likely for mammals, while there is circumstantial evidence for it in agnathan fishes.

Moreover, in recent years no compelling evidence against the general validity of this concept has been presented. Although further analysis is highly desirable, particularly in fishes and reptiles, it may be assumed for the present that in the vertebrates germ cell history is characterised by the one-time origin of

PGCs during early development and that the resulting PGC population gives rise to all the definitive gametes of the adult. Waldeyer introduced the term 'germinal epithelium' for the putative germ-cell-producing outer layer of the gonadal anlage.

The now convincingly demonstrated extra-gonadal origin of the PGCs makes this term, which is unfortunately currently used in the literature, a very confusing one. It would be better to speak of 'gonadal epithelium' when describing the proliferative layer of the somatic gonadal anlage, and this is the term we shall use in the following chapters. In the lower chordates the problem of the origin of the PGCs is still unsolved.

In the Cephalochordata the PGCs seem to arise *in situ* shortly before gonad formation. Moreover, gonadogenesis differs markedly from that in the vertebrates. Since the further history of the germ cells has not been studied, the conclusions drawn above for the vertebrates cannot yet be extended to the entire phylum of the chordates.

3

FERTILIZATION

Mature gametes are products of complex differentiation processes that prepare them to participate in establishing a new generation of organisms. During their differentiation, male and female gametes acquire distinct characteristics that define their separate roles in fertilization. The male gametes normally are motile and must travel to the egg.

To facilitate their mobility, sperm acquire specialized locomotory organelles, and they divest themselves of excess cytoplasm, thus easing the burden on the transport mechanism. They must also have the means to attach to the egg surface, penetrate it, and deliver the haploid nucleus to the egg interior.

The egg, on the other hand, need not have elaborate locomotory organelles, but it must have large amounts of cytoplasm, from which the embryo is formed. In addition, the egg possesses the physiological and morphological capacity to be fertilized by the sperm.

Successful fertilization is the culmination of mating, which is a behavioral phenomenon adapted to the conditions of fertilization for each species. The essential aspects of mating are that the sperm and egg encounter one another and that the zygote be in an environment favorable for development.

Each sex must form sufficient gametes to ensure that enough zygotes will be produced to perpetuate the species. The enormity of this task is illustrated by marine invertebrates, which release their gametes into the open sea, where they are rapidly dispersed. For this reason, vast numbers of gametes may be produced to maintain a sufficient gamete concentration in the water.

A single sea urchin, for example, may release up to 400 million eggs or 100 billion sperm during a single breeding season lasting but a few months. An improvement in the efficiency of reproduction is clearly advantageous to any species. Any adaptation that facilitates successful fertilization reduces the energy cost of reproduction. This is especially true in the female, since the production of eggs requires the synthesis of considerable cytoplasm.

Various mechanisms have evolved to facilitate successful reproduction. One kind of mechanism involves the *timing of mating*. If mating occurs when the gametes of both sexes are mature and when the zygotes are most likely to thrive, fewer gametes will be "wasted." Most species have a restricted breeding season that coincides with gamete maturity. The timing of the breeding season is usually controlled by climatic conditions.

The estrous cycle of mammals is an example of close synchrony between mating and gamete release. Another type of mechanism that facilitates fertilization is *morphological adaptation,* which increases the likelihood of sperm-egg contact. An example is internal fertilization in mammals in which the gametes of both sexes are deposited in the female reproductive tract. Containment of the gametes prevents dispersal such as occurs in water.

Furthermore, fluid movements within the reproductive tract assist in transporting the gametes to the site of fertilization. Sperm movement toward the egg may either be random or the consequence of directed movement caused by an "attractive force" associated with the egg. The latter mechanism, called chemotaxis, typifies a number of plant species, including both plants we have been discussing as examples of primitive plants-Fucus and *Marsilea.*

As we shall discuss elsewhere in this chapter, plant chemotactic substances are released by the immotile egg or by the surrounding archegonium to guide the free-swimming sperm to the immediate vicinity of the egg, where the gametes may fuse. Recent systematic searches for chemotaxis in animals by Miller have revealed that the mechanism is widespread. Chemotaxis has been demonstrated in certain hydrozoan cnidarians, ascidians, mollusks, and echinod-erms.

Mechanisms that facilitate sperm-egg encounter ensure that sufficient sperm will arrive in the vicinity of the eggs. The sequence of events after the encounter culminates in the development of a diploid zygote from the haploid gametes.

These events are critical to the survival of the species. For this reason, a number of mechanisms have evolved to ensure:

(1) that the sperm can penet-rate the egg accessory layers and fuse with the egg and

(2) that only a single sperm of the same species fertilizes the egg. We shall first discuss how gametes make contact with one another and then consider the response of the egg and the formation of a diploid zygote nucleus.

SPERM-EGG ASSOCIATION

Embryologists have been fascinated by the relationships between sperm and eggs since fertilization was first observed a century ago. What enables these cells to recognize one another? How do eggs discriminate between sperm of their own species and unrelated sperm? Why does only a single sperm fuse with an egg?

These questions have led to considerable speculation during the past 100 years, but the groundwork for the resolution of these problems has been laid only in recent years. Detailed experimental analyses of fertilization were first conducted with sea urchins and other echinoid echinoderms; these species have also been the targets of recent investigations that have greatly clarified how sperm-egg association occurs. It is therefore appropriate that we begin this section with the echinoids.

Echinoids

These common marine organisms have many advantages for the study of fertilization. They can be readily induced to shed large numbers of gametes simply by injecting the mature adult with a solution of potassium chloride. The gametes can be mixed by the investigator, who can follow the sequence of fertilization events in precise chronological order.

The echinoid egg is surrounded by an outer jelly coat composed of a polysaccharide-glycoprotein complex and by a vitelline envelope that is composed of a network of glycoprotein fibers. The vitelline envelope is attached to the egg surface by a series of short processes called *vitelline posts*.

Initial sperm-egg contact involves the jelly coat. A great deal of controversy about the jelly coat has arisen over the years, beginning with an observation made by F. R. Lillie. He described the response of spermatozoa to seawater in which unfertilized eggs had been suspended. This "egg water" causes sperm to form clusters consisting of aggregates of sperm oriented head to head.

Lillie proposed that this *agglutination* of sperm is due to a substance that he called fertilizin, which diffuses from the eggs into the seawater. It was later demonstrated that the fertilizin is derived from the jelly coat. The agglutination was described as a cross-linking of sperm heads by the fertilizin, much like the agglutination of cells by antibodies.

A receptor molecule (*antifertilizin*) was proposed to exist in sperm heads and to combine with fertilizin to cause the agglutination. When sperm and egg jelly from different species were combined, it was reported that egg jelly agglutinated sperm of its own species more readily than unrelated sperm.

These observations led to the widespread belief that the fertilizin functions as the species-specific sperm receptor during fertilization. The fertilizin-antifertilizin system has dominated thought on fertilization for several decades, and many attempts have been made to establish it as a universal system for sperm-egg interaction in the animal kingdom.

However, in recent years the fertilizin concept has come under attack. It has been shown that agglutination is not necessarily speciesspecific. Furthermore, it has recently been proposed that the clusters of sperm that form in response to jelly are caused by an increase in sperm motility induced by the jelly.

The rapidly moving sperm swarm together, forming a large cluster. The sperm are not physically held together, but are moving within the cluster. According to this hypothesis, the clustering is not actually an agglutination.

The responses of sperm to dissolved egg jelly, discussed earlier, are artifacts that are observed in the laboratory and are not normal events in fertilization. However, they do reflect specific interactions between egg jelly and sperm that are of undoubted significance in fertilization.

One of the most important effects of the egg jelly is to trigger the release of the *acrosomal process*, the sperm organ-elle that attaches and fuses with the egg surface. Release of this process is the acrosome reaction. The formation of the acrosomal process in the sand dollar, *Echinarachnius parma*.

The anterior tip of the intact sperm prior to the acrosome reaction. The major components of the acrosome are the spherical, membrane-enclosed acrosomal vesicle, which contains the acrosomal granule, and the surrounding periacrosomal *material*. The acrosomal vesicle rests in a cup-shaped depression in the apex of the nucleus, the subacrosomal fossa. Most of the periacrosomal material is located in the fossa.

Note the close association between the acrosomal vesicle membrane

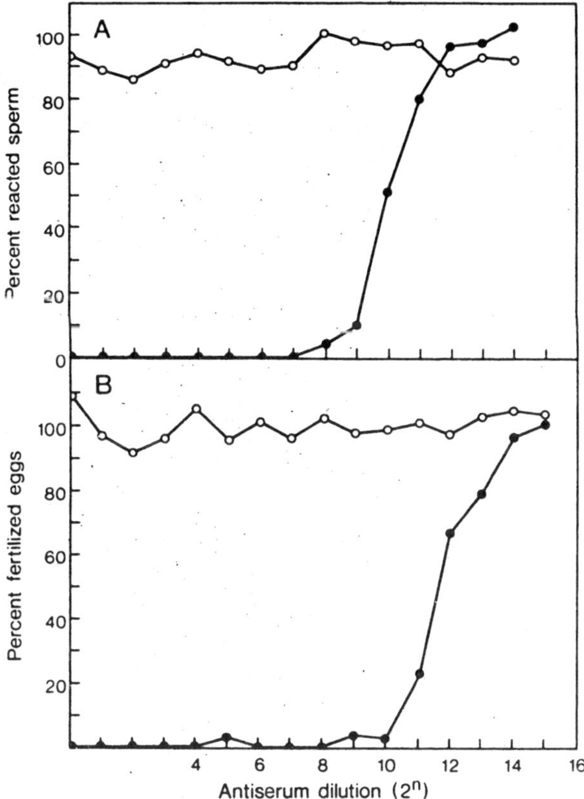

Figure 3.1 : Effects of sperm surface glycoprotein Fab on the jelly-induced acrosome reaction (A) and on fertilization (B). Lines with open circles represent responses to preimmune Fab and serve as controls. Lines with closed circles represent responses to immune Fab.

and the sperm plasma membrane in the region anterior to the two arrows. During the acrosome reaction these two membranes fuse along a line that circumscribes the acrosome at the "rim of dehiscence," which is marked by the arrows.

As a consequence, the entire anterior half of the acrosomal vesicle membrane and the overlying plasma membrane are shed. The basal half of the acrosomal vesicle membrane is everted to form the acrosomal process. The growing process is coated with the contents of the acrosomal vesicle.

The acrosomal process contains numerous microfilaments that are organized from the periacrosomal material of the subacrosomal fossa. The polymerization of the actin subunits to form microfilaments has

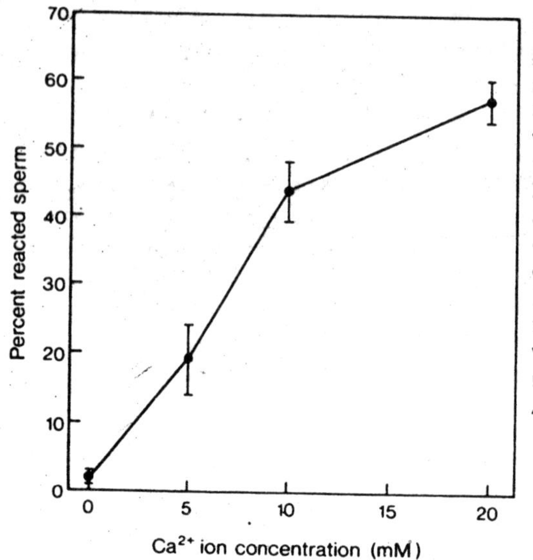

Figure 3.2 : Calcium ion dependency of ionophore induction of the sea urchin acrosome reaction. Sperm were added to artificial sea water containing A23187 and the indicated amounts of Ca". After 2.5 minutes the sperm were fixed and examined with the electron microscope to determine percentage of reacted sperm.

been studied in the sea cucumber, *Thyone briareus*. In sperm of this species, polymeri-zation is initiated in an organelle called the *actomere*, which appears to contain a small number of preformed microfilaments embedded in a dense matrix.

It will be interesting to learn whether this-or a similar-organelle also initiates microfilament formation in the echinoid acrosome reaction. The belief that the acrosome reaction is triggered by the jelly coat is based upon observations that the sperm of some echinoid species undergo the acrosome reaction while traversing the jelly coat and that a solution of dissolved jelly coat material can cause sperm to undergo an acrosome reaction.

The component of the egg jelly that induces the acrosome reaction has been identified as a sulfated fucose polymer. Treatment of sperm with univalent fragments of antibodies to a sperm surface glycoprotein prevents the sperm from responding to egg jelly. This result indicates that a component of the egg jelly, possibly the fucose polymer, interacts with a sperm surface glycoprotein to induce the acrosome reaction.

The sperm of some sea urchin species are induced to undergo the acrosome reaction only with jelly coat from the same species, whereas others will respond equally well to either a homologous or heterologous jelly coat. Since only acrosome-reacted sperm can fertilize an egg, the sperm-egg jelly interaction in the former group appears to play a role in the species specificity of fertilization. The latter group may rely upon a block to cross-fertilization, which occurs at a subsequent step.

It has been known for some time that induction of the acrosome reaction by egg jelly requires the presence of calcium ions in the medium. Dan interpreted this to mean that the acrosome reaction is triggered by an increased permeability to Ca^{++} caused by egg jelly.

In recent years a simple tool has become available for approaching this problem. This tool belongs to a group of antibiotics called *ionophores*, which are lipid soluble molecules that selectively bind to certain cations and transport them across membranes. One ionophoreA23187-is specific for transport of divalent cations such as Ca^{++} and Mg^{++}.

Thus, A23187 can be used to raise the levels of Ca^{++} in the sperm. Indeed, the acrosome reaction is triggered by the ionophore, but only when Ca^{++} is present in the medium. The Ca^{++} dependency indicates that it, not Mg^{++}, is responsible for mediating the effects of the ionophore.

Additional evidence implicating Ca^{++} uptake as a trigger of the acrosome reaction is provided by experiments in which uptake is prevented by the use of drugs. For example, two drugs that specifically block Ca^{++} transport (D600 and verapamil) inhibit the acrosome reaction.

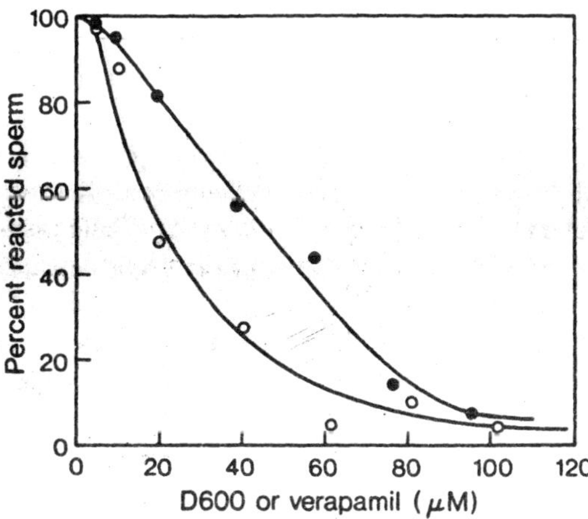

Figure 3.3 : Inhibition of the acrosome reaction by D600 and verapamil. Sperm were added to artificial sea water containing egg jelly, $CaCl_2$, and the indicated amounts of D600 (line with closed circles) or verapamil (line with open circles).

Finally, $^{45}Ca^{++}$ has been found to be incorpo-rated into sperm during the acrosome reaction.

The Ca^{++} is thought to be necessary for the fusion of the acrosomal vesicle membrane and the sperm plasma membrane, which exposes the contents of the acrosomal vesicle.

The acrosome reaction also involves uptake of sodium ions and a discharge of protons. A role for sodium in the acrosome reaction is suggusted by this evidence:

(1) The reaction fails to occur in Na^+ -free seawater; (2) $^{22}Na^+$ is incorporated into sperm during the acrosome reaction;

(3) The acrosome reaction can be induced in the absence of jelly by the monovalent cation ionophore gramicidin S. This action of gramicidin S is dependent upon the presence of sodium ions in the medium, indicating that it causes a Na^+ influx.

The proton release occurs very rapidly upon exposure of sperm to jelly and is reflected by an acidification of the medium during the acrosome reaction. The Na^+ uptake and the H^+ release both occur within 15 seconds of sperm exposure to egg jelly, which is approximately the time of the appearance of the acrosomal process.

Furthermore, roughly equal molar amounts of Na^+ and H^+ are transported in opposite directions. These results suggest that the Na^+ is exchanged for the H^+ which is then transported out of the cell. The efflux of H^+ raises the sperm intracellular pH. The rise in pH triggers the formation of the acrosomal process, possibly by inducing the polymerization of actin, which is necessary for eversion of the process.

An increase in the levels of the monovalent cation potassium in the medium has been shown to *inhibit* the acrosome reaction. The drug tetraethylammonium chloride, a potent inhibitor of K^+ transport, also inhibits the acrosome reaction.

These results suggest that K^+ must be transported out of the sperm to allow the acrosome reaction to proceed. The movement of K+ has been traced by using $^{42}K^+$ and K^+-selective electrodes. These studies have confirmed that an efflux of K^+ occurs during the acrosome reaction. A consequence of K^+ efflux could be a change in the sperm plasma membrane potential.

In fact, a decrease in K^+-dependent membrane potential upon exposure of sperm to egg jelly has been observed. The extensive ionic permeability changes we have discussed appear to play key roles in mediating the induction of the acrosome reaction by egg jelly. These events have a close parallel at a later stage in fertilization. The

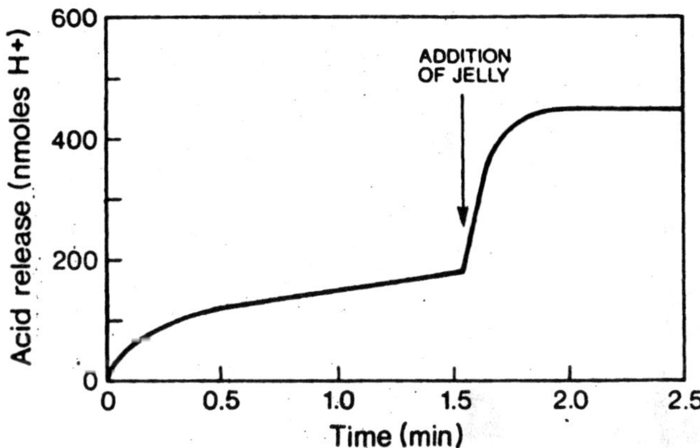

Figure 3.4 : Acid release from sperm induced by egg jelly. Sperm were added to artificial sea water and were allowed to approach a steady-state level of acid release. At this time jelly was added, and the resulting pH shift was followed.

activation of the sea urchin egg by the sperm is also mediated by ionic fluxes.

After traversing the jelly coat, sperm encounter the vitelline envelope, to which they attach in large numbers by their acrosomal processes. The acrosomal processes appear to bind to the vitelline envelope by means of the acrosomal material that coats the processes. This material has been called bindin because of its functional role in fertilization.

The attachment between the acrosomal process and vitelline envelope is species-specific; the specificity is due to a lectinlike interaction between the sperm bindin (which is a protein) and a specific polysaccharide sperm receptor on the vitelline envelope. The specificity of this interaction appears to be a significant block to cross-fertilization between species.

Further clarification of the sperm-binding mechanism could be obtained if the receptor molecule were isolated and its role in sperm binding were studied in detail. Recently, partial purification of the putative receptor has been reported.

The substance, which has a glycoprotein component, has been obtained from isolated membranes of the eggs of *Arbacia punctulata* and *Strongylocentrotus purpuratus*. The partially purified receptors will bind to sperm of their own species but will not bind sperm of the other species. If these substances are the actual vitelline envelope sperm receptors, they should compete with eggs for sperm binding and therefore inhibit fertilization in a *species-specific manner.* In fact, fertilization of *S. purpuratus* eggs by *S. purpuratus* sperm is inhibited by the homologous receptor, but *A. punctulata* receptor has no effect on *S. purpuratus* fertilization.

Thus, these substances have properties that are consistent with those of the actual receptors for their respective sperm. Further study of these substances should greatly increase our understanding of sperm-egg binding.

In spite of the large numbers of sperm that can bind to the vitelline envelope, only a single sperm normally penetrates this layer and fuses with the egg plasma membrane. Penetration of the vitelline envelope may be due to a chymotrypsinlike protease enzyme associated with the acrosome process.

A role for the enzyme in vitelline envelope penetration is indicated by experiments in which an inhibitor of chymotrypsinlike enzymes was shown to prevent specifically sperm penetration of the vitelline envelope. Fusion of the sperm with the egg plasma membrane initially involves the tip of the acrosomal process and results in the plasma memb-ranes of the two cells becoming continuous and forming a cytoplasmic bridge that eventually enlarges to allow the entire sperm (minus its plasma membrane) to enter the egg.

The egg appears to be actively involved in internalizing the sperm. This process is illustrated by the scanning electron micrographs. After fusion, microvilli adjacent to the sperm begin to elongate and cluster around the sperm to form a *fertilization cone*, which engulfs the sperm.

This interpretation of sperm incorporation is supported by experiments in which both fertilization cone formation and sperm incorporation are inhibited by cytochalasin B. These results suggest that microfilaments in the egg cortex are responsible for actively drawing the sperm into the egg and that the fertilization cone is a manifestation of this activity. The subsequent behavior of the sperm head in the egg interior will be discussed in detail elsewhere in this chapter.

MAMMALS

The study of mammalian fertilization is much more than an academic exercise, because of the social and economic consequences of reproduction of humans and of domesticated animals. Human overpopulation can be a major social problem, and infertility can be the source of much personal anxiety.

An understanding of gamete interaction might assist in solving these problems by suggesting new means of birth control or ways of improving fertility, when desired. Furthermore, the economic implications of reproduction of domesticated mammals are of mammoth proportions, and proper management of livestock reproduction is important in optimizing the food supply in any economy.

Mammals present special technical problems for the study of fertilization. Since gamete fusion occurs internally, it is difficult to observe the initial stages of gamete interaction. Investigators prefer to observe these events *in vitro*. Early attempts at *in vitro* fertilization were un successful, since sperm collected directly from the male are unable to penetrate the egg accessory layers and make contact with the egg surface.

This ability is normally acquired by sperm only after they have resided in the female reproductive tract for a number of hours. This acquisition of fertilizing competence by sperm in the female reproductive tract is called capacitation. The changes undergone by the sperm include morphological modifications to the plasma membrane and modification, redistribution, or loss of surface glycoprotein molecules.

These alterations are assumed to prepare the membrane to participate in the acrosome reaction. Sperm also become more motile after exposure to the female reproductive tract. The enhancement of motility may also be a component of capacitation. Increased motility enables the sperm to move farther up the female tract where they ultimately make contact with the egg. *In vitro* techniques for inducing sperm capacitation are now available, and capacitated sperm are routinely used for *in vitro* fertilization of mammalian eggs.

Before encountering the egg, sperm must penetrate the egg accessory layers. Penetration is usually attributed to the enzymes released from the acrosome during the acrosome reaction, which occurs in sperm during their approach to the zona shortly before making contact with the cumulus oophorus. The enzyme *hyaluronidase* is responsible for the penetration of the cumulus oophorus. Hyaluronidase digests the hyaluronic acid that binds the loosely organized cumulus cells.

Another enzyme-corona-penetrating *enzyme* is apparently responsible for allowing sperm to penetrate between the tightly bound corona radiata cells. As we shall discuss subsequently, penetration of the zona pellucida may also be due to an acrosomal enzyme.

The mammalian acrosome reaction differs from that of the sea urchin in that no acrosomal process is formed. The dehiscence of the acrosomal membrane occurs in much the same way, however. The plasmalemma and the outer acrosomal membrane fuse intermittently over the tip of the sperm head, causing the formation of numerous vesicles. This process produces gaps through which the acrosomal contents are released and leaves the inner membrane of the acrosome as the outer surface of the sperm head.

The equatorial segment of the acrosome is spared from this vesiculation process and remains intact. Like the sea urchin acrosome reaction, the mammalian reaction requires calcium ions.

After penetrating the cellular layers surrounding the egg, the head of the sperm encounters the zona pellucida, a rather thick envelope composed of glycoprotein. Sperm binding occurs between the newly exposed inner acrosomal membrane and the surface of the zona.

This binding is species-specific in some (but not all) species and is analogous to the binding of the sea urchin sperm to the egg vitelline envelope. In the mammal, the sperm bind to species-compatible sperm receptors on the zona.

This receptor system has been the target of recent attempts to develop new contraceptive techniques.

Penetration of the zona has been ascribed to the action of an acrosomal proteolytic enzyme, acrosin, which has been proposed to digest a path through the zona in advance of the sperm. Recent evidence, however, suggests that the proteolytic activity may be necessary for sperm binding to the zona, rather than penetration.

Further research is necessary to establish the function of the proteolytic enzyme and to determine whether penetration is enzymatic or mechanical. Factors released at the point of contact between sperm and the zona may block penetration by supernumerary sperm and provide an initial block to polyspermy. Subsequent polyspermy blocks are also operative and will be discussed in the next section.

Passage of the sperm through the zona is tangential to the egg surface. As a consequence, the tip of the sperm does not make the initial contact with the surface of the egg. Instead, the equatorial segment of the sperm head contacts the egg.

A process of egg cytoplasm flows around the midposterior part of the sperm head, superficially internalizing it, while the tip of the head and the tail project into the perivitelline space.

Microvilli from the egg surface seem to draw the anterior portion of the sperm head into the egg by a process similar to phagocytosis. After the head has been incorporated, the tail follows and is stripped of its plasma membrane in the process.

THE FERTILIZATION RESPONSE: EGG ACTIVATION

Sperm-egg interaction elicits far-ranging changes in the egg. These responses, collectively called *egg activation*, include a block to the entry of additional sperm ("block to polyspermy") and characteristic morphological and physiological changes, which initiate embryonic development.

Block to Polyspermy

Since polyspermy usually leads to abnormal development, it is beneficial for the egg to prevent it. In most organisms, polyspermy is prevented by mechanisms that bar additional sperm-egg fusions after the initial sperm has interacted with the egg. In the sea urchin two sequential blocking mechanisms have been proposed. The first is transient and occurs within seconds of sperm-egg contact, whereas the other is permanent and takes longer to develop.

The fast block, which was first demonstrated by Rothschild and Swann, is caused by electrical depolarization of the egg plasma membrane from a negative value to a positive value that accompanies the entry of sperm. This depolarization is the result of an influx of sodium ions that is due to an opening of Na^2 channels, triggered by the fertilizing sperm.

The positive charge on the membrane renders it unreceptive for fusion with additional sperm. A similar fast block to polyspermy has been reported for anuran amphibians, starfish, and the echiuran worm *Urechis*.

The slower block to polyspermy results from the *cortical reaction*. This change in the cortex begins at the site of sperm entry and propagates around the egg circumference. During the cortical reaction the cortical granules rupture due to fusion of their membranes with the egg plasma membrane, which causes vesiculations.

As a consequence of the cortical reaction, the vitelline envelope dissociates from the egg plasma membrane. This dissociation results from breakage of the vitelline posts that attach the vitelline envelope to the membrane. The breakage of the vitelline posts is thought to be caused by a protease enzyme that is released from the ruptured cortical granules.

Separation of the vitelline envelope from the egg surface creates a space into which the contents of the cortical granules spill. This results in elevation of the vitelline envelope from the egg surface. It is then called the *fertilization envelope*, or *activation* calyx. The area between the membranes is the *perivitelline space*. Gamete contact has occurred at some point to the right of this section.

Release of cortical granule contents has caused elevation of the fertilization envelope on the right. However, elevation is only beginning on the left. A portion of an intact cortical granule is shown at the extreme left, which is the limit of the cortical reaction. The contents of recently ruptured cortical granules can be seen in the perivitelline space.

Note the effect that the cortical reaction has upon the egg plasma membrane, which becomes a mosaic composed of the original membrane plus the membranes that formerly enclosed the cortical granules. This results in a tremendous increase in surface area. The excess membrane is accommodated by the formation of elongated and highly branched microvilli and by the formation of endocytotic vesicles, which may be the vehicles for resorption of excess membrane into the egg.

The cortical reaction results in formation of hydrogen peroxide (H_2O_2), release of enzymes that are involved in establishing blocks to polyspermy, and release of material for the formation of the *hyaline layer*. The H_2O_2 is thought to inactivate supernumerary sperm after entry of the first sperm.

This conclusion is based upon the observations that (1) treatment of sperm with H_2O_2 at concentrations similar to those released by eggs at fertilization reduces the fertilizing capacity of sperm and (2) addition of catalase (which decomposes H_2O_2 to H_2O and O_2) to the seawater causes 100% polyspermy.

Thus this polyspermy block is only effective if H_2O_2 is present. One of the enzymes released from cortical granules is a protease that breaks the bonds attaching the vitelline envelope to the egg plasma membrane. Another protease destroys the glycoprotein sperm receptors on the vitelline envelope.

Both of these enzymatic activities assist in the polyspermy block. Elevation of the fertilization envelope removes the sites of sperm binding from their proximity to the egg plasma membrane, whereas destruction of the receptors themselves causes supernumerary sperm to become detached from the membrane and prevents the binding of additional sperm.

The clear, viscous hyaline layer is formed from material that spills out of the cortical granules and covers the egg surface. This layer has an important role to play during morphogenesis.

Upon elevation, the fertilization envelope undergoes changes in its chemical and physical properties; it becomes hardened and resistant to solubilization and proteolytic digestion. Hardening of the envelope is caused by its interaction with a peroxidase enzyme released from the cortical granules, that utilizes H_2O_2 released from the eggs to promote cross-linking of the individual molecules that compose the envelope, which is thus converted to a polymer.

The hardened envelope is resistant to sperm entry and surrounds and protects the embryo until it is dissolved by the hatching enzyme released from the embryo during early development.

A cortical reaction to fertilization is not unique to sea urchins but is found in a wide variety of organisms, including vertebrates. Anuran amphibians have a cortical reaction that is remarkably similar to that of sea urchins; that is, the release of cortical granule components causes the elevation of the vitelline envelope and its conversion to a fertilization envelope that is impenetrable to sperm and has a reduced binding affinity for sperm.

Likewise, in some mammals, release of cortical granule constituents results in modifications of the zona pellucida (called the zona reaction), which in turn cause the loss of sperm-binding sites on the zona. As in sea urchins, sperm detachment apparently results from the action of a protease enzyme.

The zona also undergoes hardening, which is analogous to hardening of the sea urchin fertilization envelope. Recent evidence suggests that the mechanism of hardening of the zona is also similar to the hardening mechanism in sea urchins; that is, peroxidase released during the 'cortical reaction catalyzes cross-linking of molecules of the zona.

The simiiarity between the polyspermy blocks in sea urchins and mammals is remarkable in view of the evolutionary distance between them. An additional block to polyspermy occurs at the mammalian egg surface. In a few mammals, such as the rabbit, a zona reaction is lacking. In these species the egg surface block is apparently sufficient to prevent polyspermy.

The Initiation of Development

Fertilization of the egg initiates the developmental program. Considerable evidence has recently been obtained, indicating that activation is mediated by changes in the ionic composition of the egg.

We shall now examine this evidence, beginning with a consideration of ionic calcium, which plays a central role in activation. Early evidence that calcium undergoes dynamic changes after fertilization was provided by Mazia (1937), who demonstrated that the proportion of free calcium increases after fertilization of the sea urchin egg, apparently as a result of the release of calcium ions from a bound state within the egg. Mazia's observations could be explained in one of two ways. Either free Ca^{++} increases as *a result* of activation, or it is *a primary* cause of activation.

If it is a cause of activation, then activation should be dependent upon an elevation of intracellular Ca^{++}. This dependence is illustrated by the fact that parthenogenic activation of fish or amphibian eggs by pricking them with a needle is only possible if Ca^{++} is present in the external medium.

Presumably, the calcium ions enter the eggs at the wound and trigger development. Recently, the ionophore A23187 has been used as a means of raising internal free Ca^{++} in unfertilized eggs. For example, bathing sea urchin eggs in seawater containing A23187 will activate them.

Surprisingly, the ionophore will also activate eggs in media lacking Ca^{++} and Mg^{++}. This observation led to the proposal that A23187 causes the release of internally bound calcium or magnesium. Most calcium in unfertilized eggs is in a bound form, whereas magnesium is unbound; thus, calcium is presumably the ion involved.

In its bound form, calcium is probably in a complex with one or more proteins. It is assumed that the ionophore mimics fertilization and that the release of bound Ca^{++} is the normal response to sperm contact. Ionophore activation has been demonstrated for a wide variety of organisms, including starfish, mollusks, amphibians, fish, and mammals, suggesting that release of bound Ca^{++} is a general-ized mechanism for egg activation in the animal kingdom.

Indirect evidence in support of the central role of Ca^{++} in egg activation is provided by an experiment in which the calcium chelator EGTA was injected into unfertilized sea urchin eggs. Upon fertilization these eggs failed to activate. Since reducing calcium prevents activation, it is likely that the elevation in ionic calcium is essential for activation.

Direct observation of Ca^{++} release is now possible through a technique that utilizes aequorin, a protein extracted from jellyfish, which luminesces in the presence of Ca^{++}. The aequorin must be injected into the egg where it can interact with liberated Ca^{++}.

The initial aequorin experiments were done with eggs of the fish *Oryzias latipes* (commonly called "medaka"), since they are relat-ively easy to microinject and since they are transparent, allowing the luminescence to be seen and measured. Fertilization or ionophore treatment causes a transient burst of luminescence that develops rapidly and dissipates more slowly. Aequorin injection has now been accomplished with sea urchin eggs as well, with similar results.

Clearly, bound calcium is liberated after fertilization. Furthermore, since the aequorin experiments indicate that this effect is transient, it would appear that the Ca^{++} again becomes bound, possibly at a new site in the cell.

ME CORTICAL REACTION

One of the consequences of the elevation in free Ca^{++} in the sea urchin egg is triggering of the cortical reaction. This was first demonstrated by Vacquier, using isolated sea urchin egg plasma membranes that possess intact cortical granules.

Vacquier developed an ingenious method for isolating the membranes so that the granules can be observed. A substrate (e.g., plastic, glass) is coated with a solution of protamine sulfate. Egg vitelline envelopes adhere to the protamine-coated surface.

Treatment of the attached eggs with calcium-free seawater containing a chelating agent causes the eggs to lyse, and the egg membranes spread out on the surface "*inside-out*," that is, with the inner surface and attached cortical granules exposed. A cortical reaction can be induced on these exposed membranes by treating them with Ca^{++}.

The response is an instantaneous swelling of granules and a discharge of their contents, which form a gel network on the exposed surface of the membrane.

A means by which Ca^{++} may be responsible for triggering the cortical reaction is suggested by recent experiments of Steinhardt and Alderton. These investigators prepared antibodies to the calciumbinding protein calmodulin. When isolated sea urchin plasma membranes are treated with the antibodies, the ability of Ca^{++} to induce the cortical reaction is lost.

This result suggests that calmodulin is responsible for mediating the effects of Ca^{++} on the cortical reaction after binding to the ionic calcium. Steinhardt and Alderton have demonstrated by indirect immunofluorescence that calmodulin is present on the inner plasma membrane surface where it could perform this function.

The effect of Ca^{++} on the cortical reaction has also been investigated in the larger amphibian eggs, which are of sufficient size that Ca^{++} can be directly micro-injected into them. As with the isolated sea urchin membranes, Ca^{++} injections induce a cortical reaction. Recently,

Ca^{++} has been injected into mouse eggs, causing not only cortical granule breakdown but the initiation of cell division as well.

This result suggests that the postfertilization increase of intracellular free Ca^{++} activates the normal sequence of events that initiate mammalian development.

METABOLIC ACTIVATION

Fertilization is often considered a "trigger" that switches on the metabolically hypoactive unfertilized egg so that a predetermined program of metabolic processes can unfold. Since experimentally induced elevations of ionic calcium will mimic the effects of the sperm, we assume that the transient rise of ionic calcium that follows sperm entry mediates the effects of the sperm in activating this program.

In the sea urchin this program includes increases in particular enzymatic activities, an increase in the rate of respiration, initiation of macromolecular synthesis, and changes in the transport characteristics of the egg plasma membrane. The initial developmental program is entirely cytoplasmic and does not rely upon concurrent nuclear input.

When is this program initiated, and how is the transient rise in ionic calcium involved in its initiation? The earliest metabolic changes include activation of respiratory enzymes, followed by a large burst in the respiratory rate.

These changes are initiated within seconds of the cortical reaction. Activation of protein synthesis is a "late response," beginning five minutes after fertilization. Amino acid, phosphate, and nucleoside transport mechanisms and DNA synthesis are also activated as part of the late response.

Calcium can activate both the early and late responses. However, since the calcium release is transient and precedes the late responses by several minutes, there presumably are intervening events that mediate the effects of the calcium release. In fact, the late events can be experimentally uncoupled from a dependency upon Ca^{++} release.

This can be achieved by activating eggs with ammonia. The late events of activation occur, but the calcium release, the cortical reaction, and the respiratory changes do not occur. In contrast, the late events can be prevented by transferring eggs to sodium-free seawater after fertilization.

These effects of ammonia and sodium are both related to another event that follows fertilization-an elevation of intracellular pH. This pH rise is reflected in the release of acid from the eggs into the surrounding medium. This "fertilization acid" is due to an efflux of H^+ from the eggs, which begins at the start of the cortical reaction and continues for four minutes.

Release of H^+ is prevented if eggs are transferred to sodium-free seawater after fertilization. However, when Na' is added to the water, H^+ efflux occurs, and its rate is linearly dependent upon Na^+ concentration. Likewise, the intracellular pH rise is dependent upon the presence of extracellular Na^+. Thus, the failure of eggs to develop in sodium-free seawater is due to the inhibition of the pH rise.

The parthenogenic activation of eggs by ammonia is due to its effects on pH; that is, it mimics the sodium-dependent pH elevation. The pH rise produced by ammonia is caused by the penetration of the cell by the uncharged base NH_3, which subsequently picks up a hydrogen ion (forming NH_4^+) and directly raises the pH of the cytoplasm. Thus, although their mechanisms are different, both sodium and ammonia produce the same result, that is, an increase in pH.

What is the significance of the sodium-dependent pH shift? Johnson et al. (1976) have proposed that the low cytoplasmic pH of the unfertilized egg maintains metabolic inhibition, whereas the rise in pH relieves the inhibition and allows development to proceed. In this connection, Grainger

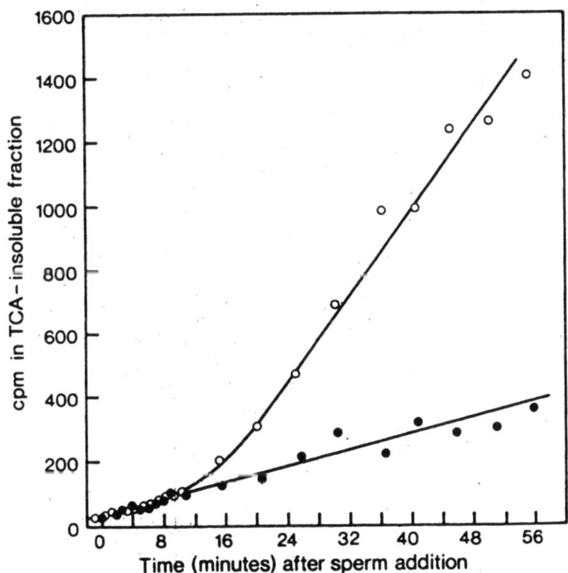

Figure 3.5 : Protein synthesis by unfertilized eggs (dots) and fertilized eggs (circles) as measured by incorporation of ¹⁴C-leucine.

et al. have recently demonst-rated that the increase and maintenance of protein synthesis at fertilization can be largely accounted for by the increase in intracellular pH. The molecular mechanisms involved in the activation of protein synthesis.

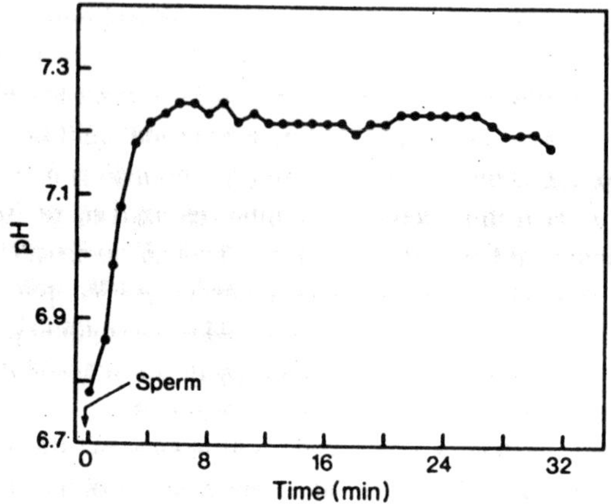

Figure 3.6 : Continuous recording of intracellular pH during fertilization.

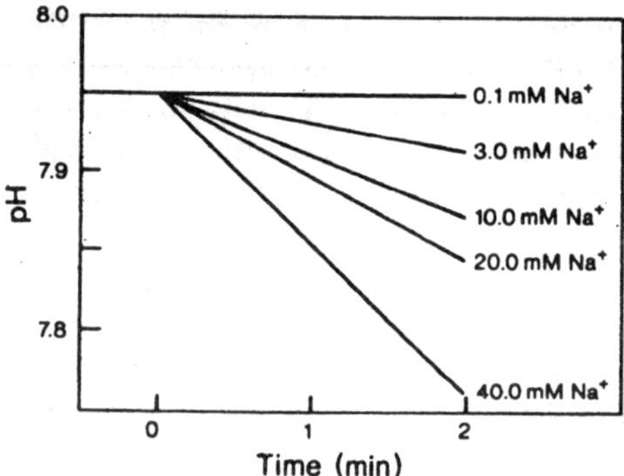

Figure 3.7 : Acid efflux from sea urchin eggs as a function of sodium concentration. Acid release is indicated by a drop in the pH in the surrounding sea water.

As we previously discussed, metabolic activation is mediated by the release of bound Ca^{++}. Calcium release, in turn, triggers the sodiumdependent increase in intracellular pH. The dual roles of calcium and sodium in the pH rise. In this experiment eggs were activated by the calcium ionophore A23187, which normally causes an elevation in pH.

However, the pH remains acidic if sodium is absent from the medium; addition of sodium allows alkalization of the eggs and activation of development. Clearly, an understanding of the interrelationships among

Ca^{++}, Na^{+}, and the pH rise is critical to our understanding of metabolic activation. Additional research should help clarify this relationship.

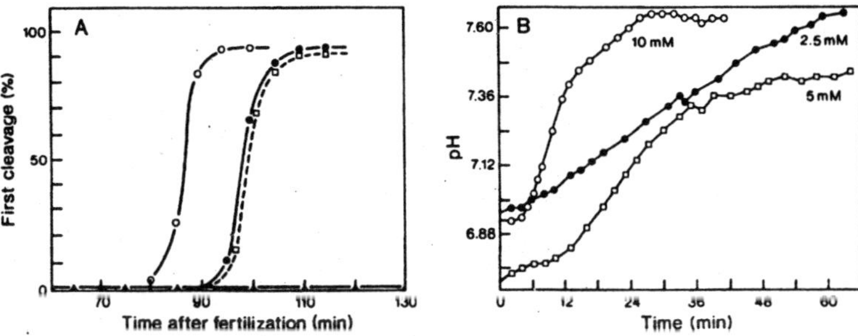

Figure 3.8 : The ammonia effect. A, Ammonia can substitute for sodium in initiating development. At 1 minute after fertilization, eggs were washed in sodium-free sea water to prevent activation. Transfer of eggs to sea water will initiate development (see solid line with open circles on graph). Likewise, transfer of eggs to sea water containing either 50 mM sodium (solid line with solid circles) or 5 mM NH_4Cl (dashed line with open squares) will initiate development Control eggs transferred to sodium-free sea water (solid line with solid triangles) will not cleave.

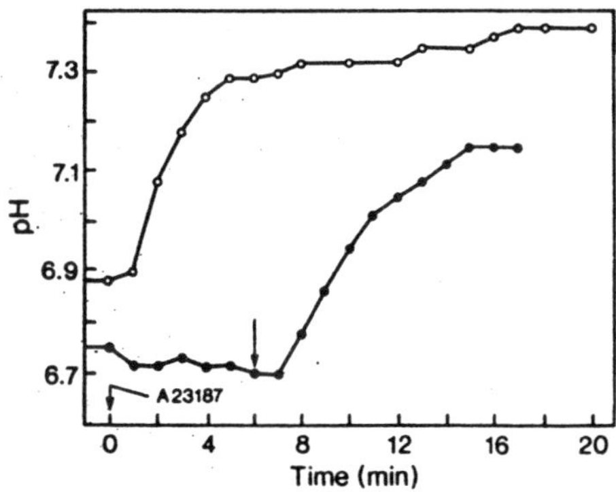

Figure 3.9 : Dual roles of Ca^{++} and Na^{+} in alkalization after activation of the sea urchin egg. Eggs were activated with A23187, and pH was monitored with an electrode. Line with open circles represents intracellular pH in eggs activated in artificial sea water. Line with closed circles represents intracellular pH in eggs activated in sodium-free sea water. At the arrow the external sodium concentration was raised to 80 mM.

How generalized are the results obtained with sea urchins? Does a similar pattern of metabolic activation under ionic control pertain to other organisms? We have only partial answers to these questi-ons. We do know that release of intracellular Ca^{++} activates devel-opment of the eggs of a variety of organisms.

However, there is reason to believe that the events after Ca^{++} release are variable. For example, the sea urchin is in a different metabolic state at fertilization from that of the vertebrates and other organisms in which meiotic maturation is under hormonal control. In the sea urchin, meiotic maturation is completed before ovulation, and the eggs become metabolically hypoactive until fertilization. In contrast, maturation of the amphibian egg (and that of most other vertebrates) is initiated at ovulation and proceeds only as far as metaphase II.

Fertilization triggers completion of meiosis so that pronuclear fusion can occur. The metaphase block in unfertilized eggs is thought to be maintained by an endogenous inhibitor that is inactivated by Ca^{++} release at fertilization. The hormonal stimulus that initiates maturation also activates protein synthesis and a number of other metabolic changes.

Thus, the unfertilized amphib-ian egg, unlike that of the sea urchin, is metabolically very active. Fertilization causes further metabolic changes, including alterations in certain enzymatic activities and an increase in the number of polysomes in the cytoplasm. An increase in polysomes after fertilization suggests that fertilization causes a further elevation in the rate of protein synthesis.

Taken together, these results indicate that metabolic *activation* in the amphibian is initiated by the hormonal stimulus, and sperm entry triggers *metabolic adjustments*. The studies on metabolic activation in other species are too meager to draw a generalized picture of the mechanisms of metabolic activation. Clearly, considerable research remains to be done on this subject.

EGG REARRANGEMENTS

Fertilization may elicit substantial reorganization of cytoplasmic components. Some of these components are *morphogenic determinants* that become segregated into particular cells during cleavage and impart specific developmental properties on the cells by affecting their patterns of gene expression.

Because of their developmental significance, rearrangement of determinants is a vital fertilization response. We shall discuss examples of rearrange-ment of morphogenic determinants and the roles of these rearran-ged components on cell.

FORMATION OF THE DIPLOID ZYGOTE NUCLEUS

Both male and female gamete nuclei must undergo a certain amount of "processing" in the egg before they are capable of forming a diploid zygote nucleus. Furthermore, they must migrate to a common point where they can make contact.

The sperm nucleus has a particularly complex history in the egg.

It is delivered to the egg as a streamlined structure with highly condensed chromatin. It must enlarge, and its chromatin must decondense before fusing with the female pronucleus. This process has been thoroughly documented in sea urchins, and this discussion will be based primarily upon the results obtained for that group.

As discussed earlier in this chapter, the sea urchin sperm is initially incorporated into an elevated fertilization cone that forms in response to sperm-egg contact. Shortly after its incorporation the nuclear envelope surrounding the sperm nucleus begins to vesiculate and dissociate.

The dissociation is not complete, however, since the apical and basal portions of the sperm nuclear envelope remain intact and are later incorporated into the male pronuclear envelope. Chromatin dispersal begins at the periphery of the nucleus after nuclear envelope dispersion and proceeds toward the center of the chromatin mass.

Before dispersion is complete, membranous vesicles begin to aggregate along the periphery of the chromatin. These vesicles fuse with one another and with the persisting apical and basal portions of the original sperm nuclear envelope to produce the male pronuclear envelope.

In addition to undergoing internal morphological and physiological changes, the pronuclei must also move toward the site in the egg where they will encounter one another. In sea urchins, movement of the male pronucleus is caused by the sperm aster, which radiates from the pronucleus and pushes against the egg cortex, causing the pronucleus to be displaced toward the center of the egg.

The astral fibers contact the female pronucleus, interconnecting the pronuclei and abruptly drawing the female pronu-cleus to the male pronucleus. Continuing action of the aster seems to displace the two adjacent pronuclei to the egg center, where they fuse to form a diploid zygote nucleus.

The migration and fusion of the pronuclei are shown in the series of photographs in figure elsewhere in this chapter. The role of astral fibers (which are microtubules) in pronuclear movement is supported by experiments in which pronuclear movement is prevented by inhibitors of microtubules.

The eggs of some animal species do not demonstrate pronuclear fusion. In these eggs, the pronuclei approximate one another and remain closely apposed until the dissolution of the pronuclear envelopes and alignment of the chromosomes on a common metaphase plate.

The rabbit zygote provides an example of pronuclear approximation. The pronuclei migrate to a central position in the egg, where they flatten and become highly convoluted on the sides facing one another. These surfaces then interdigitate and become tightly apposed. After apposition, chromatin condens-ation occurs, indicating that the chromatin has begun preparing for the first cleavage division.

Next, the pronuclear envelopes begin to break down by the familiar process of fusion of inner and outer layers followed by vesiculation.

After disruption of the pronuclear envelopes the condensing chromosomes from the two pronuclei intermix and orient on the first mitotic spindle. Thus, in contrast to the sea urchin, a zygote nucleus is not formed until the two-cell stage, when the maternal and paternal genomes are enclosed within a common nucleus for the first time.

The absence of pronuclear fusion has also been reported for the mouse and golden hamster zygotes. It will be interesting to discover whether this is a general feature of mamma-lian zygotes.

PHYSICAL AND CHEMICAL CHANGES AT FERTILIZATION

In the following, we shall be primarily interested in sea urchin eggs, whereby one should draw a distinction between early and late changes at fertilization, as pointed out by D. Epel in 1978.

Early Changes

The first recorded change at fertilization is a brief *depolari-zation of the egg membrane potential;* it is believed to be due to an influx of Na^+ ions and to be responsible for the so-called fast block to polyspermy. This fast electrical block prevents the entry into the egg of more than one spermatozoon in *Xenopus* as well as in sea urchin eggs.

A slower block to polyspermy is correlated with the elevation of the fertilization membrane which prevents the entry into the egg of supernumerary spermatozoa.

Another important early ionic change is an *increase in free* Ca^{2+}, much larger than that which follows progesterone addition in *Xenopus* oocytes (see preceding chapter). This increase is responsible for the breakdown of the cortical granules and the exocytosis of their contents. Increasing artificially the free Ca^{2+} content of unfertilized eggs, by treatment with *a* Ca^{2+} *ionophore* called A23187, is an excellent method for inducing a cortical reaction in eggs of many animal species.

Activation of sea urchin eggs (chromosome condensation. DNA replication, formation of asters) can also be obtained by treatment of sea urchin eggs with *ammonia.* Ammonia bypasses the calcium rise step: there is no cortical reaction. The cortical granules remain intact and a fertilization membrane does not develop around the egg. Ammonia penetrates easily into unfertilized sea urchin eggs and increases their *internal pH*. Normal fertilization also induces, within 1 and 4 min after sperm addition, an increase in the sea urchin egg internal pH as high as 0.3-0.5 pH units.

This rise is due to an exchange between extracellular Na' and intracellular protons (H^+) which are ejected from the egg into the outer medium. A similar increase in internal pH at fertilization has been found in *Xenopus* eggs. In both cases, the internal pH remains constant during cleavage after its initial rise shortly after fertilization.

As we shall see in more detail, the classical method for inducing *parthenogenesis* in sea urchin eggs is that of Loeb. This involves two successive steps: treatment of the unfertilized eggs first with hypertonic seawater and then with butyric acid, or vice versa.

With this two-step method, asters form around typical centrioles. With the other activation methods (treatment with the A23187 ionophore or with ammonia) asters also form, but the microtubules originate from the chromosomes and from amorphous osmiophilic foci. It seems that self-assembly of typical centrioles is a prerequisite for full parthenogenetic embryogenesis. In contrast, eggs activated by A23187 or ammonia treatment, stop developing after a few abortive cleavages.

Later Biochemical Changes

Oxygen Consumption

Warburg discovered, more than 70 years ago, that the respiration of unfertilized sea urchin eggs increases considerably (three to four times) a few minutes after the addition of sperm. However, this is not a general phenomenon, present in all eggs. If one compares the oxygen uptake of the fertilized and unfertilized eggs of a number of animals, one finds that respiration can either increase (as in the sea urchin), remain constant or even decrease at fertilization.

Fertilized eggs of all the species studied (which have about the same size and all live in seawater) have almost the same oxygen consumption; on the other hand, the respiration of the unfertilized eggs of the various species show very great variations.

The conclusion that can be drawn from these experiments is the following: fertilization does not necessarily increase the respiration of the egg, but it *regulates* this process. The respiration of the unfertilized egg, which is an inactive cell, is often abnormal; fertilization brings it back to the normal value.

That this conclusion is correct for sea urchin eggs has been shown by delicate experiments which made it possible to measure the respiration of a small number of carefully selected, perfectly synchronous oocytes. The oxygen consumption of the oocyte (which still has an intact germinal vesicle) is high; it falls considerably in the unfertilized egg (which has eliminated its two polar bodies) and remains low for a considerable time; it is restored to the initial high level by fertilization.

What is peculiar to the sea urchin egg is not so much the strong increase in respiration, which follows fertilization, but the very low respiratory rate of the unfertilized egg.

Despite many efforts, the identity of the substrate which is oxidized by sea urchin eggs at fertilization has long remained mysterious. It is only recently that it has been discovered that this substrate is simply - water. The large and sudden increase of respiration at fertilization is mainly due to the formation of hydrogen peroxide (H_2O_2) which is used by an egg *ovoperoxidase* for hardening the fertilization membrane; H_2O_2 also inactivates spermatozoa in excess and this contributes to the prevention of polyspermy.

It has been shown that at least one enzyme is activated, within a few seconds, after fertilization. This is a NAD kinase, the enzyme which phosphorylates the respiratory co-enzyme NAD (nicotinamide adenine dinucleotide) into its phosphorylated form, NADP. This synthesis or activation of NAD kinase is the fastest biochemical change that has so far been detected when sea urchin eggs are fertilized.

When sperm is added to sea urchin eggs, there is a big burst in CO_2 production, much larger than could be accounted for by cell oxidation. This burst is due to the production of a "fertilization acid", as Runnstrom first found in 1928. For many years, numerous unsuccessful attempts have been made to identify its chemical nature.

The mystery was solved when Epel discovered, in 1976, that a proton efflux takes place at fertilization; this efflux acidifies the seawater which surrounds the eggs; as a result a burst of CO_2 at the expenses of carbonates and bicarbonates present in sea-water, takes place. Thus, the Na^+/H^+ exchange which, as we have seen, occurs soon after fertilization has two consequences: increase of the internal pH of the egg and production of a CO_2 burst.

A few words about the changes in respiration after fertilization in *frog eggs*. Their oxygen consumption is not affected, but there is, as in the sea urchin eggs, an initial burst in CO_2 production. This excess CO_2 has a different origin than in sea urchin eggs. When the unfertilized eggs are accumulated in the uterus of the female frog, CO_2 cannot escape through the mucus-coated uterine walls. The unfertilized eggs, at the time of laying, are thus intoxicated by a large excess of CO_2.

The true CO_2/O_2 ratio (respiratory quotient: R.Q.) drops from the value 1, characteristic of carbohydrate oxidation found in the unfertilized egg, to the very low value of 0.65 after fertilization. This low R.Q. suggests that substrate oxidation is incomplete soon after fertilization and during early cleavage in frog eggs.

Carbohydrate Metabolism

Both sea urchin and frog eggs contain all the enzymes needed for the classical metabolic pathways of carbohydrate metabolism (glycolysis, tricarboxylic acid cycle and hexose-monophosphate direct oxidation). In sea urchin eggs, it has been reported that the glycogen content decreases during the first 10 min after fertilization and then slowly returns to its initial value.

Protein Metabolism

Protein synthesis will be discussed later in a special paragraph. Fertilization in the sea urchin induces a number of changes in the *solubility* and in the *electrical charge* of the proteins; the electrophoretic pattern changes and a new KCI-soluble protein can be isolated.

This protein, which is localized in the egg cortex and contains -SH (sulhydryl) groups, probably plays a role in the cortical changes which follow fertilization and lead to the first cleavage. The sulfhydryl-disulfide (-SH/-SS-) ratio undergoes changes in this KC1-soluble protein fraction after fertilization; these changes are probably related to the contractility of the cortex. since this protein presents all the characteristics of the "*contractile*" proteins. It will be discussed in more detail in the next chapter.

A temporary decrease in total proteins and particularly in the ribosomal proteins has been observed soon after fertilization of the sea urchin egg. Probably correlated with this proteolysis is a transient activation of several distinct proteases, which has been reported by several authors and which, as we shall see, might be physiologically important.

Lipid Metabolism

As for glycogen and proteins, fertilization of the sea urchin egg is followed by the breakdown of some of the lipid constituents. The content in *phosphatides* decreases, while the *free cholesterol* increases at the expense of the cholesterol esters. Of some interest is the fact that the sperm contains a lecithinase, an enzyme which transforms the phosphatides in the surface active lysolecithins.

NUCLEIC ACID AND PROTEIN SYNTHESIS

A sudden and important burst of protein and DNA synthesis, in the absence of appreciable RNA synthesis, makes sea urchin egg fertilization one of the most fascinating problems of molecular embryology. We shall first examine DNA synthesis.

DNA Synthesis

DNA synthesis begins, in the two pronuclei, a few minutes after insemination. It is complete, in sea urchin eggs, in about 20 min, thus before amphimixy. The DNA content of each of the pronuclei exactly doubles, so that the DNA content of the zygote nuclei is four times that of a haploid cell (4 C).

In amphibian eggs, the microinjection of labelled *thymidine* (the classical precursor used in the study of DNA synthesis) shows that replication of DNA begins shortly after fertilization and is complete before amphimixy.

We have seen, when we studied maturation, that the whole machinery for DNA synthesis is made ready for use during that period. In particular, DNA polymerase a is synthesized during maturation. This enzyme is present in the cytoplasm, in an active form, since foreign DNA is replicated if it is injected into the egg.

However, the DNA present in the egg chromosomes is not replicated until a few minutes after fertilization; DNA synthesis does not begin in the male pronucleus before it swells. In cases of polyspermy, DNA synthesis takes place simultaneously in all the sperm nuclei and in the female pronucleus.

Its initiation thus results from *cytoplasmic* changes, perhaps from a migration of cytoplasmic DNA polymerase into the nuclei. As we have seen, initiation of DNA synthesis is not related to the cortical reaction and to Ca^{2+} release since it takes place in ammonia-treated unfertilized eggs, where the Ca^{2+}-dependent step is bypassed.

It is likely that the cytoplasmic changes which induce DNA synthesis are correlated with the rise in internal pH which follows fertilization. There is no doubt that DNA polymerase a is involved in the initiation of DNA synthesis in fertilized or activated sea urchin eggs: a specific inhibitor of this enzyme, *aphidicolin* stops development at amphimixy and completely prevents further DNA replication.

RNA Synthesis

There is no sudden burst in RNA synthesis at fertilization. Very little, if any, nuclear RNA synthesis can be found before the two- to four-cell stage; the low level RNA synthesis, which can be detected in recently fertilized sea urchin eggs, is mainly due to mitochondrial RNA synthesis. Suppression of RNA synthesis by actinomycin D treatment has no effect at all on fertilization or even cleavage in both sea urchin and amphibian eggs.

This lack of RNA synthesis is probably somehow correlated with the presence of a large store of many kinds of RNAs which, as we have seen, have accumulated during oogenesis. As we know, unfertilized sea urchin eggs possess a large population of mRNAs theoretically capable of coding for as many as 10.000 to 20,000 different proteins.

Curiously, maternal mRNAs in sea urchin and *Xenopus eggs* seem to be unfinished molecules: they contain repetitive DNA transcripts and are more similar to nuclear RNAs than to typical mRNAs (E. Davidson). These maternal mRNAs are bound to proteins in the form of *ribonucleoprotein particles* (RNP) smaller than the ribosomes; more will be said about them later.

Although there is almost no net RNA synthesis at fertilization, addition of sperm or activation with ammonia are quickly followed by *polyadenylation* of pre-existing mRNAs: the average length of the (poly A) "tail" increases from 100 to 200 adenylic acid residues. Polyadenylation presumably increases the stability of the maternal mRNAs in the egg cytoplasm.

However, the biological significance of polyadenylation at fertilization remains obscure: it can be inhibited by cordycepin, an analogue of adenosine; yet, fertilization and cleavage are normal in eggs treated with this drug.

It is probable that in sea urchin eggs, the excess of ribosomes and mRNAs is stored in *heavy bodies:* these large, RNA-rich granules appear in the cytoplasm after germinal vesicle breakdown and disappear during early cleavage. They increase in size whenever cleavage is inhibited. The identity of the RNAs present in the heavy bodies remains uncertain, but recent work in our laboratory has shown that they contain large amounts of ribosomal RNA.

Protein Synthesis

Monroy's experiments were the first to demonstrate that fertilization stimulates, in sea urchin eggs, protein synthesis in the same dramatic way as oxygen consumption. In order to circumvent the impermeability barrier of the unfertilized eggs, he injected a labelled amino acid (methionine) into female sea urchins. The amino acid penetrated very well into the mature ovarian oocytes, but was not incorporated into the proteins of the unfertilized eggs.

A few minutes after sperm addition, the free labelled amino acid began to be very actively incorporated into the proteins, demon-strating that intense protein synthesis was now under way.

The first explanation proposed for this sudden burst in protein synthesis was the following: the pronuclei, in the fertilized eggs. would immediately produce new mRNA molecules, which would combine with the pre-existing ribosomes; the newly formed polysomes would be the site of protein synthesis.

But we have just seen that there is no measurable burst of RNA synthesis at fertilization and that no nuclear RNA synthesis can be detected until early cleavage. Furthermore, actinomycin D, which should block RNA synthesis, does not prevent the rise of protein synthesis at fertilization.

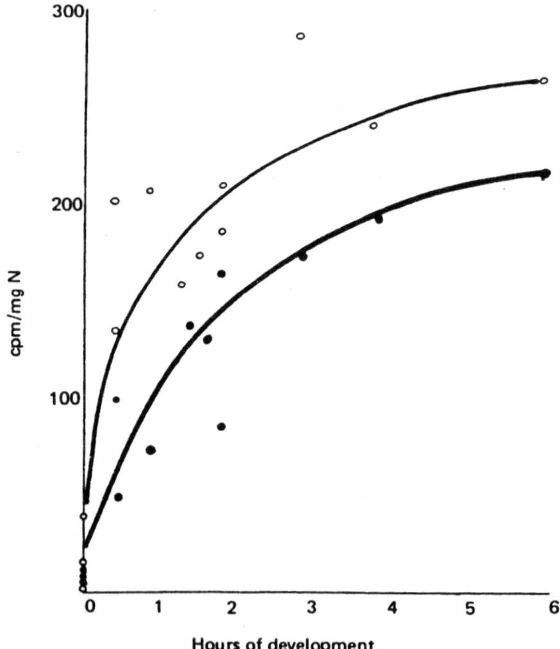

Figure 3.10 : Fast labelling of the microsomal (m) and soluble (l) protein fractions following fertilization of Paracentrotus eggs.

However, it was still possible that these negative results were simply due to a low permeability of the fertilized eggs for the RNA precursors and actinomycin. This possibility has been ruled out by experiments made on anucleate fragments of sea urchin eggs.

The potentialities for development and macromolecular synthesis of such fragments will be examined in another chapter. For our present purpose, it is enough to say that unfertilized eggs of a certain sea urchin species, *Arbacia,* can easily be cut into two halves, nucleate and anucleate, by centrifugation under proper experimental conditions.

Neither nucleate nor anucleate fragments of unfertilized eggs are capable of measurable protein synthesis. But, if these fragments, are *activated* by treatment with hypertonic seawater—which induces the cortical reaction, with formation of a fertilization membrane in the egg fragments, but does not allow further development—both nucleate and anucleate halves become capable of active protein synthesis.

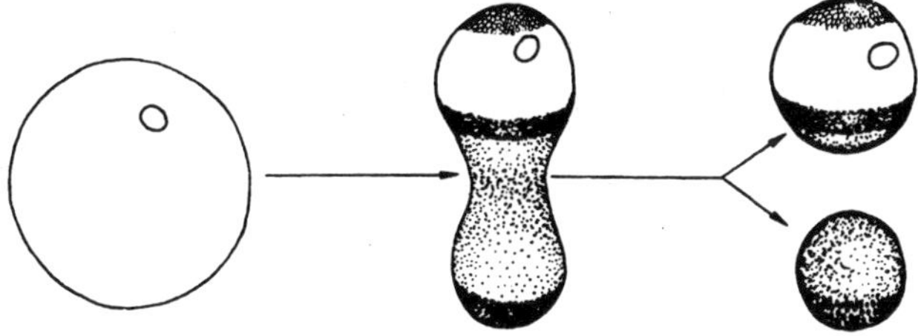

Figure 3.11 : Centrifugation of unfertilized sea urchin eggs results in the sedimentation of the various inclusions present in the egg. Finally, the eggs break into two parts; only the lighter one contains a nucleus.

In fact, incorporation of amino acids into proteins is higher in the *anucleate* than in the nucleate halves. That the anucleate halves, after activation, are the site of true protein synthesis is proven by the fact that the incorporation of amino acid is inhibited all classical inhibitors of protein synthesis.

Furthermore, physical methods show that anucleate activated fragments and normally fertilized eggs synthesize the same proteins (in particular, tub ulin). Finally, it has been demonstrated that the anucleate fragments contain polysomes, which are actively engaged in protein synthesis. Active polysomes cannot be detected in anucleate fragments which have not been activated by hypertonic seawater treatment.

These experiments definitely rule out the possibility that the stimulation of protein synthesis which follows fertilization is due to the synthesis of mRNAs by the pronuclei. An identical stimulation can be obtained in the complete absence of the nucleus. This situation reminds us of what has been said in the preceding chapter about maturation in amphibian eggs. We saw that progesterone induces the synthesis of the same proteins, whether the germinal vesicle is present or has been surgically removed. In both cases, protein synthesis is necessarily controlled at the level of *translation* and not of transcription.

There are two possible explanations left for the stimulation of protein synthesis in fertilized sea urchin eggs and in activated anucleate fragments. One of them is the existence of *masked mRNA* molecules, which would have been synthesized beforehand by the oocyte and would be stored in the unfertilized egg; these mRNAs would be unable to bind to ribosomes to form polysomes and to support protein synthesis.

The second possible explanation is that mRNAs are synthesized on a *cytoplasmic DNA* template after activation or fertilization. Such newly formed cytoplasmic mRNAs would bind to the pre-existing ribosomes and direct protein synthesis. It should be added that these two possibilities are not necessarily mutually exclusive.

Let us begin with the second possibility, that of an intervention of *cytoplasmic DNA* in protein synthesis after fertilization or activation. It is a fact that the nucleate and anucleate halves of the unfertilized sea urchin eggs (which contain about ten times more DNA than the spermatozoon) have about the same DNA content.

This proves that the major part of the DNA present in sea urchin eggs is localized in the cytoplasm. This cytoplasmic DNA is, as expected for mitochondrial DNA. made of circular molecules. It can be transcribed, and there is indeed a low level of RNA synthesis in activated anucleate fragments of sea urchin eggs.

But all attempts to demonstrate that mitochondrial RNA can bind to cytoplasmic ribosomes and direct the synthesis of a large number of proteins have failed. The number of proteins synthesized by sea urchin eggs exceeds 400; the information contained in mitochondrial DNA would allow, at best, the synthesis of about a dozen different proteins.

It should be added that unfertilized sea urchin eggs synthesize, but at a very low rate, the same set of 400 proteins as the fertilized eggs. It is thus the *rate* of protein synthesis which increases dramatically at fertilization and there are no major qualitative changes in protein synthesis at that time.

We can conclude from this analysis that protein synthesis directed by cytoplasmic, mitochondrial DNA and RNAs probably occurs in the fertilized sea urchin egg, but that it has a minor importance and is probably limited, at best, to the synthesis of a few mitochondrial proteins.

The bulk of the proteins synthesized at fertilization must have another origin; the only possibility is the intervention of *masked mRNA molecules of maternal origin.*

What is the evidence in favor of this masked mRNA hypot-hesis? Monroy has shown and Spiegelman has confirmed that unfertilized sea urchin eggs contain large amounts of "template" RNA. This means that if RNA extracted from unfertilized eggs is added to ribosomes isolated from bacteria or mammalian liver, it supports protein synthesis in this in vitro, acellular system.

In order to support protein synthesis, it must bind to the bacterial or liver ribosomes and form polysomes. This is, of course, one of the fundamental properties of the mRNAs. There is no big change in the template activity of the RNAs extracted from sea urchin unfertilized eggs and larvae *(plutei);* the same is true when one compares the template activity of the RNAs isolated from unfert-ilized frog eggs and tadpoles.

This means that the amount of template RNA present in the unfertilized eggs is high enough to support, at least in theory, the intense protein synthesis which exists in plutei and tadpoles. That this template RNA is a mixture of true m RNAs has been demonstrated in Gross' laboratory. It could be shown that unfertilized sea urchin eggs (and even anucleate fragments) contain, in masked form, the specific mRNAs for tubulin, actin and histories.

The masked mRNAs or template RNA present in the unfertilized eggs probably originate in the nucleus of the oocyte during oogenesis. We have seen that about 3% of the RNAs synthesized at the lampbrush stage by the amphibian oocyte is of the mRNA type.

It is likely that a large part of the mRNAs synthesized during oogenesis is *stored* in an inactive form and is used at later stages of development only; their utilization begins at fertilization. As already mentioned, the maternal mRNAs are bound to proteins in the form of 60S ribonucleoprotein (RNP) particles smaller than the ribosomes.

The mRNAs associated with these particles cannot be translated in an in vitro system unless the protein moiety of the RNP particles has been removed. Evidently, the maternal mRNAs of the unfertilized sea urchin eggs are in *a masked,* inactive form so long as they are hound to the proteins of the RNP particles.

How are these maternal mRNAs *unmasked* at fertilization? It is only known with certainty that the stimulation of protein synthesis, in sea urchin eggs at least, is closely linked to the *increase in internal pH* which takes place shortly after fertilization. This has been shown by experiments in which the internal pH value has been manipulated by addition of either ammonia or acetate to the eggs.

While ammonia, as we have seen, stimulates both DNA and protein synthesis. acetate, which lowers the internal pH of the eggs, inhibits protein synthesis and arrests development at amphimixy (like the classical and potent inhibitors of protein synthesis, puromycin and emetine).

There is thus some evidence for the view that, at least in sea urchin eggs, unmasking of maternal mRNAs is a pHdependent process. But there are so many pH-dependent processes occurring constantly in living cells that such a statement does not lead us very far. More work is obviously needed for a better understanding of the changes that the protein-synthesizing machinery undergoes at fertilization.

One possibility, which was proposed by Monroy in 1971, is that the *rihosomes* of unfertilized sea urchin eggs might be abnormal in the sense that they would be unable to bind maternal mRNAs; mild digestion with trypsin indeed restores the ability of unfertilized egg ribosomes to bind to an artificial messenger, polyuridylic acid (poly U).

Since there is some evidence for the release of a protease at fertilization, one can imagine that this enzyme removes some inhibitory proteins from unfertilized egg ribosomes; this would allow these rihosomes to bind more efficiently to the maternal mRNAs.

Such a protease could also play a role in the unmasking of the mRNAs stored in the RNP particles. It is only to be expec-ted that the activity of this proteolytic enzyme is pH- dependent. However, one must admit that the data, which have accumulated on the possible differences between the ribosomes of unfertilized and fertilized eggs, are contradictory.

A recent publication reported that there is a 15 min lag before ribosomes from unfertilized sea urchin eggs become active in in vitro protein synthesis, while there is no lag for ribosomes from fertilized eggs. It has also been shown that, a few minutes after fertilization, one of the ribosomal proteins (called S6) is phosphorylated.

Since phosphorylation of this S6 protein also takes place during *Xenopus* oocyte maturation and when mammalian cells cultured in vitro are induced to multiply at a faster rate, it is likely that this phosphorylation process has some biological importance. But there is no evidence so far that phosphorylation of the S6 ribosomal protein affects the rate of protein synthesis.

It seems therefore unlikely that changes in ribosome conform-ation or composition, resulting from phosphorylation or mild proteolysis of the ribosomal proteins, are the *only* factors involved in the stimulation of protein synthesis at fertilization.

Evidently many *control mechanisms* operate when the rate of protein synthesis increases sharply at sea urchin egg fertilization. Negative, inhibitory controls must exist in the unfertilized eggs in order to keep protein synthesis at a very low level.

However, the suppression of protein synthesis in these eggs is not complete: they already contain a few polysomes and, as we have seen, they synthesize trace amounts of hundreds of proteins. Fertilization, probably by raising the internal pH of the eggs, suppresses the negative controls which were operating in the unfertilized eggs.

Thus, its main biochemical effect is to increase the availability of translatable maternal mRNAs and to allow them to bind to the ribosomes. Another factor involved in the increase in protein synthesis, which follows insemination, is a faster elongation rate of the growing polypeptide chains.

Can the findings obtained with sea urchin eggs be generalized to other cells and eggs? Very similar facts have been found in the *germinating seeds* of many plants: the ribosomes are inactive in the dry seeds, which contain, like sea urchin eggs. a store of masked mRNAs.

When the seeds are soaked in water in order to induce germination, the preformed mRNAs bind to the ribosomes; polysomes are formed and protein synthesis begins. The only difference is that the stimulation of protein synthesis takes place later in germinating seeds than in sea urchin eggs: it takes 1.5 to 2 h in seeds, but only 4-5 min in sea urchin eggs before a large increase in the rate of protein synthesis can be observed.

When one examines what is known about protein synthesis in eggs other than those of the sea urchin, one comes to the conclusion that the latter represent an extreme case. Unfertilized eggs of the starfish, the mollusk *Mactra* and the worm *Cerebra-tulus* are active in protein synthesis; the latter is increased after fertilization.

But not in the same dramatic manner as in the sea urchin egg. We have already seen that the sea urchin egg is also a special case insofar as oxygen consumption is concerned. One may wonder whether protein synthesis would not *decrease* after the fertilization of eggs which show a diminution of oxygen consumption at fertilization.

Experiments on the polychaete worm *Chaelopterus* have shown us that there is a temporary arrest of protein synthesis when the eggs are activated by the addition of an excess of KCI to the seawater; such a treatment, like fertiliz-ation, decreases the oxygen consumption of the eggs by almost 50%. Therefore, it is possible that *energy production,* resulting from the oxidations and phosphorylations in the mitochondria, might be another control mechanism of protein synthesis at the time of fertilization.

Frog eggs also deserve mention. Their protein-synthetic activity decreases as they move through the genital tract (oviduct, uterus) of the female. At fertilization, protein synthesis reverts to a level equivalent to that of the ovarian egg after maturation. But it is not at all certain that this effect on protein synthesis is due to fertilization itself.

It could be due to the removal of the excess CO_2, which has accumulated when the eggs were packed in the uterus of their mother, as already mentioned. Since an excess of CO_2 in the eggs should lower their internal pH, it is possible that this factor, as in sea urchin eggs, controls the rate of protein synthesis in frog eggs.

We think that additional work, on eggs from a variety of animal species, is required before one can generalize the hypothesis that internal pH controls the rate of protein synthesis. For instance, one may wonder whether, in activated *Chaetopterus* eggs. in which there is a transient decrease in protein synthesis after activation, there is a concomitant decrease in the internal pH; only new experiments, which are now technically feasible, will provide an answer to this question.

In conclusion, it seems that the magnitude of the increase in both protein synthesis and respiration, which occurs at fertilization, largely depends on the stage of repression which prevails in the unfertilized egg. The main effect of fertilization is to remove the inhibitions which make the unfertilized egg a "sleepy," almost "dormant" cell.

In other words, the main effect of fertilization would be a derepre-ssion rather than an activation; but this derepression does not operate, as usual, at the genetic, but at the post -transcriptional level.

PARTHENOGENETIC ACTIVATION

The experimental embryologists of the beginning of this century were not as sophisticated as the molecular embryologists of today. They made no distinction between derepression and activation, although, as we shall see, one of them at least was very well aware of the fact that the unfertilized egg is inactive because it is "intoxicated (we would say, in our modern terminology, repressed).

Parthenogenetic stimulation of development after treatment of unfertilized eggs by chemical or physical agents was therefore considered as an activation of the "sleeping" egg.

We shall briefly present here the two main methods of *parthenogenesis* (which were discovered, early in this century, by Loeb and Bataillon) because they still retain a good deal of interest for molecular embryologists.

Loeb was working on sea urchin eggs and discovered that they undergo *activation* when they are treated with *hypertonic seawater*. The fertilization membrane is elevated in the normal fashion, but only a single aster, a "monaster" (of course, of maternal origin) can form. At best, the chromosomes can undergo a few cycles of replication around this monaster; since there is only one aster, no typical mitotic apparatus, with two asters and a spindle, can form.

The consequence is that the egg cannot divide and will not develop further. But Loeb found that if the activated eggs are submitted, after a few minutes, to a second treatment, with *butyric acid* this time, a second aster will appear and the egg will develop *parthe-nogenetically* (that is, without the intervention of the spermatozoon) until the larval pluteus stage is reached. As we have seen, the two asters form around typical centrioles in such eggs.

Bataillon worked on frog eggs and found that they can be activated by pricking them with a clean needle. They form a fertilization membrane and undergo a series of monasterial, abortive division cycles. If, however, the needle is not clean and is contaminated with blood cells, so that a nucleate cell is introduced into the egg at the time of pricking, a second aster will form around the injected cell. Mitosis will then be normal and development will proceed until the tadpole stage.

The cell which has been introduced in the unfertilized egg thus contains a "second factor" necessary for the formation of a second aster and further development. Since the introduction of an anucleate mammalian red blood cell had no effect, Bataillon concluded that the second factor must somehow he linked to the cell nucleus and that a "karyocatalysis" is required for parthenogenetic development.

This "*karyocatalysis*" theory is no longer held, since there is now good evidence that Bataillon's second factor is a *centriole*, not a nucleus. The lack of activity, as a second factor, of mammalian red blood cells is due to the fact that these cells do not possess centrioles.

It has been shown that injection of cell extracts enriched in centrioles has the same effects as injection of a nucleate cell, i.e. induction of complete parthenogenetic development. Treatment with colchicine, which inhibits the formation of astral microtubules, suppresses the activity of these cell extracts.

However, this activity is enhanced by agents, which, like heavy water (D_2O), stabilize the microtubules. The second factor of Bataillon is thus a *"microt-ubule organizing center"* (MTOC); it is necessary for the building up of a bipolar mitotic apparatus in eggs which have been activated by pricking with a needle.

It is important to note that parthenogenetic activation, even if it is not followed by further development, induces the same metabolic changes in the egg as fertilization. Treatments with hypertonic seawater, even in the absence of a second treatment with butyric acid, is sufficient to increase both respiration and protein synthesis in sea urchin eggs. The part played by the spermatozoon in these early metabolic events thus appears as negligible.

This statement is no longer true for later development. Parthog-enetic embryos are, as a rule, *haploid* and die during development. The *haploid syndrome* can very easily be demonstrated in frog eggs which have been fertilized with sperm previously irradiated for a short time (1 to 2 min) with an UV lamp.

The spermatozoa remain perfectly motile and the same percentage of fertilization can be obtained as with normal, nonirradiated sperm. But the chromatin of the spermatozoon has been irreversibly damaged; the paternal chromosomes are eliminated into the cytoplasm, where their DNA undergoes complete degradation, during the first cleavage of the egg.

Development is at first normal; but a few eggs already fail to gastrulate properly. The others may reach, at best, a very abnormal tadpole stage. They display strong oedema, have a reduced head and very small gills, and are unable to digest their yolk completely. This haploid syndrome is always *lethal.*

A few eggs, however, may escape death and become adults. Cytological observation shows that they are always diploid or polyploid. They contain *2n, 4n, 6n* or *8n* chromosomes. This regulation of the number of chromosomes, which was first obse-rved by Bataillon with his pricking method, is due to the fact that, in a few eggs.

The chromosomes undergo one or more monasterial cycles of replication before a second aster forms around the injected nucleate cell. The biochemical reasons for the haploid syndrome remain totally unknown.

MOLECULAR EMBRYOLOGY AND CLASSICAL THEORIES OF FERTILIZATION

Both Loch and Bataillon formed theories of fertilization based on their own work on sea urchin and frog eggs, respectively.

Loeb proposed that the two factors needed for inducing parthenogenetic development in sea urchin eggs (hypertonic seawater and butyric acid, or vice versa) could have two opposite effects. Incipient cytolysis would first he induced and would lead to the death of the egg unless restoration to normal conditions was engendered by the second parthenogenetic agent.

By analogy. the spermatozoon would, during fertilization, first induce lytic changes and then restore the normal situation. There is some biochemical evidence supporting Loeb's theory at least in the case of the sea urchin egg. We have mentioned the initial breakdown and later resynthesis of macromolecules such as glycogen. certain proteins, phosphatides and cholesterol esters.

The presence of a lecithinase in spermatozoa, as well as the synthesis or activation of proteases very soon after fertilization, can be taken as arguments in favor of the incipient cytolysis hypothesis. However, the breakdown of the cortical granules is a normal response to a local increase in calcium ions; the exocytosis of its contents, can hardly he taken as a lytic process.

For Bataillon, the egg is an intoxicated cell and the elimination of the perivitelline fluid would lead to a "purification" of the cell *(reaction d'epuraiion)*. Although this part of the theory is probably no longer true, since certain eggs show very little, if any, cortical changes at fertilization, molecular embryology provides ample demonstration for the view that the unfertilized egg is intoxicated (or repressed) by the products of its own metabolism.

This intoxication is probably related to the very low permeability of the unfertilized egg. As we have seen. the studies made on the respiration of invertebrate eggs have shown that oxygen consumption is abnormal in unfertilized eggs and is brought hack to normality by fertilization.

The fact that the cytoplasm of frog oocytes, which have finished their maturation, blocks cell division when it is injected into a cleaving egg is further evidence in favor of Bataillon's theory.

Bataillon has placed emphasis on the role of CO_2 in the intoxication of the unfertilized egg. In fact, he could show that if frog eggs are treated with an excess of CO_2 they undergo activation (monasterial cycles of chromosomes) instead of cleavage. If CO_2 is removed, normal cleavage becomes possible.

Some of the experiments described in this chapter have shown that the CO_2 content of the eggs might indeed play a role by controlling the intracellular pH in the regulation of protein synthesis in both sea urchin and amphibian eggs.

When we look today at the theories proposed long ago by Loeb and by Bataillon, we come to the conclusion that these two great embryologists showed remarkable insight. At present, their theories no longer contradict each other, but simply complement each other in a harmonious way.

4

KARYOTIC REGULATION

A living cell is a partnership between the nucleus and the cytop-lasm. Now, although this concept has its limitations, it will serve as a basis on which to divide our subject matter. This chapter concerns the nucleus. The remainder of the book deals mainly with the cytoplasm.

The nucleus, with its store of hereditary factors, the genes, is the conservative member of the partnership. It controls the syntheses which go on in the cell. Hence it directs that most amazing of all syntheses, the course of development.

Hence also it determines the species and individual , characteristics which are the outcome of development. On the other hand, it is primarily the cytoplasm which develops and is the progressive member of the partnership. It begins in relative simplicity and then, by differen-tiation and molding, becomes the mature organism in all of its structural and functional complexity.

Both the nucleus and the cytoplasm are needed for the continued life and functioning of the cell. Many experiments have been performed in which the nucleus of a cell has been removed. For example, when an amoeba is cut into two parts, the part containing the nucleus heals and continues to live.

The part which lacks a nucleus may carry on for a time, but ultimately it wears itself out and perishes, for it cannot continue to rebuild its substance. It is reported that a sea urchin egg from which the nucleus has been removed may cleave several times, but it will not continue to develop.

It is even more obvious that a nucleus cannot live and carry on its functions without the cytoplasm. In particular, the nucleus depends upon the cytoplasm for its raw material and energy. It is probable that only in the cytoplasm do the oxidations take place by which energyrich molecules (ATP) are synthesized.

THE LIFE CYCLE OF THE NUCLEUS

The story of the nucleus as it is revealed by the microscope is a repetitive story, yet one that is full of interest. It begins at fertilization when the chromosomes of the two germ cells (gametes), an egg and a sperm, come together in the nucleus of the fertilized egg or zygote.

The story ends and is ready to start again, at least as far as the contribution of the individual to the race is concerned, when the new individual has matured and has produced gametes of its own. Between fertilization and the maturing of the germ cells, the story of the nucleus is one of alternating growth and division by mitosis.

Fertilization

From the standpoint of the nucleus, fertilization is the coming together of the chromosomes of the egg and the chromosomes of the sperm in the same nucleus, the zygotic nucleus. One set of chromosomes, the haploid number, is present in the egg.

Another set is present in the sperm. This makes two sets, the diploid number, present in the zygote. In man the haploid number is 23. Hence the human diploid number is 46.

Each chromosome of a haploid set has its own individuality; that is, it is of its own kind. It becomes visible at the beginning of cell division and disappears at the close of cell division in its own characteristic manner.

Very significantly, its genes are arranged within it in a definite sequence. Moreover, it affects development in its own specific way. Each chromosome of a diploid set usually has a mate. Yet, although the two mates are within the same nucleus, with rare exceptions each remains independent of the other throughout the cell divisions (mitoses) of development.

Mitosis

The coming together of the male and female nuclei in fertilization is followed *by mitosis,* the process by which one cell becomes two cells in so complicated but precise a manner that each daughter cell possesses just the same two sets of chromosomes that the mother cell possessed.

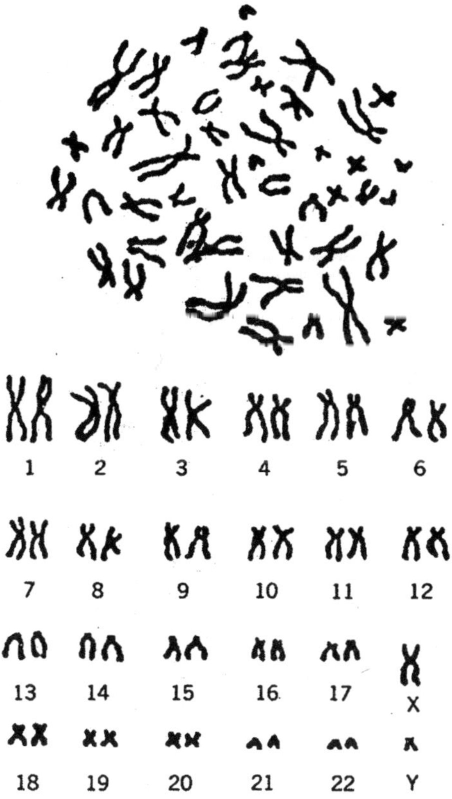

Figure 4.1 : The 46 chromosomes of man, seen in polar view at the metaphase of a somatic mitosis.

Several phases of the process are now described.

Interphase

Let us begin our account with the interphase, that is, with a cell that is not dividing. Such a cell has commonly been called a "resting cell." Actually it is a busy cell doing everything which a cell does except divide. It is during the interphase that the genes self-duplicate and carry on their function of supervising syntheses.

Following sclf-duplication, thc chromosomcs continuc to bc strctchcd out. Usually they are so attenuated that their boundaries cannot be seen with the light microscope. The extended chromos-omes are able to carry on their chemical functions in an efficient manner.

"Nuclear sap" or karyoplasm fills the interstices between the chromosomes. One or more rounded bodies, the *nucleoli,* are usually

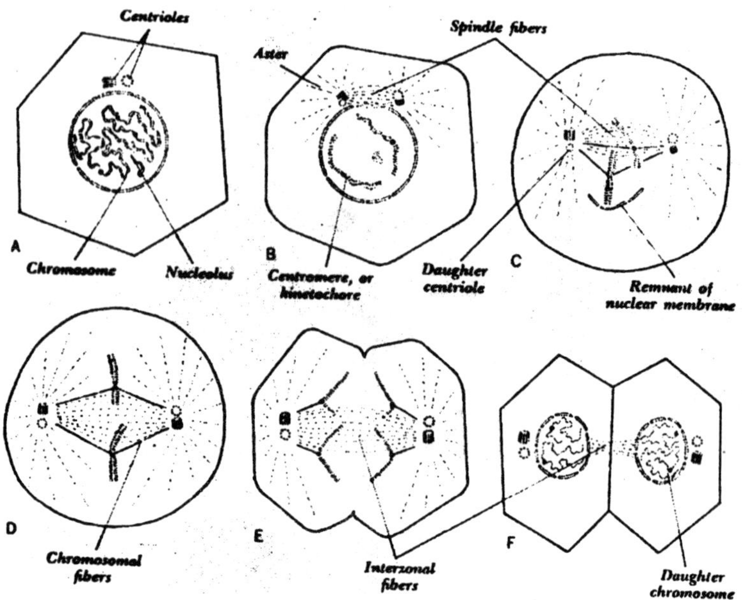

Figure 4.2 : Diagram of mitosis showing one pair of chromosomes. A. Interphase after replication. B. Early prophase. C. Late prophase. D. Metaphase. E. Anaphase. (The telophase is not shown.) F. Early interphase before the chromatids, or "daughter chromosomes" have replicated. The centrioles have been drawn larger in order to show their structure. Replication of the centrioles is shown here beginning in the early prophase, but in some cases it begins as early as the preceding anaphase.

present in the interphase nucleus. The visible boundary between the nucleus and the surrounding cytoplasm is the *nuclear membrane.*

In the cytoplasm adjacent to the nucleus there is a body, the central body, which consists of two granules or, after each granule has replicated, of two pairs of granules, the *centrioles.*

Prophase

The events in the nucleus and in the cytoplasm which lead to the division of the cell constitute the prophase. The following is a generalized account:

1. In the cytoplasm the pairs of centrioles move apart, and a sphere of gel rays, the *aster,* grows around each pair. Together the two asters constitute the *amphiaster.* It is of interest that the centrioles are short, cylindrical bodies, each consisting of nine rods (or double or triple rods).

   In this they are structurally comparable to a cilium or flagellum. (They lack the two central tubules of a cilium.) In a typical case, the centrioles of a pair lie at right angles to each other and to the axis through the pairs.

2. The chromosomes condense into tight spiral coils. Under the light microscope these appear to be rods. But when the material is suitably fixed and stained, each chromosome can be seen to be a double structure, each half of which is a chromatid.
3. A fusiform body, the *mitotic spindle,* now forms between the pairs of centrioles, sometimes in the cytoplasm, sometimes in the substance of the nucleus. The spindle's poles are anchored in the region of the centrioles of the cytoplasm. The electron microscope shows that it consists of hollow fibrils (microtubules).

   It elongates as the centrioles move apart. Possibly this is a result of imbibition of water, or incorporation of other molecules, or change in the shape of its constituent proteins. The amphiaster and the spindle together form the "*achromatic figure*," so-called because it does not stain deeply with basic dyes.
4. According to some accounts, chromosomal fibers grow out from particular locations in each chromosome until they reach the poles of the spindle. The locations from which they grow out are known as *centromeres* or kinetochores. The chromosomal fibers seem to consist of a contractile fibrous gel.
5. Toward the end of the prophase, the nuclear membrane disappears.
6. The nucleolus also disintegrates, and its substance, along with the nuclear sap, passes into the cytoplasm.

Metaphase

The chromosomal fibers appear to contract and pull the chromosomes to the equator of the spindle (prometaphase). When thus spread out and seen in polar view, their number can be counted.

Anaphase

Continuing, or resuming, what appears to be contraction, the chromosomal fibers seem to drag each chromatid toward one pole of the spindle. Actually two processes may be at work: a continuing enlargement and elongation of the spindle and a shortening of the chromosomal fibers. Since the fibers shorten without becoming thicker, the process

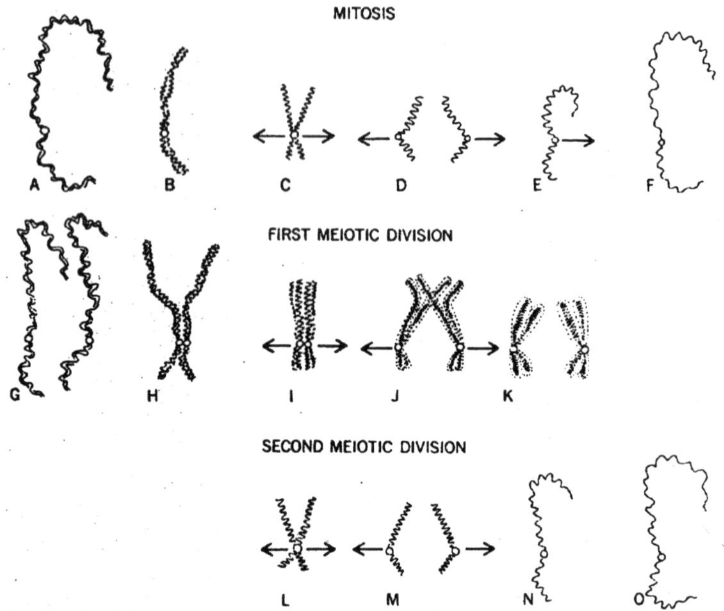

Figure 4.3 : The behavior of chromosomes in mitosis and meiosis. A to F. Mitosis, illustrated by a single chromosome: A, interphase after replication has taken place; B, prophase; C, metaphase; D, anaphase; E, telophase; F, interphase before replication has occurred. G to K. First meiotic division, illustrated by a homologous pair of chromosomes: G, leptotene stage; H, synaptene (zygotene) stage; 1, pachytene stage; J, diplotene (at which time "crossing over" between adjacent segments of homologous chromosomes may take place) followed by diakinesis; K, anaphase. L to O. Second meiotic division, illustrated by one chromosome: L, metaphase; M, anaphase; N, telophase; O, interphase.

probably involves the removal of water or other molecules from the fibers. As the chromatids move apart, they are often Vor J-shaped depending on the location of the centromeres to which the chromosomal fibers are attached.

A band of "interzonal fibers" is often seen for a time after the separation has been accomplished, connecting the chromosomes which have pulled apart, and often including a remnant of the spindle.

Telophase

Having reached the poles of the spindle, the chromatids swell and disappear from view, as seen with the light microscope. Alth-ough usually called "daughter chromosomes," they are actually single chromatids. The nuclear membrane reappears, and shortly one or more nucleoli re-form. Of the mitotic apparatus, little more than the centrioles of the cytoplasm remain.

Interphase

The daughter cells have now returned to the interphase, or "resting stage," except for one feature: each chromosome still consists of only one chromatid. Investigations by a staining technique known as the Feulgen reaction (specific for DNA) have made it clear that the restoration of chromosomal material takes place during the interphase. At that time each chromatid becomes two chromatids and the double nature of each chromosome is restored.

Studies employing radioactive isotopes as tracers have added some important information. It is not the chromatids which replicate. It is half-chromatids, also termed chromonemata, which undergo self-duplication.

Taylor placed bean seedlings in a solution containing radioactive (tritiated) thymidine (a substance which the cell uses to synthesize DNA). He left them in this solution until those cells which were in early interphase had each replicated once. He then cut the roots off and transferred them to a nonradioactive solution.

In this they replicated a second time. Then, having killed the cells, he determined by autoradiography whether or not the chromosomes were radioactive. Now, if whole chromatids had replicated when they were in the first or radioactive thymidine solution, they would have made for themselves radioactive mates.

After cell division half of the daughter chromosomes would have been radioactive, and half, namely, the original chromatids, would have been nonradioactive. But Taylor found that *all* the daughter chromosomes were radioactive! Why? The answer is that half chromatids replicated, so that each chromatid now consisted of a new radioactive half-chromatid and an original nonradioactive half-chromatid.

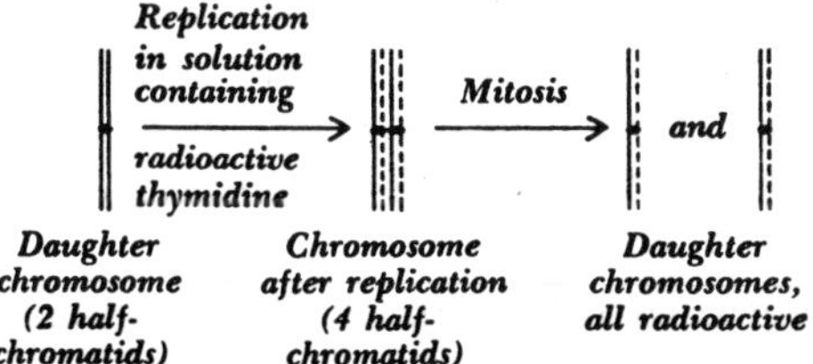

At the second replication, this time in a nonradioactive solution, each half-chromatid, whether radioactive or nonradioactive, made for itself a nonradioactive mate. Therefore, following the next cell division, the daughter chromosomes (chromatids) were half radioactive and half nonradioactive.

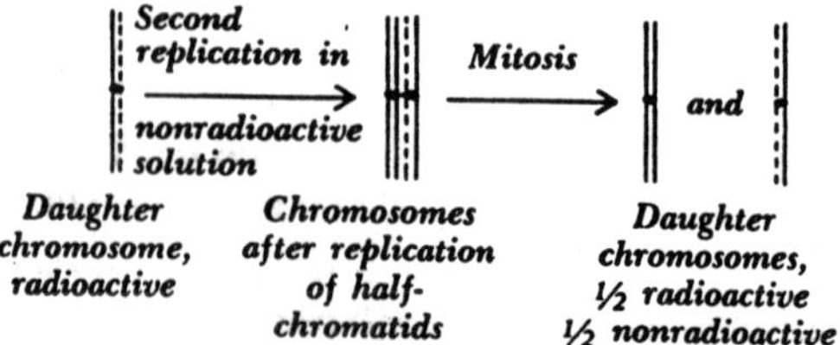

Similar results were obtained by Prescott and Bender, using cells of hamster tissues raised in tissue culture and also human leucocytes.

Experiments of this sort emphasize the remarkable stability of the half-chromatids. Although most of the macromolecules within the cell are in a state of flux, continually breaking down and being replaced, the half-chromatids are passed on intact from one cell generation to the next. And this has been going on for countless generations!

There is evidence that the centrioles may self-duplicate even earlier than the chromosomes, in some cases as early as the preceding anaphase. At this time each centriole (there are two at each pole) makes for itself a partner, not usually by longitudinal splitting, however, as might be expected, but apparently by a process of lateral budding.

The introduction of mitosis in the history of life on the earth was one of the grand innovations of all time. Possibly it took place only once, for, with variations, it occurs in all plants and animals except bacteria and blue-green algae.

Even in these there must be a process akin to mitosis, for without some such process the transmission of hereditary characters to each new generation in a balanced fashion would not be possible.

Meiosis

In 1884, Van Beneden described the processes of maturation, fertilization, and cleavage as they take place in the parasitic round worm of the horse, *Ascaris megalocephala*. He demons-trated that the egg and sperm contribute an equal number of chromosomes to the offspring. This led, a few years later, to the discovery of *meiosis*, the process by which the number of chromosomes is reduced to onehalf when the germ cells ripen.

Each body cell and each unripe germ cell or *gonium* has two sets of chromosomes, the *diploid number*. Each of the chromo-somes has a mate (except in the case of the unpaired sex chromo-somes). When meiosis takes place, an oogonium or spermatogonium, as the case may be, gives away one set of its two sets of chromo-somes; that is, it gives away one chromosome of each pair of chromosomes. The result is that each ripe egg or sperm has left only one set of chromosomes, the *haploid number.*

Now, meiosis, unlike mitosis (1) takes place only in the gonads, (2) occurs only in those unripe germ cells which at the time are in the process of ripening, and (3) although it begins earlier, reaches completion only during the period of sexual maturity of the plant or animal.

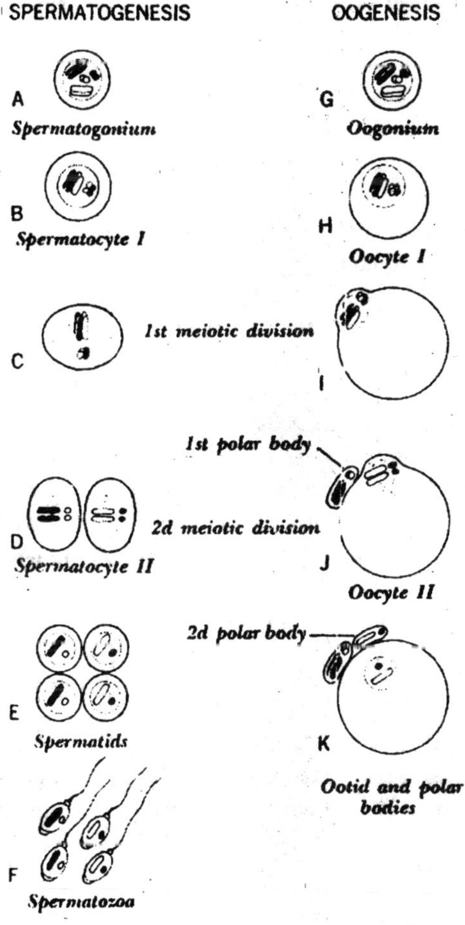

Figure 4.4 : Spermatogenesis and oogenesis. The black ovals represent paternal chromosomes (those derived originally from the sperm). The white ovals stand for maternal chromosomes (derived from the original egg).

In most cases some unripe germ cells (residual gonia) remain in the gonad where they multiply by mitosis and so produce more gonia. In female birds and mammals, however, the period of multiplication of oogonia comes to an end about the time of hatching or birth. At this time several million gonia may be present; yet only a very few ever ripen into ova—about 400 in the case of women.

The gonia are thus lineal descendants, by mitosis, of the original fertilized egg or zygote. Hence they are cousins of the body cells, and, generally speaking, they possess the same two sets of chromosomes which the body cells possess.

Meiosis has been compared to two mitotic divisions, for two successive spindles are formed, and two separations of chromosomes take place. Yet there is only one complete prophase, and this differs from a typical prophase in that it involves a pairing of the chromosomes side by side, each with its mate.

The process of pairing is known as *synapsis*. In a sense this pairing of homolo-gous chromosomes completes the coming-together of the germ cells at fertilization; for all through the mitotic cell divisions of development the chromosomes of the egg and those which came from the sperm have remained separate. Now each chromosome joins its mate.

The stages of the prophase of meiosis have been given names: leptotene, when the chromosomes become threadlike; synaptene (or zygotene), when they come together in pairs; pachytene, when they contract and become tightly coiled; diplotene, when they begin to separate, except for locations where "*crossingover*" between synapting chromatids has taken place; and diakinesis, when again they become compacted prior to separating.

Note that at the beginning of meiosis each chromosome consists of two chromatids as in ordinary mitosis, so that, when pairing takes place, the result is bundles of four chromatids . These are known as *bivalents* because each consists of two chromosomes. They are also known as tetrads because each is four chromatids.

During this period preparatory to the first meiotic division, the ripening germ cell is called *a primary gametocyte,* primary spermatocyte or oocyte, as the case may be. The diplotene stage of oocytes is of particular interest, not only because it is long drawn out (it may be as long as 40 years in the human female), but because it is a period of great growth. The nucleus enlarges greatly due to the accumulation of nuclear sap and is often referred to as the *germinal vesicle.* The chromosomes become *fuzzy* objects which have been compared to lampbrushes.

First meiotic division

The stages of the prophase which have just been described end as the metaphase approaches. The bivalents move to the equator of the first meiotic spindle. At the metaphase they split; and at the anaphase each bivalent divides, and *a monovalent chromosome* or dyad (pair of chromatids) goes to each pole of the spindle.

The monovalents are actually the original chromosomes which united in synapsis, except that some crossing-over has usually taken place. The products of the first meiotic division are known as *secondary gametocytes.*

Second meiotic division

Usually the telophase of the first meiotic division is brief and is quickly followed by the much reduced prophase of the second meiotic division. In some instances, the anaphase of the first meiosis leads directly into the metaphase of the second meiosis.

In any case, the monovalent chromosomes (dyads, two chromatids each) take a position at the equator of the second metaphase spindle. At the anaphase the chromatids or *daughter chromosomes* (monads) move apart to the poles of the spindle. The resulting cells are known as *tids,* ootids or spermatids, as the case may be.

Note that a reduction in the number of chromosomes has taken place. The gonia had two sets (diploid number) of chromosomes. The primary gametocyte had one set (haploid number) of bivalent chromosomes (tetrads). The secondary gametocytes had one set of monovalent chromosomes (dyads).

Now the tids have one set of daughter chromosomes, that is, one set of chromatids (monads). But this is important: the set which each tid has is a complete set with one chromosome present to represent each pair of chromosomes of the original diploid set.

In the case of spermatogenesis, each primary spermatocyte divides twice equally and produces four equivalent spermatids. In oogenesis, however, although the meiotic divisions of the nucleus are equal, those of the cytoplasm are grossly unequal.

At each division almost all the cytoplasm goes to only one of the daughter cells and thus is conserved to supply the substance of the embryo. The result is that the first meiotic division gives rise to one secondary oocyte and one small *first polar body.* The second meiotic division similarly produces one ootid and a small *second polar body.* The spermatids undergo metamorphosis and become sperm cells or *spermatozoa.*

The ootids, on the other hand, as a rule need only to burst from the ovary to cause them to ripen and become ready for fertilization.

It is a strange fact that the stage in meiosis which is attained before the fertilizing sperm enters the egg is not the same for all species. In sea urchin eggs meiosis is complete, and both polar bodies have been formed before the egg is receptive to the sperm.

This is rather rare. In vertebrates the first polar body has been given off and the second meiotic division has progressed to the metaphase before fertilization takes place. The extreme case is the parasitic roundworm, *Ascaris,* in which the sperm enters the cytoplasm of the egg before even the first meiotic spindle has formed. It remains inactive in the center of the cytoplasm while the egg completes meiosis.

In most animal species, therefore, the meiosis which closes the nuclear cycle of the egg overlaps to a greater or lesser extent the entrance of the sperm which begins the nuclear cycle of the new individual.

Why does this complicated process of meiosis exist? The answer was pointed out in 1903 by Walter Sutton, who showed the parallelism between the story of hereditary factors (later termed genes), as worked out by Mendel and other students of plant and animal breeding, and the behavior of chromosomes in fertilization, mitosis, and meiosis.

Meiosis, Sutton showed, is nature's way of reducing the number of chromosomes in each germ cell so that, at fertilization, the normal diploid number of chromosomes will be restored. But it is more than this: It is nature's way of seeing to it that the germ cells are all different.

Let us assume that the chromo-somes of different pairs are different and also that each chromos-ome has some mutant genes which make it unlike its mate. In nature this is probably always true. Since all the cell divisions during development are made by mitosis, every gonium will have the same two sets, or diploid number, of chromosomes and genes.

Then, when meiosis takes place, each ripe germ cell retains only one chromosome of each pair of chromosomes, one gene of each pair of genes. Moreover, it is a matter of chance which chromosome of a given pair of chromosomes the cell retains.

Thus, from the standpoint of heredity, there can always be as many different genetic kinds, or *genotypes,* of ripe germ cells as there are possible combinations of chromosomes taking one chromosome from each mating pair of chromosomes.

If there were just one pair of chromosomes in an unripe germ cell,

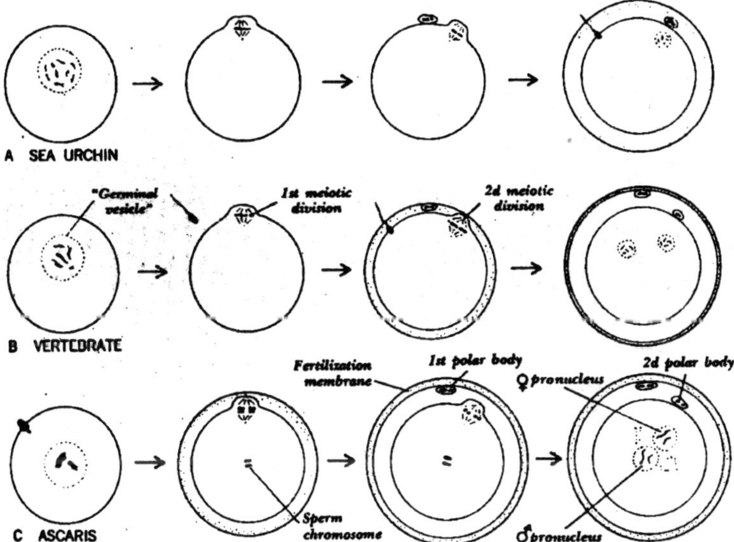

Figure 4.5 : Fertilization (entrance of sperm) takes place at different stages of meiosis in different animals: A, after meiosis is complete in the sea urchin; 8, midway in the second meiotic division in most vertebrates; and C, before meiosis in the parasitic roundworm Ascaris.

there would be two kinds of ripe germ cells. If there were two pairs to begin with, there would be four possible kinds of germ cells. If there were three pairs, the number would be eight.

In general, if there are N pairs of chromosomes before meiosis, there will be 2N possible kinds of genotypes among the ripe germ cells. Now, man has 23 pairs of chromosomes. 2^{23} is 8,388,608. In other words, there is no more than one chance in over eight million that a man or woman will produce two germ cells with identical genotypes. (Crossing-over decreases the chance tremendously.)

What is the maximum chance that two identical offspring will be born to the same parents by ordinary sexual reproduction? It is one chance in 8,388,608 times one chance in 8,388,608, or one chance in 70,368,744,177,664. Crossing over at synapsis, mutations, and chromosomal aberrations greatly increase this variability.

THE ROLE OF THE NUCLEUS

The account of fertilization, mitosis, and meiosis which has just been given is similar to that usually found in textbooks of biology. It stops short, however, of explaining how the nucleus controls development.

Indeed, it avoids the problem, for the compacted chromosomes of fertilization, mitosis, and meiosis are not active in development.

It is only when they are expanded, as at the interp-hase, that they carry on their functions in growth and development. (The chromosomes of oocytes are expanded also during the long prophase.) Furthermore, the chromosomes presumably are alike in every cell of the body of an embryo.

Now, how can chromo-somes which are everywhere the same account for the origin of differences within the embryo? In order to approach this problem we must consider the chemistry of the nucleus.

The Chemistry of the Nucleus

The distinctive chemical substances of the nucleus are long chain molecules known as *deoxyribonucleic acid,* or DNA for short. Each such molecule is a gene; more likely it is a string of genes, the material units of heredity.

A closely related type of nucleic acid is *ribonucleic acid,* or RNA. Both DNA and RNA are present in the nucleus. DNA is in the chromosomes. RNA is in the chro-mosomes, but especially in the nucleoli, and it is also present in the cytoplasm.

Now, nucleic acids have three unique and remarkable properties:

1. They are self-duplicating molecules. Possibly they are the only truly self-duplicating molecules. Centrioles and plastids are morphologically self-duplicating, but it is likely that, in this case, nucleic acids are involved.
2. Nucleic acids supervise the synthesis of proteins within the cell. The proteins are the principal structural compounds of protoplasm and are the chemical basis of its functioning.

   They include the enzymes which control the innumerable chemical reactions of the cell and also structural proteins. In fact, a gene *(cistron)* has been defined as the template for the production of one polypeptide chain. It is probably true that, in the absence of nucleic acids, no proteins are ever produced.

   These statements are oversimplified. The self-replication of a nucleic acid requires the presence of specific enzymes (polyme-rases) and numerous other molecules in the protoplasm. The process is therefore a circular one: nucleic acids bring about the synthesis of proteins, and proteins are needed to make possible the replication of the nucleic acids. It is the entire protoplasmic system which endures.

Genes are not indivisible. Crossing-over of parts of genes may take place between synaptic mates during meiosis. The smallest part of a gene which may interchange thus with a mate has been termed *a recon.*

3. Nucleic acids (genes) occasionally mutate; that is, a short segment of a gene may change in composition. When this occurs, the gene replicates according to its new or mutant form. If this were not so, life on earth would never have evolved, and the ever-increasing adaptation of living things to their environments would not have been possible. The shortest segment of a gene which may mutate is probably a single nucleotide unit. It is known as a *muton.*

Great progress has been made in the last few years in the knowledge of the structure of nucleic acids and of the manner in which they duplicate. According to the Watson-Crick hypothesis, each DNA molecule is a long double chain of units known as nucleotides.

It may be compared to a rope ladder twisted into a helix. Each single nucleotide consists of a sugar, deoxyribose (S), a phosphate group (P), and one of four different "bases." The bases are adenine (A, a purine), guanine (G, also a purine), thymine (T, a pyrimidine), and cytosine (C, a pyrimidine).

The single chain consists of alternating sugars and phosphates, with a base attached as a side group to each sugar. The precise sequence in which the nucleotides are arranged in the long nucleic acid chain is the basis of the coded information which is passed on by heredity and which guides the course of development.

In the double chain of the DNA molecule, each base of one chain is linked crosswise, by weak hydrogen bonds, with its mate in the other chain. The relations are precise.

Adenine is always bonded with thymine, and guanine with cytosine. There are no other combinations. Hence the two chains fit each other exactly like a mold fits a model.

When a nucleic acid (DNA) molecule selfduplicates, the chains supposedly separate from each other. Each single chain then makes for itself a replica of its former partner by attracting and assembling to itself appropriate nucleotides from the surrounding protoplasm.

The result is that there are now two identical DNA molecules (double chains of nucleotides) where previously there was one. One notes at once the comparison between the replication of the half-chromatids (chromonemata), which takes place during the interphase of

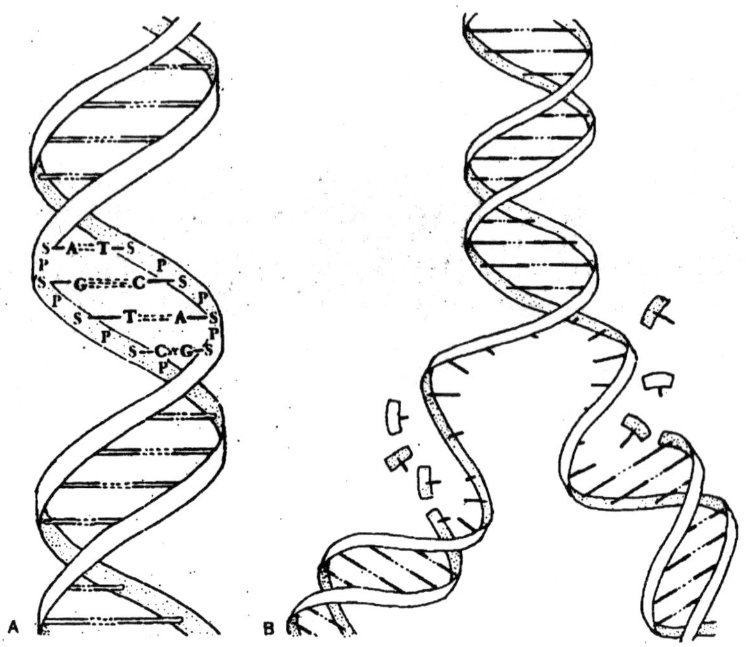

Figure 4.6 : The structure of DNA and the manner of its replication.

mitosis,, and the replication of the single chains of DNA molecules. The nucleus also contains proteins of two sorts: histones (proteins rich in basic amino acids), which are associated with the genes; and nonhistones, which compose the larger amount of protein and may be a part of the metabolic and mitotic machinery of the cell.

The relation between histones and DNA is loose, yet it is so consistent that the combination is usually referred to as nucleoprotein. A small amount of nonhistone protein, termed residual protein, presumably holds the genes in order and organizes them into the definite bodies, the chromosomes. Lipids are also present in the nucleus.

The Influence of the Nucleus on Development

The genes (DNA) do not directly influence development. Instead, they "transcribe" their coded information to RNA, which then collects around the chromosomes and in the nucleoli. Apparently one of the two strands of a DNA molecule assembles ribonucleotides and makes for itself a complementary strand of RNA. (The other strand possibly self-duplicates.)

The nucleotides (ribonucleotides) which compose the RNA, however, differ from those which compose DNA. The sugars are ribose instead of deoxyribose, and one of the four nucleotides is uracil instead of thymine (it lacks the methyl group of thymine). Although these differences from a chemical standpoint appear to be slight, yet they account for important differences in function.

The RNA molecules are single chains and as such are not self-replicating. (RNA is self-replicating in some viruses.) They are dependent on DNA for their production. Moreover, the RNA molecules do not remain in the nucleus.

From time to time they migrate into the cytoplasm where they function in the manufacture of proteins. Some of them serve as templates for the production of proteins, notably for the production of the enzymes which catalyze the chemical processes of the cell. Possibly this release of RNA is not a continuous process.

An impressive release takes place at the prophase of the first meiotic division of oogenesis when the swollen oocyte nucleus (the so-called germinal vesicle) breaks down and its nuclear sap and the material of the large nucleolus mingle with the cytoplasm. A similar release occurs at the prophase each time a cell divides.

In very active cells, such as growing oocytes and certain cells of larval insects, the nuclear membrane has been shown by the electron microscope to be perforated. Nuclear material has been found to "bleb" through pores into the cytoplasm where it supervises the syntheses of proteins.

There may be a delay between the time of the formation of the RNA in the nucleus and the time when its influence becomes apparent in cytoplasmic processes. The clearest evidence that this is so comes from experiments on species hybrids, that is, on eggs which have been fertilized by sperm of another species.

In this case the early events of development, for example, the pattern and rate of cleavage, are determined by the egg cytoplasm alone, presumably by the RNA that was synthesized while the oocyte grew in the mother's ovary and under the influence of the mother's genes.

Generally speaking, the effect of the genes (DNA) brought in by the sperm is not apparent until about the time of gastrulation. But from then on, the genes of both the egg and the sperm control the course of development.

Even more crucial are certain experiments in which eggs were enucleated before they were fertilized by the sperm of another species, or in which the egg nuclei were destroyed immediately after the foreign sperm had entered. In such haploid eggs (termed androgenetic hybrids), the cytoplasm belongs to one species and the nucleus to another. Such combinations have not lived beyond the early embryonic stages, but the evidence indicates that, from the time of gastrulation onward, the influence of the chromosomes of the stranger nucleus is pronounced.

Do Nuclei Undergo Differentiation?

In the early days, Weismann and Roux believed that the nuclei of body cells become different during cleavage as a result of what were thought to be unequal divisions of the nuclei. The organization of the embryo, so they theorized, is the expression of the inherited organization within the chromosomes of the zygote nucleus.

This view has long since been given up. There is organization within the cytoplasm of the uncleaved egg. There is strong evidence that the nuclei of every cell of the body contain initially the same set of chromosomes and genes, and that, during cleavage at least, mitosis is equal cell division in so far as the chromosomes are concerned.

This fact was originally demonstrated in an experiment by Spemann in which he constricted an uncleaved amphibian egg by tightening a hair noose around it. The egg was pinched into two halves, with only a narrow neck of protoplasm between them.

One half contained the nucleus; the other half lacked a nucleus. After several cleavages had taken place in the nucleated half, one of the descendant nuclei, migrated across the isthmus of cytoplasm into the non-nucleated half (delayed nucleation). Cleavage followed, and both halves (provided the needed cytoplasm was present) became whole embryos of half size.

This proves that a cleavage nucleus can do everything that the original zygotic nucleus can do. It is an exact copy of its original. The fact that the daughter nuclei are equal during early development has been clinched by some remarkable experiments by Briggs and King. They removed the nucleus of an uncleaved frog's egg, and substituted a nucleus from a cell of a blastula or an early gastrula.

With refined techniques, many such operations were performed and were successful. Normal embryos developed. Here again, it is demonstrated beyond question that at this early stage of development when the future fates of the regions of the embryo are being determined, the nuclei, i.e., the genes of the cells of the different tissues, are equivalent. Since the nuclei are alike, it therefore must be the cytoplasms which are different.

Gurdon and others have extended this work and have found that

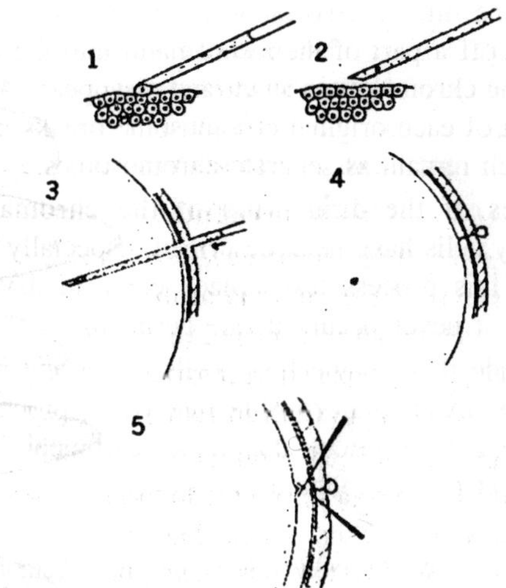

Figure 4.7 : Method used by Briggs and King when they substituted nuclei from older embryos for the nuclei of uncleaved eggs of the frog.

some nuclei of stages as late as swimming larvae, when transpl-anted to enucleated eggs, are capable of supporting a complete development. Some of the tadpoles even live to metamorphose into frogs and become sexually mature.

This, however, is not the whole story. It is a common observation that the nuclei of cells of different tissues do differ in size and shape. Somehow during development nuclei do change.

The cytoplasm, which supplies them with substance, energy, and necessary enzymes, also affects the visible characteristics which distinguish them. It does more than this: it affects their activity; and, at least in some cases, it affects their composition.

The nuclear changes which arise during development are of three sorts, and they arise in three distinct ways:

(1) by the diminution of the chromatin,

(2) by stable changes in the chromos-omes, and

(3) by the epigenetic control of gene activity by the cytoplasm.

1. Diminution of the chromatin : This is a rare phenomenon. Boveri observed many years ago that a reduction of the substance of the chromosomes (diminution of the chromatin) takes place during the early cleavages of the parasitic roundworm, *Ascaris*.

At the two-, four-, eight-, and 16-cell stages only one of all the blastomeres retains the typical four chromosomes of the zygote. In each case this cell is in the line of descent of the germ cells (germ line). The other cells cast off a part of their chromatin into the cytoplasm; that is, the tips of the chromosomes swell and disappear. At the same time the central part of each original chromosome breaks into several small segments which remain as separate chromosomes.

Other examples of the diminution of the chromatin in the development of body cells have been described, especially among the insects. Sometimes this process takes place early in develo-pment, sometimes late. But it never occurs in the germ line.

We may conclude that, in the body cells of these embryos, the chromatin (DNA) has already performed its function by producing RNA. The RNA is stable and functions during development. The DNA (chromatin), no longer needed in the development of the body cells, is discarded.

Experiments have shown in *Ascaris* that a certain substance localized in the cytoplasm prevents the diminution of the chromatin. Normally, this material is passed on to only one blastomere during each of the first four cleavage divisions.

But if, as a result of manipulation (centrifuging), this material becomes divided between two daughter cells, then diminution of chromatin does not take place in either of them. This is a clear case in which substances in the cytoplasm control the nucleus.

2. Changes in the chromosomes : Diminution of the chro-matin is not known to occur in vertebrate development; yet there is evidence that changes in the genome or sum total of genes may take place. In their studies of transplanted nuclei, Briggs and King found that, when a nucleus from a late gastrula or neurula of an amphibian is substituted for the nucleus of an uncleaved egg, normal development does not usually take place, although it *may* take place.

When examined under the microscope, the cells often do not have the normal chromosomal equipment. It is not clear whether changes of this sort take place in normal development. Possibly the nuclei of differentiating cells become increasingly sensitive to ,experimental manipulation. Possibly the transplanted nuclei are forced by the mitotic apparatus to divide before they are fully ready.

Possibly, also, when cells differentiate, the mitotic cell divisions become slovenly. In any case, in these experiments the daughter cells did not always receive the normal complement of chromo-somes.

The nuclear changes which take place in this manner are stable. Briggs and King have demonstrated that when the nucleus of a defective embryo is transplanted to an uncleaved egg, the result is an embryo with a defect similar to the defect of its nuclear parent. This transfer of nuclei may be repeated over and over with the same result (nuclear cloning).

For a time it was suggested that the changes in the genome which Briggs and King had discovered might be related to the normal differentiation of cells. For example, the nuclei derived from the endoderm seemed to favor the development of endodermal organs.

But this interpretation has not been borne out. Defects of a similar sort have also been obtained when the nuclei are taken from the neural tube. On the other hand, some nuclei are capable of supporting a normal development even when taken from the gut of a swimming tadpole.

However, one thing is certain: Any changes which may take place in the genome of differentiating cells always occur after the cytoplasm has already become specialized and committed to its fate.

3. The epigenetic control of the genes by the cytoplasm : Whether or not nuclear substance differentiates during development, it is certain that the same genes are not equally active in every cell and at every place, nor are they equally active at all times. Genes represent potentialities (in the sense of possibilities rather than powers) which may or may not have opportunities to express themselves.

Which potentialities are realized depends upon the conditions within the cytoplasm which surrounds a given nucleus; these conditions change epigenetically as development progresses. It has been said that the cytoplasm turns the genes off and on like faucets. To change the metaphor, a nucleus is like a stockroom of a foundry full of patterns.

From its vast store, patterns are requisitioned when, where, and as they are needed. No casting can be poured at the foundry for which a pattern is not provided, and, of course, the form of the casting depends upon the shape of the pattern.

In the same way, no synthetic chemical process can take place in a cell unless a gene is present to supply the pattern (i.e., template) for the enzyme which mediates the process. Genes are templates in stock. But the decision as to which genes are active at a given time and place-this depends upon cond-itions in the cytoplasm. And these conditions change epigenetically as development goes on.

It is a mistake to pass over lightly the epigenetic factors which turn genes on and off, for they, and not the genes, ultimately control the time and place of developmental processes.

Direct evidence that this is true is found in the behavior of the giant chromosomes of certain insect larvae as they approach metamorphosis. Such chromosomes consist of multiple strands of chromatids (so-called polytene structure).

They are the result of self-duplication of the chromatids without accompanying separations. For example, the chromatids of the salivary glands of fruit fly larvae multiply nine or ten times without separating, with the result that each giant chromosome consists of 512 to 1,024 chromatids.

These compound chromosomes show bands (possibly representing genes) of DNA which stain darkly and are large enough to be seen under the light microscope. They also show swellings, or "puffs," at definite loci. These seem to be the result of RNA and protein accumulating and forcing the strands apart.

The puffs may expand until they become what are known as Balbiani rings. Beermann has studied the puffs and rings in the larvae of the midge, *Chironomus,* and has shown clearly that they occur at different locations in the chromosomes of different tissues and at different times.

The explanation seems to be that the puffs and rings are centers of intense synthetic activity in particular tissues, in preparation for the role which that tissue will play in the metamorphosis which is soon to follow.

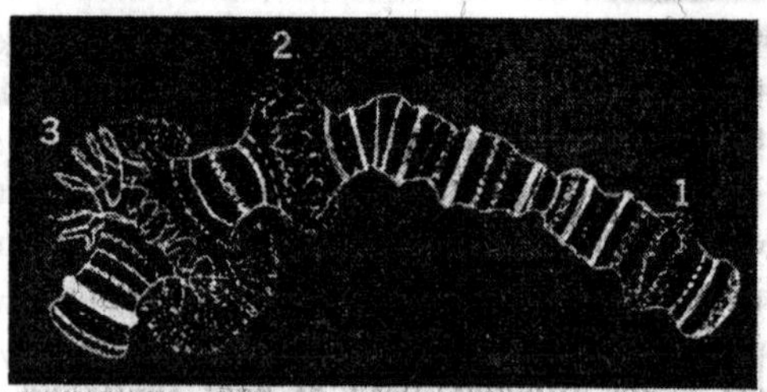

Figure 4.8 : Polytene chromosome of the salivary glands of the midge, Chironomis, showing stages of "puffing."

How are the Genes Controlled?

The analogy of a nucleus to a stockroom of patterns at a foundry has one obvious flaw. A stockroom has a stockkeeper who delivers the patterns as they are called for. The nucleus of a cell, however, is automated so that, cued by conditions in the surrounding cytoplasm, it delivers the templates when they are needed.

Lately developmental geneticists have been much interested in the problem of how the genes are activated. We shall make reference to the closely reasoned concepts of Jacob and Monod of the Pasteur Institute, who studied enzyme induction in the bacterium, *Escherichia coli*. Normally this microbe produces no enzyme to metabolize the nutrient sugar lactose.

Yet, when lactose is supplied in the nutrient medium, the enzyme b-galactosidase, which is able to metabolize it, appears in abundance. The presence of the substrate (substance acted upon) serves as an *"inducer"* and leads to the production of the enzyme.

The authors reasoned that no enzyme could have been produced if a "structural gene" for its production had not already been present in the germinal material. (Genes which determine the structure of a protein are called structural genes.)

If this is so, then why was the enzyme b-galactosidase not produced until lactose was added? Genetic studies by Jacob and Monod on mutant strains of *Escherichia coli* supplied them with the answer: The gene for the production of b-galactosidase is inactive because another gene, so-called "regulator gene," is present which produces a *repressor substance* that inhibits the activity of the structural gene.

When the substrate lactose is supplied, however, it inactivates the regulator gene (or the repressor substance which it produces). The result is that the structural gene, now freed from repression, gives rise to the enzyme b-galactosidase. The structural gene is thus activated by a process of *derepression*.

This statement is oversimplified. The repressor substance produced by the regulator gene acts upon an "operator gene" within a complex of structural genes. It is the operator gene which directly controls the structural genes, turning them on and off.

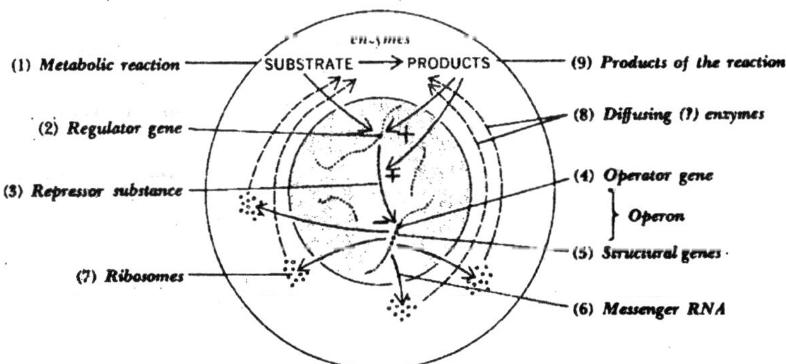

Figure 4.9 : Schema illustrating Jacob and Monod's concept of nucleo-cytoplasmic interaction.

An operational complex of genes of this sort was called an "operon" by Jacob and Monod. Such a complicated mechanism would be unbelievable if its several steps had not been experimentally verified. Generally speaking, a substrate acts as an inducer.

Its presence leads to the production of the enzyme which metabolizes it. In the language of modern communication theory, this is a case of "positive feedback." In a similar manner, the products of a reaction are inhibitors which repress the production of the enzymes which produce them. This is a "*negative feedback.*"

Embryologists are much interested in concepts such as this one of Jacob and Monod because they suggest how epigenetic factors in the cytoplasm (represented in the above case by the substrate lactose) can control the activity of the genes. There is, however, an important difference between the actions of genes in bacteria and in the development of higher organisms. The genes of bacteria are immediately responsive to their environment, and the mRNA which they produce by transcription is immediately active in protein synthesis. It is also short-lived. The genes which control the development of higher organisms, on the other hand, commonly produce "masked" mRNA, which may be stored for a time until the time and place for its activity have arisen.

5

Cytoplasmic Regulation

The genes of the nucleus control the course of development, but it is mainly the cytoplasm which develops. The genes are the "physical basis of heredity," but principally it is the cytoplasm which expresses this heredity insofar as it is expressed. There is interaction between the genes and the cytoplasm, but their roles are different.

The genes largely control the species characteristics of the new organism and to a great extent its individual characteristics as well; but the characteristics which distinguish one tissue from another, or one organ from another *in the same individual,* are decided initially by processes which take place in the cytoplasm.

The history of the cytoplasm is indeed a true epigenesis; that is, it is a straightforward coming-into-being of molecules, structures, and functions which did not at first exist. It starts in the relative simplicity, or at least plasticity, of a single cell, the unripe egg.

It moves forward irreversibly in ever-increasing complexity through the stages of the life cycle: embryo, larva (if any), juvenile, adult, and finally the senescent individual. The ultimate end of development is death unless, as in some of the lower organisms which are capable of regeneration, the story is able to begin again.

Embryology is therefore, for the most part, a recital of processes which take place within the cytoplasm. What is this cytoplasm which, synthesized under the direction of the nucleus, does and becomes so many different things? It is, to be sure, a colloidal aqueous system of proteins, lipids, sugars, and numerous other organic molecules plus various mineral salts. But cytoplasm is more than a chemical system, no matter how complex.

It has a structure, an intimate, precise anatomy such as no system of a merely chemical nature ever had. Its proteins and lipids are organized into membranes, filaments, and granules, at the surfaces of which chemical reactions take place.

These are the organelles which carry on the many processes of living. The precise patterns of the protein surfaces are the principal basis of the specificity of the cell's chemical reactions, for the proteins include the enzymes which catalyze and control the rates of chemical processes. They also include nonenzyme proteins.

A very few years ago cytoplasm was described as a homog-eneous, transparent gel or sol which was optically "empty." The various formed bodies which were seen under the microscope were spoken of as being within the "ground cytoplasm"; or they were interpreted as artifacts produced by fixation and staining.

Now this picture has changed. The phase-contrast microscope, the electron microscope, and the ultracentrifuge have revealed an intricate morphologic structure in cytoplasm.

Although the cytoplasms of different kinds of cells have specialized functions (no single type of cell can be said to be "typical"), yet the cytoplasms of most cells posses certain organelles in common. These include: a plasma membrane (cell membrane or plasmalemma, the outer layer of the living cytoplasm); inner membranes, which form vacuoles, tubules, and lamellae; a nuclear membrane (actually a part of the cytoplasm); and formed bodies such as ribosomes and mitochondria.

Each of these organelles plays a role in the physiological life processes of the cell. Some of them presumably play a part in development, but it is not always clear what that part may be. Specialized cells have filaments and granules of various kinds which perform distinctive functions.

Muscle cells have contractile fibrils; nerve cells have conductile fibrils; secretory cells have secretory granules; certain other cells produce pigment granules; while still others store reserve food materials such as yolk platelets, glycogen granules, and fat globules. The list could be multiplied.

The embryologist sees these various structures appearing in the cytoplasm as development proceeds, each at the proper time and in the proper place. He has reason to believe that they result from the synthesi-

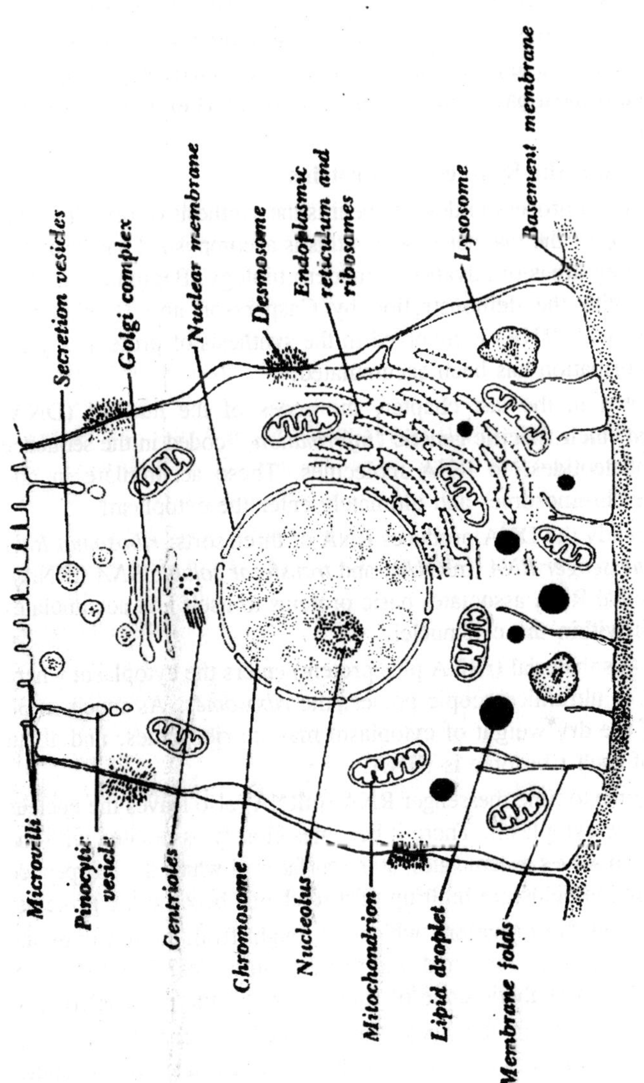

Figure 5.1 : Diagram of an animal cell. The size of the finer structures has been exaggerated.

zing activities of distinctive enzymes which are released in the cytoplasm, and that in turn each enzyme obtains the "information" (specific pattern) for its production from a particular gene or genes in the nucleus.

But why a particular gene is active producing a particular enzyme at a particular time and in a particular cell, he cannot as yet say. The answer to this most basic question of embryology is to be sought in the cytoplasm.

Ribosomes and the Synthesis of Proteins

A principal process in development is the synthesis of proteins. Our understanding of the manner in which this is accomplished has been one of the exciting areas of advance in modern biology. Beginning in 1941 and 1950 with the demonstration by Caspersson and Brachet that ribonucleic acid (RNA) is involved in the synthesis of protein, a great deal of information has been accumulated.

As noted in the last chapter, the genes of the nucleus (DNA) "*transcribe*" their specific genetic "*information*," coded in the sequence of their nucleotides, to RNA molecules. These accumulate in and around the chromosomes and ultimately enter the cytoplasm.

Apparently the DNA produces RNA of three sorts: *ribosomal RNA* (rRNA), *messenger RNA* (mRNA), and *transferor soluble RNA* (tRNA). The ribosomal RNA associates basic proteins to itself and accumulates in nucleoli within the cell nuclei.

Later this material (rRNA plus protein) enters the cytoplasm where it appears as ultramicroscopic bodies, the *ribosomes*. As much as 30 percent of the dry weight of cytoplasm may be ribosomes, and about one-half of each ribosome is rRNA.

From time to time messenger RNA (mRNA) also leaves the nucleus and enters the cytoplasm. There it becomes closely associated with the ribosomes and serves as templates, at the surfaces of which the polypeptide chains of amino acids are built up. These chains become the proteins.

How is this "information" which is brought from the nuclei by the mRNA transferred ("translated") to protein molecules? It is carried in the form of an alphabetic code of four "*letters*," the ribonucleotides.

These differ from the nucleotides of DNA in that the sugar is ribose, instead of deoxyribose, and the bases are adenine, guanine, cytosine, and uracil (which substitutes for thymine). The precise sequence of the nucleotides is the most important feature, for it determines the sequence of the amino acids in the polypeptide chain.

But since there are only four kinds of nucleotides and there are

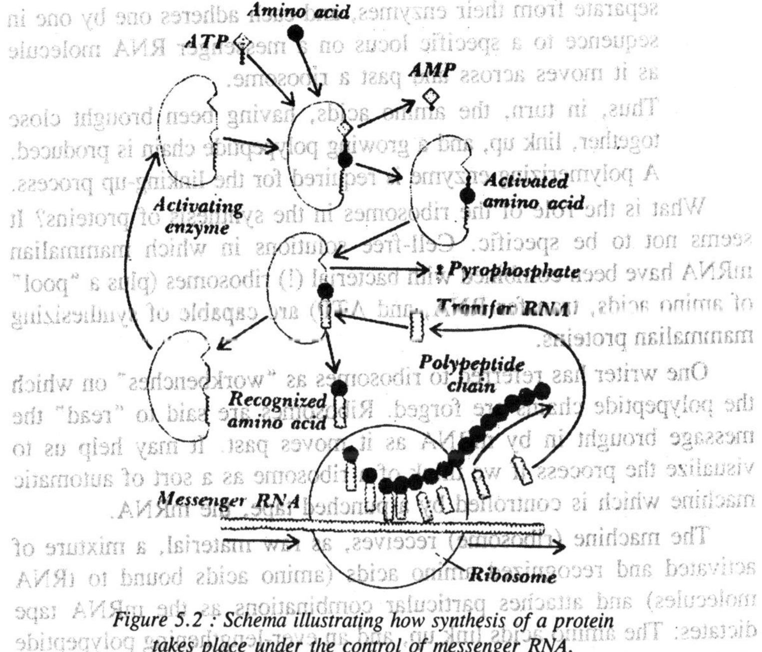

Figure 5.2 : Schema illustrating how synthesis of a protein takes place under the control of messenger RNA.

twenty or more kinds of amino acids in a polypeptide chain, it must take a "word" or *codon* of at least three nucleotides to pick or choose one amino acid.

The third sort of RNA is transfer or soluble RNA (tRNA). It consists of smaller molecules, each of which bears, possibly at one end, a triplet of nucleotides which corresponds specifically to one amino acid. Since there are twenty kinds of amino acids, there must be at least twenty specific kinds of tRNA.

According to present evidence, the "translation" of information from mRNA to the polypeptide chain takes place mostly in the cytoplasm in the following manner:

(1) First the amino acids in solution in the cytoplasm become activated with the aid of energy-rich molecules, such as ATP, and activation enzymes.

(2) Then the activated amino acids, still attached to the enzymes, are "recognized" by specific tRNA molecules. By recognition we mean that each amino acid becomes attached to its proper tRNA molecule.

(3) Having accomplished this, the tRNA-amino-acid combinations

separate from their enzymes, and each adheres one by one in sequence to a specific locus on a messenger RNA molecule as it moves across and past a ribosome.

Thus, in turn, the amino acids, having been brought close together, link up, and a growing polypeptide chain is produced. A polymerizing enzyme is required for the linking-up process.

What is the role of the ribosomes in the synthesis of proteins? It seems not to be specific. Cell-free solutions in which mammalian mRNA have been combined with bacterial (!) ribosomes (plus a "pool" of amino acids, transfer RNA, and ATP) are capable of synthesizing mammalian proteins.

One writer has referred to ribosomes as "workbenches" on which the polypeptide chains are forged. Ribosomes are said to "read" the message brought in by mRNA as it moves past. It may help us to visualize the process if we think of a ribosome as a sort of automatic machine which is controlled by a punched tape, the mRNA.

The machine (ribosome) receives, as raw material, a mixture of activated and recognized amino acids (amino acids bound to tRNA molecules) and attaches particular combinations as the mRNA tape dictates. The amino acids link up, and an ever-lengthening polypeptide chain is produced. (The tRNA is released to be used again.) When the tape runs out, the polypeptide chain is complete and moves away.

The entire process takes but little over one minute. The machine is also free to start over again with the same or another tape. Since the tape is long, it may be moving through or past several ribosomes at the same time, each of them in the process of producing a polypeptide chain.

Such a series of ribosomes, connected by a thin thread of mRNA molecules, can be demonstrated with the electron microscope. They are known as *polyribosomes* (polysomes). (There is some evidence that ribosomes are not all the same in their nucleotide composition, even that they differ in the cells of different tissues. Still, they may all be just workbenches.)

The "code" by which mRNA transfers its information to polypeptide chains has been broken. This was accomplished first by students of microorganisms. Bacteria were homogenized without destroying the ribosomes.

The resulting cell-free system was supplied with amino acids, a source of energy (ATP), and artificially synthesized RNA chains of known composition. The polypeptide chains which were produced were then analyzed. The breakthrough came in 1961 when Nirenberg "fed" his system an artificial RNA consisting of the nucleotide urocil only. He obtained polypeptide chains composed of the amino acid phenylalanine.

From that point on it was only a matter of time until the entire code was broken. At first the ribosomes in the cytoplasm are free and scattered. Then, about the time that newly synthesized proteins begin to be secreted away from the cell, the ribosomes become attached to the endoplasmic reticulum within the cytoplasm.

This work on RNA and the ribosomes is of great interest to embryologists because it is possible that differentiation manifests itself first quantitatively in the distribution of the ribosomes. At any rate, gradients in the arrangement of ribosomes exist early in development, and any manipulation which disturbs their distribution, such as centrifuging, results in abnormalities.

Mitochondria and the Energy of the Cell

Whence comes the energy which makes these reactions possible? Proteins are highly ordered molecules, and their synthesis is an uphill (endergonic) process which requires a continuous source of free energy to make it go. If free energy were not provided, the living system would quickly revert to a more probable, that is, a more mixed-up state, namely, death.

Now, the development of an organism is, beyond anything else in the known universe, an improbable sequence of events. Only free energy, that is, ordered energy, can cause it to progress. Furth-ermore, an organism, and especially a developing organism, is a "dynamic system" like a flame or cyclone.

Within it are gradients of substance and energy which would immediately flatten out and disappear if it were not that a continuous expenditure of free energy sustains them. Where shall the needed order and energy be found? Heat cannot provide it, for it is only when heat is guided from a hotter source to a reservoir at a lower temperature, as in a heat engine, that some of it can be made to do work. Even then, a heat engine is inefficient.

Since temperature differences do not exist in the cell, the cell cannot be a heat engine. Green plants possess remarkable machinery, the chloroplasts, which capture the free energy of light and store it in the form of highly ordered organic molecules, primarily sugars. These molecules, built initially by green plants, are the ultimate source of the free energy which an animal uses in its development.

If the free energy of glucose (about 690,000 calories per gram-molecule) were to be released all in one step, as when sugar burns, most of the energy would become heat, and would be lost. But in the living cell, the energy is released stepwise in a complex series of controlled chemical reactions.

In this way, a large part of the energy is transferred to the high-energy phosphate bonds of a ubiquitous substance known as adenosine triphosphate or ATP. Adenosine diphosphate (ADP) plus energy (approximately 12,000 calories per mole) becomes adenosine triphosphate (ATP).

The energy-rich ATP can then migrate to where the energy is needed and, like a charged battery, give off its energy in the form of highenergy phosphate bonds, and so revert to ADP.

Perhaps the thing which, more than anything else, distinguishes living from nonliving matter, is this ability to transform and transfer energy in a controlled and useful manner. It is accomplished with the aid of enzymes, or rather, organized brigades of enzymes. One enzyme controls each step of the complex process.

Three successive chains of reactions are involved:

Chain 1 The glycolytic sequence (cf. fermentation)

Glycolysis consists of a series of 14 chemical reactions by which glucose, glycogen, or other carbohydrate, is split and partially oxidized into a simpler chemical molecule, pyruvic acid. It takes a little energy (supplied by ATP) to start the process, much as it takes energy to start a stone rolling down a hillside; but the ATP which is used to start the process is more than restored by the ATP produced in the process. To control the 14 steps of glycolysis, at least 12 different enzymes are involved.

In the case of anaerobic organisms (those which do not use molecular oxygen), this is the end of the story, except that the resulting pyruvic acid is changed either to lactic acid, to acetic acid and CO_2, or to alcohol and CO_2, without much gain or loss of energy. But the process as it occurs in anaerobic organisms is incomplete, and over 90 percent of the energy of the glucose still remains in the pyruvic acid or related molecules.

Chain 2 The citric acid cycle

Aerobic organisms, on the other hand, namely, those which ultilize oxygen (and this includes most developing eggs), continue the metabolic process by breaking down the pyruvic acid into CO_2 and H atoms. This

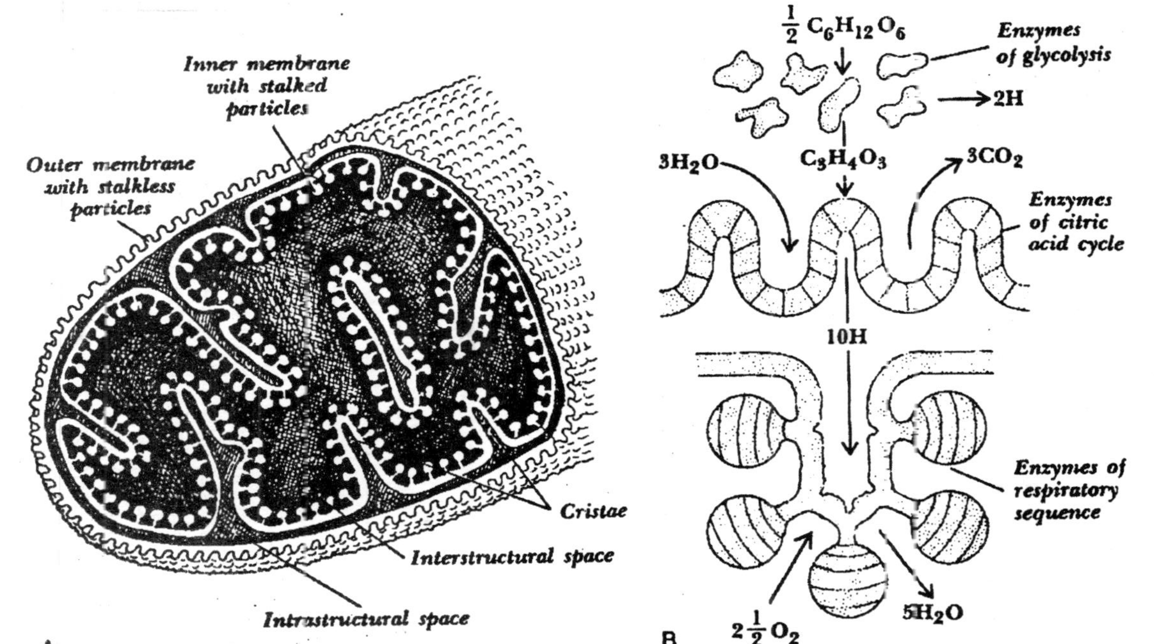

Figure 5.3 : The mitochondrion as a converter of energy. A. The finer structure of a mitochondrion. B. Detail of the outer and inner membranes. The enzymes of gly colysis are probably in the cell solution. Those of the citric acid cycle may be in the outer membrane of the mitochondrion. The enzymes and coenzymes of the respiratory sequence probably occupy the inner membrane.

is done in a closed series of re actions known as the citric acid cycle (of Krebs). It has appropriately been called "the metabolic mill." It is as though pyruvic acid were fed into a hopper and CO_2 and H atoms were ground out.

Again specific enzymes and involved, and these act in organized manner. Fats and fragments can also be fed into this same metabolic mill. The CO_2 which comes out is waste and must be discarded, but the H atoms are high in energy. Indeed, a single H atom (gram atom) represents some 52,000 calories of potential energy, four times the energy of an energy-rich phosphate bond.

Chain 3 The respiratory sequence

The third chain of reactions is the one by which the H atoms that were produced in the first two chains are oxidized to water. The energy which is released is transferred to ATP. As before, the process does not take place in one step, for, if it did, a large part of the energy would be lost as heat.

Instead, it takes place by a series of steps. At each step a special molecule accepts a pair of H atoms (or electrons-a H atom equals a H^+ ion plus an electron) at a higher energy level., It then transfers them to another molecule at a lower energy level. At some of the steps, ATP molecules are synthesized. The final hydrogen acceptor is oxygen, and the end result is water.

The overall consequence of the three chains of chemical reactions is that, for each sugar molecule which is consumed, 6 molecules of CO_2 are given off, 6 molecules of H_2O are produced, and 38 molecules of ATP are restored. These represent 456,000 calories, or 66 percent of the potential energy of the glucose which was consumed. The remaining 34 percent of the energy is lost as heat.

The enzymes and coenzymes of the last two chains of metabolic reactions are organized in the form of certain ubiquitous organelles which are visible with the light microscope and which are known as *mitochondria*. It is their business to oxidize pyruvic acid and restore ATP. This they apparently do after the manner of a bucket brigade at a fire; that is, they pass on pyruvic acid and its products from locus to locus until it is completely oxidized.

Mitochondria are present in abundance in all cells which are capable of utilizing oxygen in the release of energy. They are most concentrated in the more active cells, such as nerve, muscle, and gland cells, and notably in those regions of these cells where metabolic activity is greatest.

A single cell may contain a thousand or more of these bodies. They may make up 20 percent of a cell's dry weight. As pictured by the electron microscope, each mitochondrion consists of a double membrane surrounding a semifluid "matrix". The inside layer of the double membrane is thrown into internal folds or cristae which partially divide the cavity within.

The enzymes of the citric acid cycle (chain 2) are thought to be in the outer layer, while those of the respiratory sequence (chain 3) are localized in the inner layer of the membrane of the mitochondrion. Motion-picture microphotography shows the mitochondria to be in incessant motion, especially at the time of cell division.

Lysosomes

Lysosomes are cell organelles which contain powerful digestive enzymes. Each is surrounded by a semipermeable membrane to prevent the enzymes from escaping and digesting the rest of the cell. Christian de Duve, a biochemist, was the first to recognize lysosomes as such. Later they were identified under the micro-scope.

De Duve noted that when he homogenized rat liver with a Waring blender a very severe treatment-enzymes were released which digested the organic molecules of the cell. When, however, he used a gentler method of homogenization, very little digestion took place. The enzymes were present, but they were held within the membranes of the bodies which enclosed them.

Lysosomes are especially important in white blood corpuscles and other phagocytic cells (cells which devour). They constitute many of the granules of the cytoplasm. Cells of this sort take in foreign particles much as an amoeba engulfs a food particle. The process is called *phagocytosis*.

The vacuoles containing the food particles and the lysosomes move together, fuse, and their contents mingle. After a short time the vacuole becomes a digestive vacuole filled with the products of digestion. Absorption takes place, and the residual vacuole, now loaded with undigested debris, is ejected by the cell.

When, under certain circumstances, the membranes of the lysosomes become permeable, digestive enzymes are released into the cytoplasm, and the cell dissolves. This is known as *autolysis*.

A controlled autolysis is not uncommon during development. For example, at the time of metamorphosis, the tail of a tadpole undergoes autolysis through the activity of enzymes released from lysosomes.

The Endoplasmic Reticulum

The electron microscope reveals the presence of a net of tubules and hollow lamellae within the cytoplasm of many cells; a light microscope is not powerful enough to reveal this. This net is the *endoplasmic reticulum.*

The membranes of the net separate two liquid phases: an outer phase (the ground substance of the cytoplasm), which bathes the net; and an inner phase, which is contained within the net. In some cases, perhaps in all, the membrane of the net is continuous with the plasma membrane and with the nuclear membrane.

Thus the fluid which is within the net has connection with the fluid outside the cell. Following cell division, the nuclear membrane seems to be restored from the endoplasmic reticulum.

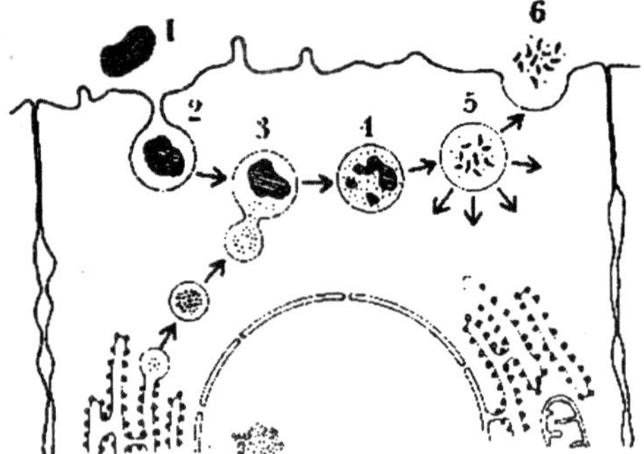

Figure 5.4 : Phagocytosis and intracellular digestion. 1 and 2. A food particle is ingested. 3. It is joined by a lysosome bearing digestive enzymes. 4 and 5. Digestion and absorption take place. 6. Final ejection of waste. The origin of the lysosome from the endoplasmic reticulum is hypothetical.

Endoplasmic reticula are of two sorts: "rough" (granular) and "smooth" (agranular). The roughness of the former is due to numerous ribosomes which adhere to and coat the surface of the net. Because the ribosomes stain with basic dyes, this material of the rough reticulum shows dark with the light microscope. It has been called *ergastoplasm.*

Not all cells show a well-developed endoplasmic reticulum. In early development embryonic cells may possess an abundance of ribosomes, but these are mostly free, and the reticulum is limited in extent and largely smooth.

Later, especially in the case of those cells which secrete protein outwardly, the endoplasmic reticulum increases greatly and becomes granular. The implication is that the reticulum is an organelle which receives the proteins that the adherent ribosomes have synthesized, concentrates them, and ultimately delivers them as secretion granules outside the cell. Examples are the gland cells, which secrete zymogen granules, and the fibroblasts, which produce the structural proteins of connective tissue.

The Golgi Complex

Most cells possess in their cytoplasm a region of specialized membranes known as the *Golgi complex* or Golgi apparatus. In thin sections the membranes appear as irregular stacks of flattened vesicles (cisternae) often located not far from the centrioles. The membranes are smooth; that is, they are devoid of a coating of darkly staining ribosomes.

Products of cytoplasmic synthesis accumulate in the Golgi apparatus, possibly having entered from the endoplasmic reticulum. Here condensation and further synthesis takes place, with the result that secretory granules are produced which have carbohydrate (mucopolysaccharides) as well as protein in their composition.

The granules collect in vacuoles which, in the case of gland cells, move to the surface of the cell and are discharged to the outside. The process may be thought of as phagocytosis in reverse. Thus the membranes of the Golgi complex act as semipermeable barriers to the free diffusion of molecules, and they may be compared in this respect to the membranes of the contractile vacuoles of protozoa.

The Cell Cortex

The form and activity of a cell are intimately dependent on the properties of the outer zone (or zones) of the cytoplasm, namely, the *cell cortex*. The outer layer of the cortex is the *plasma membrane*. It directly relates the cell to its environment. Nutrients must enter through it, and unwanted substances must be kept out.

Wastes are eliminated by it. Stimuli, except perhaps those of light, affect it first. The plasma membrane is of the order of 100 *A* thick (= 100×10^{-7} mm). It has properties of semi permeability, but it must not bethought of as a passive sieve.

On the contrary, it is an extremely active membrane which produces and retracts minute processes (microvilli) from its external surface, and forms microtubules and vesicles that push inwardly into the interior of the cytoplasm. By this means the cell feeds (phagocytosis) and drinks (pinocytosis).

The particles and droplets which thus enter the cell do not directly penetrate the plasma membrane. Rather, the membrane moves before

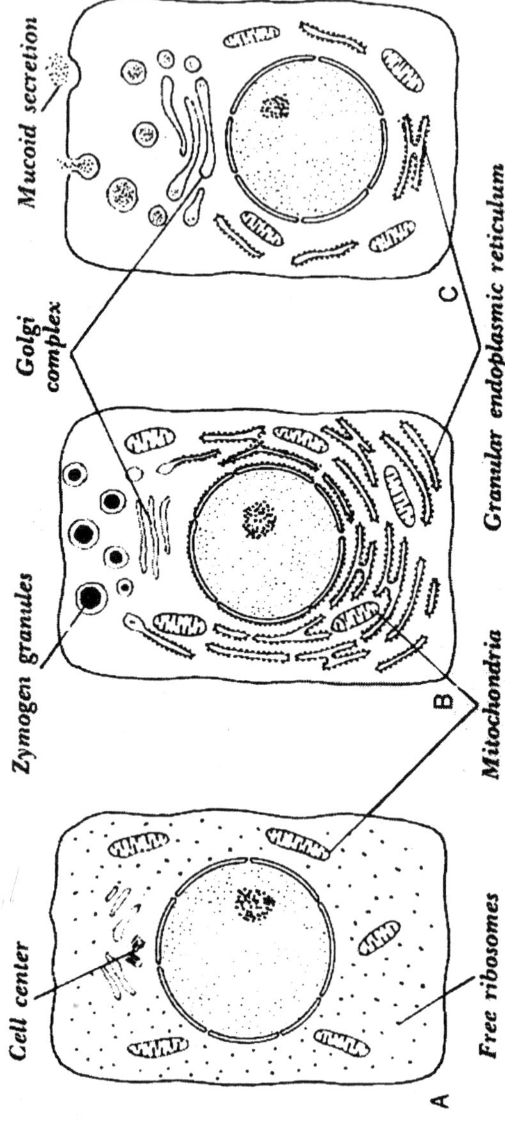

Figure 5.5 : Three types of cells engaged in synthesis (idealized). A. A protein-retaining cell with free ribosomes. B. A protein-secreting cell with an extensive endoplasmic reticulum coated with ribosomes. C. A mucus-secreting cell with a welldeveloped Golgi complex.

them and surrounds them as they advance into the cell. Later the membrane may dissolve, or the contents of the vacuole may be digested and absorbed into the cytoplasm.

In addition to the plasma membrane, the cell cortex commonly includes a somewhat thicker layer of cytoplasm which is from 1 to 5 IL thick (= 1 to.5 × 10^{-3} mm). This layer, sometimes referred to as ectoplasm or plasmagel, is usually gel-like in its physical structure, although it may change its state from gel to sol and back again.

When photographed with the electron microscope, the plasmagel often differs very little, if at all, from the more liquid internal cytoplasm (endoplasm, plasmasol). Indeed, it may not be present at all. But in other cases it is specialized by the presence of filaments or granules which are not found elsewhere.

The various responses by which a cell changes its shape are often cortical reactions. Development, also, is to a remarkable degree a matter of cortical response. This is obvious in the case of the extrusion of the polar bodies, the rising-up of the fertilization membrane, and the division of the cytoplasm during cleavage.

It is not so obvious that the folding of the cellular layers by which the gut and later the nervous system are formed are also to a large extent cortical reactions.

Protozoologists long ago became aware that the form and much of the activity of one-celled animals are properties of the cell cortex. Consider, for example, the ciliate, *Stentor*. The body of this animal is conical in shape. The sides are marked by a hundred or so longitudinal rows of cilia separated by intervening pigmented stripes.

The base of the cone (anterior surface or "head") is a disc with spiral rows of cilia. The head is surrounded by an almost complete circle of membranelles. At the clockwise end of the arc of membranelles are the structures used in feeding, namely, the oral pouch, cytostome, and gullet. Nearby is the anal pore.

The apex of the cone (posterior end) is specialized as an organ of temporary attachment, the holdfast. All these structures are specializations of the cell cortex.

Tartar has performed numerous surgical operations on *Stentor* and has repeatedly observed its extensive capacity for regeneration and reorganization. The controlling feature, he finds, is the presence and orientation of the cortex.

A tiny bit of *Stentor*, only 1/123 the size of a normal animal, is capable of regenerating a whole *Stentor* provided that a little of the cortex and some nuclear material are present. But a far larger piece of *Stentor,* lacking a bit of cortex, does not regenerate even though adequate nuclear material is included.

Undoubtedly the nucleus is necessary to control the processes by which substances of the cell are synthesized; but, Tartar concludes, the polarity and asymmetry of the body are transmitted from one generation to the next solely by the cortex.

In a somewhat similar fashion, the cortex of a metazoan egg transmits the pattern which largely controls the polarity, symmetry, and asymmetry of the future embryo. For a time these features are labile to varying degrees and subject to modification by the environment.

Raven has studied the role of the cortex in the eggs of mollusks. When such eggs are centrifuged, the visible materials of the cytoplasm are displaced either toward or away from the axis of rotation. They are displaced toward the axis if they are less dense than the fluid cytoplasm which surrounds them, and away from it if they are more dense.

Development proceeds nor-mally except that the granules and droplets remain unnaturally distributed. In other cases, the displaced visible stuffs move back to their original locations. Now, since the internal cytoplasm is quite fluid and subject to flow (witness its movements during meiosis, fertilization, and cleavage), it must be the cortex which remains fixed and so transmits the pattern of the developing organism. (Extreme centrifugation liquefies the cortex and results in abnormalities.)

The Adhesion of Cells

The cells of an embryo adhere to one another in varying degrees and in manners which are characteristic of the different tissues. At first the adhesions are weak. Cells easily dissociate when early embryos (blastulas and gastrulas) are placed in calcium-free solutions.

As development proceeds, however, the adhesions between most cells become stronger. This is especially true in the case of epithelial cells, that is, in the case of cells which form layers and either bound the outside of the embryo or line cavities within it.

Adhesion is especially, close and strong immediately adjacent to where the contacts between cells border a cavity. In effect, an epithelium presents a continuous outer surface-Holtfreter thought of it as a "coat"— which is semipermeable and which serves to separate the external medium which surrounds an embryo (or fills its cavities) from the internal medium which bathes its inner cells.

Actually, it is not a continuous coat, but a network of close contacts between cells (often referred to as *terminal bars).* Except at these outer

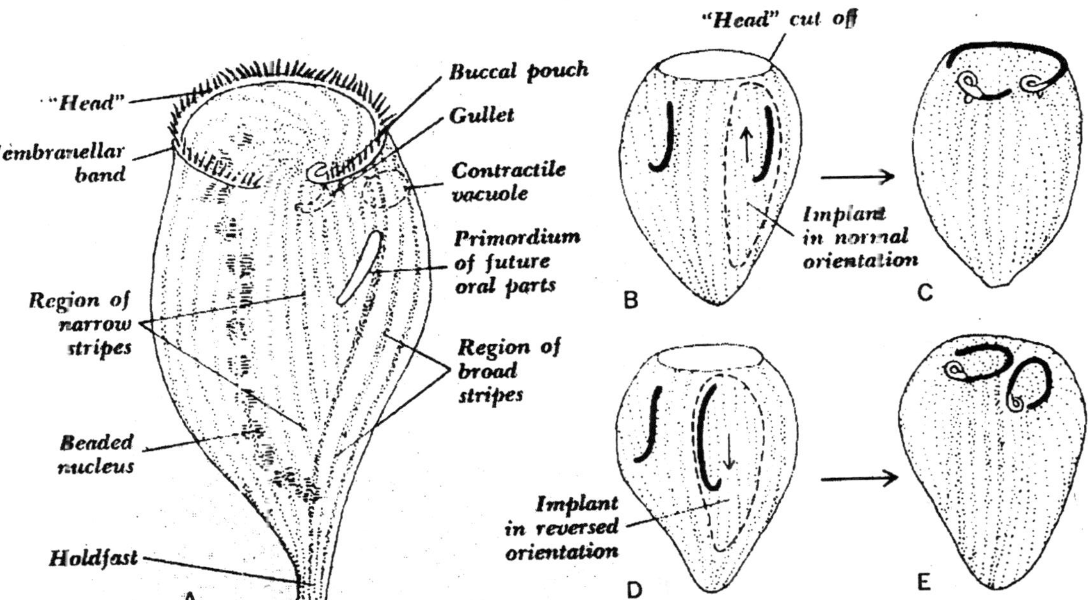

Figure 5.6 : A. Schematic diagram of the ciliate protozoan Stentor. B and C. Tartar's experiment in which a region of the cortex with narrow stripes was implanted in a region of broad stripes in normal orientation. Result-a "doublet" Stentor. D and E. A similar experiment but with the implant revolved 180°. The implant developed a set of oral parts with reversed asymmetry.

contacts pockets of liquid may separate the plasma membranes. As development proceeds, adherent patches known as *desmos-omes* develop between adjacent cells and strengthen the bonds between them. The electron microscope sometimes shows filaments extending from each desmosome into the cytoplasms on both sides.

Other cells of the embryo, notably those known as mesenchyme, are less adherent to each other. Apparently they are free to wander independently; but they do not wander haphazardly. On the contrary, they move to appropriate positions in the embryo, and there they remain as if trapped by their surroundings.

What guides mesenchyme cells? What finally brings their wanderings to an end? In general, what holds embryonic cells to their places in the embryo?

Holtfreter has attempted to account for the direction of cell migration in terms of "contact guidance." Movement on a surface, he supposes, is controlled by the oriented macromolecular structure of the surface. He explains the final location of cells by the principle of "selective adhesion."

That is, cells (or rather, cell surfaces) adhere most strongly to other cells for which they have an "*affinity*" (possibly a comparable surface pattern). When a cell makes contact with a congenial cell, its wanderings cease. Abercrombie terms the behavior "*contact inhibition.*"

These responses of the cell surfaces are markedly specific. In some cases they appear to be species-specific, as when an egg cell reacts to sperm cells of its own species, or when lymphocytes react to foreign proteins (antigens) by producing antibodies.

In other cases—and this is especially characteristic of embryos-the specificity of the cell surface seems to be tissue-specific. When an embryo is experimentally dissociated into its constituent cells and then the cells are permitted to reaggregate, they first adhere to each other and then tend to sort out and reform the tissues from which they were taken.

Superficially, it seems as though each cell, when it contacts another cell, recognizes whether the other cell is of its own or another kind. Steinberg offers a less mysterious explanation of contact guidance and selective adhesion. He explains segregation in terms of quantitative differences in the strength by which cell surfaces adhere.

The principle is much the same as that which causes oil and water to separate; namely, the mean of the cohesive forces between like molecules (water to water and oil to oil) is greater than the adhesive

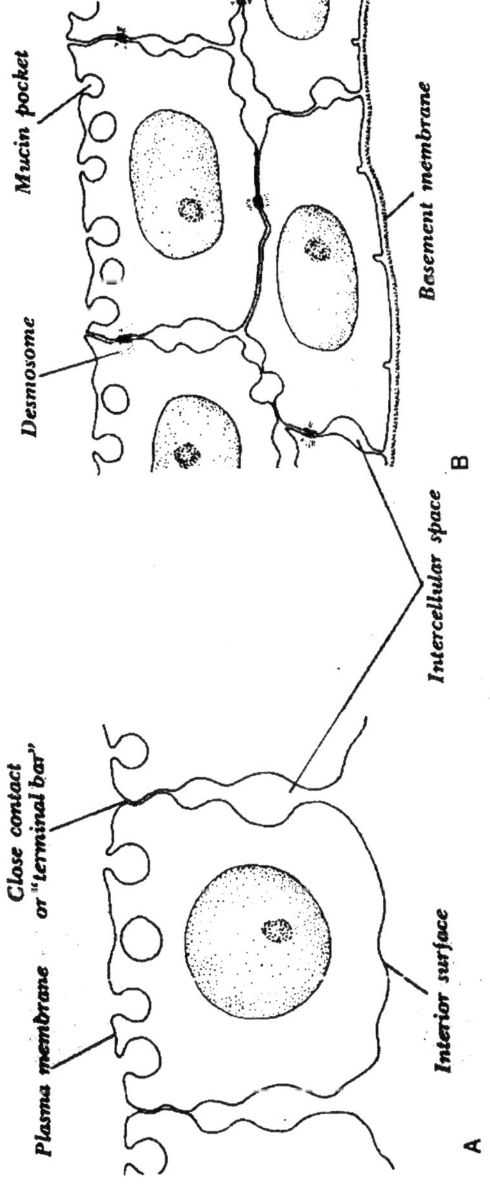

Figure 5.7 : Adhesion of epithelial cells. A. Early one-layered epithelium in which there is close contact next to the external surface. B. Later two-layered epithelium with desmosomes and a basement membrane.

forces between unlike molecules (water to oil). Similarly, 'intermingled cells sort out because like cell surfaces adhere to each other with more average strength than unlike surfaces adhere. Sorting out, therefore, is not so much a property of cells as individuals, as it is a property of cells in populations.

A corollary of this is that those cells of a mixed population which have the strongest mutually adhesive properties tend to pull together and become encapsuled within an outer mass of cells with less adhesive tendency.

The foldings and moldings of tissues during development (about which we shall have much to say in the following pages) have been interpreted in terms of programmed changes in the adhesive characteristics of cell surfaces.

For example, a thickening of an epithelial layer of cells involves an increase in the areas of contact between cells. Presumably this results from an increase in the strength of their cohesiveness. A thinning of an epithelium is associated with a decrease in the areas of contact.

The Role of the Cytoplasm in Development

Many embryologists today are engaged in describing in detail the chemical and cellular changes which take place in an embryo as development proceeds. This is the problem of *cytodifferentiation*. Notable progress is being made with the aid of sophisticated physical and chemical techniques such as *electron microscopy*, *radioactive tracers*, and *immunological* methods.

As a result, today we know something about the sequence of chemical events in the embryo and when and where new cellular structures appear and disappear. But it is one thing to *describe* the physical and chemical changes which take place during development, and quite another thing to give a *causal account* as to how these changes are brought about. The key question is: What controls differentiation?

There is no doubt today that the cytoplasm controls the course of cell differentiation. But there are two "levels" at which this control may be exerted:

(1) It has been generally assumed that control takes place at the level of "transcription"; that is, that it takes place in the nucleus when DNA becomes activated and transfers its information to messenger RNA. Some genes are "turned on" at any given time and place, while other genes (probably most of them) remain repressed.

(2) There is, however, increasing evidence that some control by the cytoplasm takes place at the level of "translation," namely, when

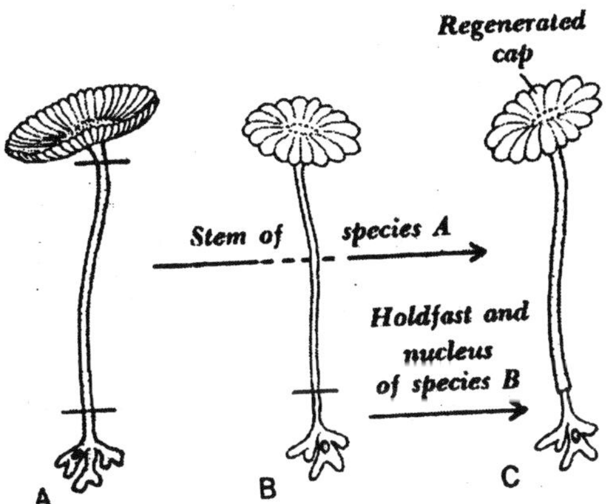

Figure 5.8 : Regeneration in the unicellular alga; Acetabularia. *A. Acetabularia mediterranea ("species A"). B. Acetabularia crenulata ("species B"). C. Result of grafting a decapitated stem (no nucleus) of species A to a holdfast (with nucleus) of species B. The first cap which regenerated was cut off. The second cap to regenerate had the characteristics of species B.*

mRNA supervises the synthesis of protein. In this case the mRNA produced in the nucleus remains "masked" until it reaches the proper time and place in the living system.

The classic example of control of differentiation at the level of translation is seen in the marine alga, *Acetabularia*. This is a giant cell, some two inches long, and shaped like a mushroom or parasol. Its apical end spreads out radially like a hat or cap. Its basal end forms rhizoids (holdfasts), one of which contains the large nucleus.

The entire life cycle of *Acetabularia* does not concern us now. Our present interest is in its capacity to regenerate a new cap when the old cap is cut off. Surprisingly, the regeneration of a cap in *Acetabularia* takes place even when the nucleus has been cut away two or three weeks previously. Clearly the presence of nuclear DNA is not directly responsible for the restoration of lost parts.

Hammerling, who studied these matters intensively, interpreted his findings in terms of gradients in the distribution of morphogenetic substances, i.e., a cap-producing substance with its highest concentration at the apical pole, and a holdfast-producing substance centered at the basal pole. His concept is that these gradients determine the location of the new parts.

The role which the nucleus plays in regeneration must be an indirect one. Presumably the genes of the nucleus produce the morphogenetic substances which then migrate within the cytoplasm and concentrate at the apex or base, as the case may be.

There they stimulate *regeneration and at the same time control the* character of that which regenerates. This interpretation is borne out by Hammerling's observation that, when a piece of the stem of *Acetabularia mediterranea* (we shall call this "species A") is grafted by its basal end to a holdfast of *Acetabularia crenulata* ("species B"), the cap which regenerates resembles the caps of species A.

He interprets this to signify that, at the start, regeneration is under the control of the morphogenetic stuffs which were emitted by nucleus A before it was removed. But if this first cap is cut away, the second cap which regenerates has the character of species B. This confirms the view that the nucleus, which is located at a distance from the apex both in time and space, is the source of the morphogenetic stuffs.

In terms of present theory we may identify the morphogenetic stuffs (as Brachet does) *with messenger RNA. They are produced by* transcription in the nucleus. Under the influence of the cytoplasm, the cap-forming stuff migrates and concentrates at the apical end of the cell. Here translation takes place; that is, the information brought by the mRNA is transferred *in situ* to newly synthesized enzymes and structural proteins. No theory of turning genes on and off is sufficient to account for this localized reaction.

What brings about the concentration of morphogenetic stuff at the apex? Since the inner cytoplasm of *Acetabularia* is engaged in active streaming, some factor in the cortex must be responsible. But it must be admitted that, as yet, we know very little of how this influence of the cortex originates or is exerted.

6

Nucleo-Cytoplasmic Interactions

We have repeatedly seen that the nucleus and the cytoplasm continuously interact during normal egg development. These interactions can be experimentally modified in two different ways. The most radical experiment is to divide the egg into two parts (merogony) and to compare the morphogenetic and biochemical potentialities of the *nucleate* and *anucleate* halves.

Another approach is to modify the *nucleus*. One of the simplest ways to attain this goal is to introduce a foreign nucleus into the egg by hybridization or nuclear transplantation.

Biochemical work on anucleate fragments of eggs and on hybrids will be the main topic of this chapter. We shall, however, start with another object, the unicellular giant alga *Acetabularia,* since this organism presents a number of advantages for those who are interested in the role of the nucleus in morphogenesis and in the biochemical interactions between the nucleus and the cytoplasm.

BIOLOGY AND BIOCHEMISTRY OF THE ALGA ACETABULARIA

Biological Cycle. Regeneration

The giant green alga, *Acetabularia mediterranea,* which is 3 to 5 cm long, has a single nucleus during the major part of its life cycle. This nucleus, which is large and contains giant RNA-rich nucleoli, but very little chromatin, is localized at the basal part of the alga (which is called the *rhizoid*). At the apical end of its stalk, the still uninucleate alga forms an umbrella (or *cap)* which will later serve for its sexual reproduction (formation of cysts, which will produce motile gametes). The morphology of the cap is species-specific and allows taxonomists to classify the different species of *Acetabularia.*

The most important of these species, for biochemists, are *A. mediterranea* and *A. crenulata,* which can easily be cultivated in large amounts in the laboratory.

Morphogenesis is affected, in *Acelabularia,* by the amount of *light* it receives. If the light supply is abundant, caps form quickly on short stalks. Reduced light supply results in the formation of very long stalks and delayed production of small caps.

Absence of light completely stops growth and induces a number of degenerative changes in the nucleus (decrease in size, reduction of the dimensions and RNA content of the nucleolus); all these changes are reversible if light is again supplied to the algae.

These simple experiments clearly show that the structure and chemical composition of the nucleus depend on the energy supply in the cytoplasm derived from photosynthesis.

Acetabularia has become a fascinating subject for biologists and biochemists since the fundamental work of Hammerling around 1930. He discovered that if the alga is cut into two pieces, both the nucleate and anucleate fragments can *regenerate* and form a typical cap. Anucleate fragments can survive for several weeks and form caps, provided that the apical part of the stalk is present.

The algae thus contain *morphogenetic substances,* which are distributed along an *apico-basal gradient;* basal fragments, although they are close to the nucleus, hardly regenerate and never form caps. Nevertheless, the morphogenetic substances are *gene products* and originate from the *nucleus.*

This has also been demonstrated by Hammerling, who succeeded in grafting nucleate fragments of A. *mediterranea* on anucleate stalks of A. *crenu-lata* and vice versa. The result is, in general, the formation of a "hybrid" cap, which soon degenerates and is replaced by the typical cap of the nucleate fragment.

There is a kind of "struggle" between the pre-existing morphogenetic substances present in the anucleate half of one species and the newly formed substances produced by the nucleus present in the nucleate half. The battle is ultimately won by the nucleus. The reasons why the gradient in the distribution of the morphogenetic substances decreases from the apical toward the basal end of the stalk and not, as one would

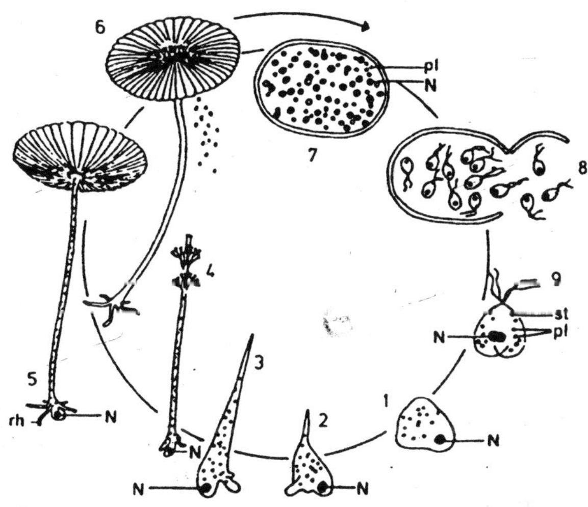

Figure 6.1 : Acetabularia mediterranea. Upper left: Life cycle of the alga. 1 zygote; 2 young growing cell; 3 slowly growing alga; 4 alga in its rapid growing phase; 5 alga with a young cap; 6 alga with a mature cap; the vegetative nucleus N has broken down and cysts disperse out of the cap. 7 Enlarged resting cyst; 8 germinating cyst giving rise to the gametes; 9 conjugation between two gametes. N nuclei; pl chloroplasts; st stigma; rh rhizoids.

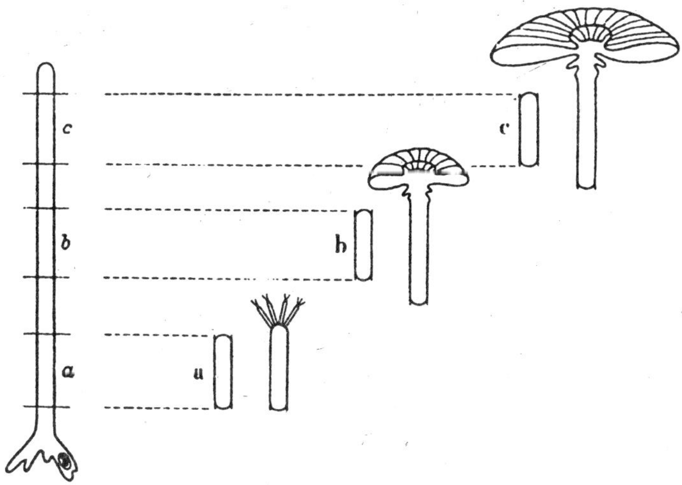

Figure 6.2 : Distribution of morphogenetic substances according to the experiments of J. Hammerling.

expect, from their nuclear origin, in the opposite direction, are not clear. Experiments by Russian workers have shown that it is possible to displace, by centrifugation, almost the whole content of the alga toward the apical or the basal end of an anucleate fragment.

Nevertheless, the polarity remains unchanged and the cap forms at the initial apical end. Thus, it is probable that polarity is linked to the properties of the cell membrane and cell wall, which are different at the apical end, where growth occurs.

There is an accumulation of RNA, proteins, glycoproteins and polysaccharides at this apical end. It is possible, but this remains a hypothesis, that the membrane contains specific receptors for the morphogenetic substances; such receptors would be accumulated at the apical end of the stalk.

Morphology of the Cytoplasm and the *Nucleus*

The algae are protected by a thick extracellular wall made of polysaccharides; it is in close contact with the plasma membrane which limits the cytoplasm. The center of the alga is occupied by a large, turgescent, acidic vacuole.

The cytoplasm is reduced to a thin sheet, compressed between the extracellular membrane and the central vacuole. Golgi bodies, which are accumulated at the growing apical end of young algae where they participate in the formation of the polysaccharide cell wall, and mitochondria are abundant in the cytoplasm.

But the most obvious cell organelles are the *chloroplasts,* they are ovoid and often contain inclusions of the polysaccharides, which they have synthesized by photosynthesis. If one observes an *Acetabularia* under the light microscope, one can see that the chloroplasts are moving continuously on cytoplasmic strands; this movement is called *cyclosis.*

Cytoplasmic contractile proteins (actin, myosin) are involved in cyclosis, since chloroplast movement quickly stops if the algae are treated with the actin-binding drug, cytochalasin.

The large (300.tm in diameter) *"vegetative" nucleus* of the alga contains, according to the *Acetabularia* species, a single large ribbon-shaped nucleolus or several smaller nucleoli. The nucleoli are very rich in RNA and are the site, as we shall see, of intensive rRNA synthesis. It has been impossible for a longtime to demonstrate cytochemically the presence of DNA in the huge vegetative nucleus.

On the other hand, DNA is easily detectable in the much smaller nuclei of the gametes and zygotes. Recent studies have shown that the vegetative nucleus contains *lamp brush chromosomes* similar to those of amphibian oocytes: 20-μm-long, DNA-containing loops extend in the abundant nuclear sap from condensed chromomeres.

The nucleus, despite its large size, is diploid and contains a little more DNA than the theoretical diploid value, which is due, as we shall see, to amplification of extrachro-mosomal (nucleolar) genes. When the cap reaches its full size, the nucleus shrinks, the nucleoli regress and a huge intranuclear spindle is built up. Meiosis occurs at this time.

The numerous small daughter nuclei, which move toward the cap and invade it, are haploid. In the cap, groups of such "secondary" nuclei, which have been repeatedly dividing, become surrounded by a polysaccharide membrane: the cap is now subdivided in cysts; each of them contain many haploid nuclei. Finally, the cysts are released in seawater and give rise to a swarm of flagellated haploid gametes.

As one can see, there is a remarkable similarity between *Acetabularia* and eggs from animals so far as the behavior of the nucleus is concerned: the vegetative nucleus is, like the oocyte germinal vesicle, in meiotic prophase.

In both cases, the nucleoli are RNA-rich and well-developed; the chromosomes are in an extended lampbrush state and the nuclear sap is abundant. However, the life cycles diverge after meiosis: in *Acetabularia,* the repeated cell divisions of the secondary nuclei precede the formation of gametes and zygotes; in eggs, cleavage follows fertilization.

Biochemical Studies

Morphogenetic Substances and mRNAs: Experiments with Inhibitors

The main problem is, of course, the *chemical nature of the morphogenetic substances*. There are many good reasons for believing that since they are synthesized in the nucleus and carry the information necessary for the production of the many proteins which are certainly needed for cap formation, the morphogenetic substances are *a family of stable mRNA molecules.*

No experim-ents contradicting this hypothesis have so far been published and there is, as we shall see, ample circumstantial evidence in its favor. However, direct proof, such as isolation of the mRNAs produced by the nucleus and demonstration that they have biological activity, is still missing.

In favor of the mRNA hypothesis are a number of facts. *UV-irradiation,* at wavelengths which affect the integrity of nucleic acids, inhibits regeneration in anucleate halves, in particular when the apex of the stalk, where morphogenetic substances are accum-ulated, is irradiated.

Irradiation of nucleate halves has only a transitory effect. Regeneration is stopped and then resumes when the nucleus has been damaged in a reversible way only. Similar results are obtained when living fragments of the algae are treated with the enzyme *ribonuclease*.

Loss of regeneration is irreversible after treatment of anucleate fragments, reversible after action of ribonuclease on the nucleate ones. Ribonuclease apparently destroys the pre-existing mRNAs; new mRNAs cannot be produced unless the nucleus is present.

The mRNA hypothesis predicts that, on the contrary, actino-mycin should not .inhibit the regeneration of the anucleate halves (since the drug does not bind the pre-existing RNAs), but would arrest that of nucleate halves (because actinomycin will inhibit the synthesis of new mRNA molecules by the nucleus). These expec-tations are fulfilled by experiments.

In actinomycin-treated anucleate fragments, an unexpected secondary effect is, however, observed after a few days. While the initiation of cap formation was as good as in the controls, the growth of the anucleate caps is definitely depressed.

The probable explanation of this secondary effect is the following: the *chloroplasts,* which exist in very large numbers in the algae, contain their own DNA. Actinomycin binds to chloroplastic DNA and this binding results in alterations of chloroplast activity which lead to a slowing down of the outgrowth of the cap.

Autoradiography brings further evidence for the view that the morphogenetic substances are stable mRNAs. If a normal alga is treated for a short time with labelled uridine, this precursor is very quickly incorporated into nuclear RNA.

If the algae are cultivated, after this uridine "pulse", in normal seawater, one can follow the migration of the RNAs which have been synthesized in the nucleus. They first move into the cytoplasm, then accumulate at the apex of the stalk (where, as we know, the morphogenetic substances are also concentrated). We shall soon see that more recent biochemical evidence also support the stable mRNA hypothesis.

Energy Production

Oxygen consumption and *photosynthesis* remain normal in anucleate fragments for a few days. Afterward, the rate of photosynthesis somewhat decreases, as well as the ATP content, in the anucleate halves. Of particular interest is the existence of a *circadian rhythm* in photosynthetic activity.

If the algae are submitted to alternate periods of light and darkness, each period lasting about 12 h, they "remember" this photosynthetic rhythm for a certain time when they are cultivated in continuous light. Anucleate fragments retain their circadian rhythm of photosynthesis for many weeks.

If however, nucleate fragments with a given rhythm are grafted on anucleate halves which have a different periodicity in their rhythm, the nucleus imposes its rhythm on the chloroplasts of the anucleate halves. It is even possible to combine, in such grafting experiments, nucleate halves which have lost their rhythmicity with anucleate fragments which possess a normal rhythm or vice versa.

Only when the nucleate half has retained its rhythm will the graft be the site of a circadian rhythm. This, as well as other experiments made with inhibitors of RNA and protein synthesis, has shown that the circadian rhythm of photosynthesis is not completely autonomous in anucleate halves.

A nuclear control, probably exerted by mRNAs, superimposes itself on cytoplasmic rhythmicity and controls it. Similar results and conclusions have been obtained for several other circadian rhythms found in *Acetabularia*. They deal with the shape of the chloroplasts, the RNA, polysaccharide and ATP content, etc.

Protein Synthesis

The protein content of both nucleate and anucleate halves approximately doubles during the 2 weeks which follow the sectioning of the algae. Since many of the newly synthesized proteins are specific *enzymes* and since we know that enzyme synthesis is controlled by the genes, it follows that the information which emerges from the genes must be stored in the cytoplasm of the anucleate fragments.

The only explanation that molecular biology can offer for such a paradox is the intervention of stable mRNA molecules, which have been synthesized in the nucleus and stored in the cytoplasm.

Certain enzymes, for instance, the phosphatases and the enzymes involved in the production of the cell walls, markedly increase in activity when the caps form. This increase occurs in anucleate as well as in nucleate fragments.

Some of these enzymes are distributed along the apico-basal gradient as morphogenetic substances, a fact which suggests that the corresponding mRNAs might have the same spatial distribution. Regeneration, in anucleate fragments, is quickly and irreversibly blocked by *puromycin* and *cycloheximide,* the two classical inhibitors of cytoplasmic protein synthesis.

On the other hand, *chloramphenicol* and *tetracycline*, which inhibit chloroplastic and mitochondrial protein synthesis, have very little effect on regeneration. Since the effects of puromycin and cycloheximide on regeneration are reversible in the case of nucleate fragments, one can safely conclude that proteins synthesized on cytoplasmic polysomes, under the direction of mRNAs of nuclear origin, play a more important role in morphogeneis than the newly synthesized chloroplastic proteins.

The chlorophasts contain their own enzymes and it would seem likely, at first sight, that this synthesis is controlled by chloroplastic rather than nuclear DNA.

Experiments by Schweiger, in which nucleate halves of A. *mediterranea* were grafted into anucleate halves of *A. crenulata* and vice vers, have shown that this is not true for the lactic and malic dehydrogenases. The enzymes present in the chloroplasts of the "hybrid" progressively change and are replaced by others, which are characteristic of the species which provided the nucleus.

The same nuclear control has been found for the synthesis of the insoluble proteins which form the chloroplastic membrane and of the chloroplastic ribosomal proteins.

Inhibitors which discriminate between chloroplastic and cytoplasmic protein synthesis have been very useful for the analysis of enzyme synthesis in *Acetabularia.* As we have seen, puromycin and cycloheximide inhibit the functioning of the cytoplasmic 80S ribosomes, while chloramphenicol blocks protein synthesis in the chloroplastic 70S ribosomes and in the still smaller mitochondrial ribosomes.

Use of these inhibitors has disclosed an unexpected fact: one would have expected that the activity of the enzymes involved in DNA synthesis increases when the vegetative nucleus breaks down and the daughter nuclei multiply at a fast rate. This is indeed what happens for thymidine kinase and other enzymes important for DNA synthesis; but a surprise came when it was found that these enzymes are encoded by chloroplastic DNA, not by nuclear genes.

Subtle interactions between the nuclear and chloroplastic genomes thus take place during the life cycle of the alga, in particular at the time of cap formation. In order to throw some light on these interactions, a direct approach has been taken recently by H.G. Schweiger and colleagues. They injected mRNAs into the algae and studied their in vivo translation on 80S cytoplasmic ribosomes. The main conclusion was that during the *Acetabularia* life cycle, translation of some proteins is not regulated, i.e. it proceeds for months at a constant rate; but translation of other proteins is turned off at a given stage.

In contrast, at the same stage, translation of still another group of proteins is enhanced. Continuation of these experiments should throw more light on the complex pattern of protein synthesis regulation during growth and morphogenesis in *A cetabularia.*

RNA Synthesis

The total RNA content, like the protein content, doubles within 2 weeks when caps are formed in either nucleate or anucleate fragments. This large increase is mainly due to the synthesis of *chloroplastic* RNAs. Only recent work, in the laboratories of H.G. Schweiger and W. Franke, has allowed precise studies on RNA production by the nucleus.

Amplification of the *ribosomal* (nucleolar) genes occurs, as in amphibian oocytes, in the *Acetabularia* vegetative nucleus. Electron microscopy of spread nucleoli and estimations of their DNA content have shown that there are about 4000 copies of the 28 S and 18 S rRNA genes in the vegetative nucleus.

As in the oocyte nucleus,. this number of copies does not increase markedly during the growth period of the vegetative nucleus, thus, amplification apparently takes place soon after zygote formation.

The general organization of the ribosomal genes is very similar in the nucleolar organizer of *Acetabularia* and *Xenopus* oocytes. Tandem repeated transcription units are separated by untranscribed spacers of varible lengths.

However, these spacers are, on the average, smaller in *Acetabularia* an in *Xenopus* and the size of the nucleolar rRNA precursor, in the alga, is not uch larger than that of the final products (28 S+ 18 S rRNAs). The rate of rRNA synthesis by the vegetative nucleus is very high (4 × 10' nucleotides/per second per nucleus); it is increased 20 times a few days after removal of part of the stalk.

The ribosomal RNAs are quickly transferred from the nucleus to the cytoplasm, where they display a great stability (half-life of 80 days); they move from the rhizoid to the apex of the stalk with a speed of 2-4 mm day. As one can see, the *Acetabularia* nucleus is, like the oocyte nucleus, a machine for the large production of cytoplasmic ribosomes.

Fortunately, the chloroplastic mRNAs have no polyadenylic "tail" and this has given the possibility of selectively "catching" the *messengers* produced by the vegetative nucleus. As expected, these mRNAs of nuclear origin are synthesized only by nucleate fragments of the alga and are heterogeneous: their molecular weights range between 0.5 and 3×10^6.

They migrate from the nucleus to the apex of the stalk independently of the rRNAs at a speed of 5 mm day⁻' (the growth rate of the stalk itself is only 1-3 mm day'), thus independently from the ribosomal RNAs.

The increase in the rate of mRNA synthesis (two- to threefold) after cutting the alga into two halves, is much smaller than for the rRNAs. These data suggest that transcription of chromosomal DNA plays only a limited role in the control of genetic activity and morphogenesis in *Acetabularia.*

The major role is played (as in maturing *Xenopus* oocytes and fertilized sea urchin eggs) by the selective translation of mRNAs of nuclear origin stored at the apex of the alga.

Little is known, unfortunately, about the nature of the factors which regulate the translation of the mRNAs accumulated at the apex of the stalk. Inhibitors of *proteolytic enzymes* prevent cap formation in both nucleate and anucleate fragments: it might be that protease activity is required for the "*unmasking*" of the mRNAs of nuclear origin which are linked to proteins in the form of ribonucleoprotein particles.

Another regulatory factor is the *polyamine* content: inhibitors of ornithine decarboxylase (ODC) prevent cap formation in both kinds of fragments. They also prevent the formation of cysts, if they are added to whole algae after the breakdown of the vegetative nucleus. Since the ODC inhibitors decrease the polyamine content of treated cells, it can be concluded that the putrescine, spermidine and spermine levels control both RNA and DNA synthesis in the alga.

In polyamine-depleted algae, the decrease in RNA synthesis is followed by inhibition of growth and morphogenesis; the slowing down of DNA synthesis in the secondary nuclei prevents normal cyst formation.

These experiments with ODC and proteolytic enzyme inhibitors have allowed an estimation of the stability of the morphogenetic substances in anucleate fragments of *Acetabularia.* If such fragments are treated with the inhibitors during different lengths of time and then cultured in normal medium, reversibility becomes partial after a 10-day treatment and completely disappears after 2-3 weeks.

This is in good agreement with the estimated half-life of the *Acetabularia* messengers (about 10 days) and provides further evidence for the view that Hammerling's morphogenetic substances are a mixture of mRNAs synthesized by the nucleus and stored in the apical cytoplasm.

DNA Synthesis. Relative Autonomy of the Chloroplasts

As we have seen, there is no replication of nuclear DNA during the whole vegetative life of the alga (which lasts several months). Only when the cap has reached its maximal size does the big vegetative nucleus break down and give rise to daughter nuclei which replicate their DNA, divide and are transferred passively to the cap by *protoplasmic streaming*.

In contrast, *chloroplastic* and *mitochondrial* DNAs are very actively synthesized by both nucleate and anucleate fragments of *A cetabularia.* The total DNA content increases two to three times in both kinds of fragments during the 10 days which follow the sectioning. This came as a big surprise 20 years age, since the very existence of DNA in chloroplasts seemed very doubtful at the time.

Now we know that chloroplastic and nuclear DNAs are very different molecules. In *Acetabularia,* chloroplastic DNA is a mixture of large circular and long linear molecules. In addition, small circular molecules, corresponding to replicating chloroplastic DNA, have been observed under the electron microscope (review by Luttke and Bonotto).

The use of highly sensitive and very specific stains for DNA allows us to see chloroplastic DNA with a fluorescence microscope, which is possible because the DNA content of the *Acetabularia* chloroplasts is exceptionally high as compared to that of other unicellular algae.

Fluorescence microscopy shows that the DNA content of all *Acetabularia* chloroplasts is not the same, i.e. the small, young chloroplasts, which are accumulated at the tip of growing algae, contain much more DNA than the large chloroplasts which surround the nucleus. In fact, many of these basal chloroplasts, which are heavily loaded with carbohydrate reserves, have probably no DNA at all.

There is thus a progressive loss of chloroplastic DNA along the apicobasal gradient. Chloropastic DNA can, in *Acetabularia* as elsewhere, be trans-cribed and translated. If they are given light, isolated chloroplasts synthesize all kinds of RNAs and a number of proteins (in particular, proteins involved in the formation of chloroplastic membranes and lamellae).

However, as we have seen, the synthesis of enzymes such as lactic and malic dehydrogenases and that of the chloroplastic ribosomal proteins requires the co-operation of the nucleus: the *autonomy* of the chloroplasts toward the nucleus is thus imperfect. Counts of the chloroplasts, in nucleate and anucleate fragments of the alga, have shown that they increase in number even in the absence of the nucleus.

However, chloroplast multiplication is more active in the nucleate than in the anucleate halves, where many chloroplasts are large and have a dumbbell shape. It is thus probable that cytoplasmic factors, which are synthesized under nuclear control, are required for fission of the chloroplasts and that they are missing in the anucleate halves.

Since even an apparently specific chloroplastic function such as the circadian rhythm of photosynthesis is regulated on cytoplasmic (80S) ribosomes, we must again conclude that the chloroplasts are not fully autonomous toward the nucleus.

In whole algae, the chloroplastic and nuclear genomes co-operate in order to specify the protein composition of the chloroplasts and to allow their multiplication.

Complexity of Nucleocytoplasmic Interactions

It is clear that the nucleus provides, in *Acetabularia,* the various mRNAs which are required for morphogenesis (cap formation), i.e. the nucleus is the source of the morphogenetic substances and therefore exerts a very important *positive control* on the cytop-lasm.

It should be emphasized that although anucleate fragments can utilize their stored mRNAs of nuclear origin and form full-sized caps, these caps always remain sterile: neither cysts, nor gametes form in anucleate caps which are therefore useless for reproduction purposes.

But it would be a mistake to believe that the vegetative nucleus of *Acetabularia* exerts only positive effects on the cytoplasm. Already 50 years ago, Hammerling demonstrated that anucleate fragments of *Acetabularia* form caps faster than control whole algae.

Similarly, we found that protein synthesis is accelerated when the nucleus has been removed. There is no doubt that the nucleus exerts *negative,* as well as positive controls over the cytoplasm.

In addition, the *cytoplasm* controls nuclear activity. If the photosynthetic function of the chloroplasts is suppressed by culture in darkness, the nucleus shrinks. The nucleolus undergoes vacuolization and stops synthesizing RNA. Thus, energy production in the cytoplasm is absolutely required for the maintenance of nuclear morphology and synthetic activity.

If, as shown long ago by Hammerling, one cuts repeatedly an alga into halves when it begins to form a cap, the breakdown of the large vegetative nucleus is inhibited, thus there will be no formation of daughter nuclei, no nuclear DNA synthesis. Repeated amputation of the cap thus leads to immortality. As in amphibian oocytes undergoing maturation, the choice between DNA transcription and replication is decided by cytoplasmic factors.

This conclusion is reinforced by more recent experiments in which an old vegetative nucleus was introduced into the cytoplasm of a young alga; this nucleus did not break down and one can speak of "*rejuvenation*" of the old nucleus by the surrounding young cytoplasm.

As we shall now see, although a giant green alga seems very different from an amphibian or sea urchin egg, there are remarkable similarities in the nucleocytoplasmic interactions in all cases. We are again faced with the paradox if unity and diversity in living organisms.

BIOCHEMISTRY OF ANUCLEATE FRAGMENTS OF EGGS

Amphibian Eggs

The huge germinal vesicle plays only a passive role in the biochemical changes (increase in protein synthesis with preferential histone synthesis, stimulation of protein phosphorylation, production of the maturation promoting factor) which take place when maturation is induced in fullgrown *Xenopus* oocytes. All these changes occur even if the oocytes have been enucleated prior to progesterone addition.

It has also been recently shown that a "*heat shock*" (heating for a few minutes above 30 °C) induces the synthesis of a few new proteins in *Xenopus* oocytes. These heat shock proteins are the same whether the oocyte possesses its nucleus or not. Enucleated *Xenopus* oocytes contain the cytoplasmic factors required for "*reprogramming*" injected nuclei.

The cytoplasm of these oocytes contains factors which inactivate certain genes and reactivate others. The outcome of the experiments in which protein or 5 S RNA synthesis has been followed after injection of adult nuclei is the same whether the oocytes have a nucleus or not. Finally, enucleated oocytes undergo the same permeability changes and the same cortical reaction after parthogenetic activation as intact progesterone-treated oocytes.

However, we have seen that the mixing of the nuclear sap and the cytoplasm is required for swelling of injected nuclei, condensation of chromatin and aster formation. It has been reported that enucleated amphibian oocytes undergo repeated cleavages after injection of sea urchin mitotic apparatuses: however, it is possible that in such experiments, sea urchin chromosomes take part in the cleavages.

Cleavage in the absence of the nucleus has been obtained repeatedly in amphibian eggs. In these experiments, the eggs were fertilized and the egg and sperm nuclei were destroyed by pricking or irradiation. In general, partial blastulae, in which only the animal pole has cleaved, are obtained; gastrulation never takes place.

The morphogenetic potencies of these *"achromosomal"* blastulae have been tested by grafting nonnucleated "cells" from arrested blastulae onto inducing sites of normal gastrulae. The grafts survived up to 4 days, but showed no sign of differentiation: nonnucleate ectoderm cells are thus totally unable to respond to the inducing stimulus of a normal organizer.

These experiments show that the morphog-enetic potencies of enucleated amphibian eggs are much lower than those of anucleate fragments of *Acetabularia:* they are limited to the capacity to divide repeatedly. We shall now see that the same is true for anucleate fragments of sea urchin eggs, which have been better studied from the biochemical viewpoint.

Sea Urchin Eggs

Biological Observations

The only materials that can be obtained in sufficient quantities for easy biochemical analysis are the anucleate fragments of sea urchin eggs. The unfertilized eggs of *A rbacia* can be cut into two halves by centrifugation in a density gradient. The heavy red pigment accumulates to the centripetal end of the egg, which successively elongates, takes a dumbbell shape and separates into two halves.

The light, transparent half contains the nucleus, fat droplets, mitochondria and a clear, ribosome-rich layer, the hyaloplasm. The red heavy half is anucleate; it contains the red pigment granules and most of the yolk. Electron microscopy shows that the light halves also possess some of the yolk platelets and that mitochondria, as well as lipid droplets, are present in both halves.

All kinds of egg inclusions are therefore presen in the two halves, but they are in different proportions; the only specific cell organelles are the nucleus, which is always in the light half, and the pigment granules, which are found only in the heavy halves.

Both kinds of fragments can be fertilized or parthenogenetically activated by the classical method of Loeb (hypertonic seawater and butyric acid treatments). Nucleate halves develop very well after *fertilization* and give rise to normal plutei. The development of fertilized

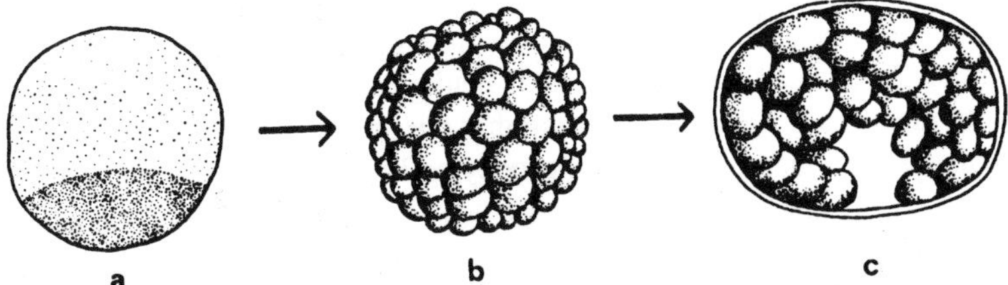

Figure 6.3 : a Anucleate half of a sea urchin egg. b Anucleate morula after 13 h. c Anucleate blastula after 3 days.

anucleate halves is less normal and plutei are not easy to obtain. This rather poor development is probably not so much due to haploidy as to the excess of yolk, which makes cleavage difficult.

Parthenogenetic development of the nucleate halves is almost as good as that of normal eggs. But the development of the anucleate halves, after parthenogenetic treatments, is completely abortive. Asters can repeatedly form in the cytoplasm of the activated anucleate halves and irregular furrows can cleave the egg to form a kind of morula.

But the typical cleavage pattern of the sea urchin eggs, with micromeres, mesomeres and macromeres is always missing. The blastula-like larvae neither hatch nor form cilia.

As one can see, there is no true morphogenesis and we are very far indeed from the formation of the elegant, species-specific cap of *A cetabularia.* Only the formation of asters and furrows displays some autonomy toward the nucleus in activated anucleate sea urchin eggs.

The only other anucleate materials that have been obtained from eggs and have been the subject of rather limited biochemical studies are the *polar lobes,* which can be separated at the "trefoil" stage from the eggs of the mollusks *Mytilus* and *Ilyanassa.*

They never form asters and do not cleave; all one can see are localized contractions of the cortex, which confirm that contractility of the cortical proteins displays some autonomy toward the nucleus.

Biochemical Studies

These results can be presented briefly, since the essentials have already been related in preceding chapters. We shall first consider energy production, then synthesis of macromolecules.

The *respiration* of nucleate and anucleate halves of unfertilized sea urchin eggs is, like that of the whole eggs, very low. If the fragments are fertilized, or if they undergo parthenogenetic activation, their oxygen consumption quickly increases.

The respiration of the anucleate halves becomes even higher than that of the nucleate ones. This proves that the part played by the female pronucleus, in cellular oxidations, is negligible. This conclusions agrees with the already mentioned finding that the oxygen consumptionn of the isolated germinal vesicle of frog oocytes is hardly measurable.

In fact, enucleated and normal frog oocytes have about the same respiratory rate. The higher oxygen consumption of the anucleate halves is probably related to the fact that, as shown by electron microscopy, the nucleate and anucleate halves contain, after their separation by centrifugation, two different populations of mitochondria.

They differ in size, structure and probably in respiratory enzyme content. In fact, the anucleate, heavy halves contain a larger proportion of the reducing mitocho-ndrial respiratory enzymes (dehydrogenases) than their nucleate counterparts.

Isolated polar lobes of the mollusk *Mytilus* have a low respiratory rate, which does not increase with time, in contrast with the blastomeres where the nuclei multiply at a fast pace.

Coming now to the synthesis of macromolecules, we have already seen that *protein synthesis* is independent of the presence of the nucleus in sea urchin eggs. Activation of anucleate fragments is followed by a strong stimulation of protein synthesis and is even stronger in activated anucleate halves than in nucleate ones.

This stimulation is accompanied by an increase in the number of polysomes in anucleate as well as in nucleate halves; these polysomes result from the binding of "unmasked" maternal mRNA molecules to the pre-existing ribosomes.

A whole spectrum of proteins are synthesized after activation in both nucleate and anucleate halves, but it is not yet known whether the neosynthesized polypeptides are identical or slightly different in the two halves. Among the stored mRNAs, which are translated by both kinds of activated fragments, those coding for *histones* and for *tubulin* deserve special interest for the following reasons.

In anucleate halves, there is of course no replication of nuclear DNA and no assembly of DNA in nucleosomes; yet histones are synthesized, apparently uselessly. This definitely proves that there is not necessarily a close link between histone production and nuclear DNA synthesis. Synthesis of tubulin in activated anucleate halves also raises an interesting question: Is translation of tubulin mRNA involved in the formation of asters in activated anucleate fragments?

It is well known that unfertilized sea urchin eggs have a large pool of unpolymerized tubulin molecules. The formation of asters and spindle, in the fertilized sea urchin eggs, results from their polymerization around microtubule organizing centers (MTOC), such as the kinetochores of the chromosomes and the centrioles.

If anucleate halves are treated with heavy water (D_2O), many cytasters appear, which are centered around small centrioles or precentrioles which have apparently been assembled de novo. It is not known whether translation of tubulin mRNA is involved in the formation of asters in activated anucleate sea urchin egg fragments.

But it is a striking fact that, despite tubulin synthesis, anucleate activated halves never form cilia; anucleate amphibian blastulae also lack cilia. Something necessary for the assembly of cilia is thus missing in nonnucleate cytoplasm.

Whatever this might be, it seems clear that its production, in blastulae, requires transcription of nuclear genes. Anucleate parthenogenetically activated sea urchin fragments are also unable to hatch: they apparently cannot synthesize the "hatching enzyme". It would be interesting to know whether they possess a maternal mRNA encoding this proteolytic enzyme.

The *distribution* of the maternal mRNAs in the nucleate and anucleate halves of sea urchin eggs (obtained by high speed centrifugation) has been recently studied. It was found that total RNA, the bulk of the poly A containing mRNAs, actin and tubulin mRNAs are accumulated in the heavy, anucleate halves.

This might explain why protein synthesis is stronger, after parthenogenetic stimulation, in anucleate than in nucleate halves. In contrast, histone mRNA is accumulated in the nucleate halves (even in uncent-rifuged eggs cut into halves with a glass needle). However, isolated nuclei contain little histone mRNA.

The distribution of the various mRNAs is thus heterogeneous in sea urchin eggs, which is probably due to the organization of the cytoplasm in gradients. If we compare amphibian and sea urchin eggs with *Acetabularia,* a striking fact emerges: although all anucleate fragments possess a large store of mRNAs, only the alga is capable of true morphogenesis.

The reasons why, in eggs, the maternal mRNA store (theoretically capable, as we have seen, to synthesize more than 10,000 different proteins) does not allow even such early events as hatching and ciliation, are not known.

It might be that the mRNAs are less stable in anucleate than in nucleate fragments. However, it is known that polyadenylation of the same mRNAs occurs in both kinds of fragments and that the enzyme responsible for polyadenylation is accumulated in the heavy anucleate halves. As we know, one of the functions of poly A addition is to increase mRNA stability in the cytoplasm.

Transcription of nuclear genes during and after cleavage is thus an absolute prerequisite for further development. In molecular terms, the maternal mRNA store corresponds to *preformation,* the neosynthesized mRNAs to *epigenesis:* that development is an epigenetic process is, of course no surprise for embryologists.

Development stops, in enucl-eated eggs, at an early cleavage stage probably because they are unable to synthesize the mRNAs and the specific proteins which are required for further developmental events.

Both nucleate and anucleate fragments of sea urchin eggs possess a large population of *mitochondria.* Since they contain DNA, the DNA content of nucleate and anucleate halves is about the same (the haploid female pronucleus represents only a small proportion of total DNA). After activation, both kinds of fragments synthesize small amounts of mitochondrial RNAs (in particular, mitochondrial ribosomal RNAs).

The mitochondria of activated anucleate fragments synthesize as RNA species which can diffuse into the cytoplasm, but, since it is unable to bind to the cytoplasmic ribosomes, it cannot play an important role in the strong stimulation of protein synthesis which follows activation of anucleate halves.

An unexpected fact has recently come to light: *mitochondrial* DNA, RNA and protein syntheses are more strongly stimulated by *parthenogenetic* activation in anucleate than in nucleate halves of sea urchin eggs. It seems that removal of the nucleus induces a multiplication of the mitochondria, as well as an increase in mitochondrial DNA transcription and translation.

If so, the often neglected female pronucleus would exert a *negative* control on the mitochondrial genome. This is reminiscent of the negative control exerted by the *Acetabularia* nucleus on the chloroplastic genome. In this alga, removal of the nucleus speeds up DNA, RNA and protein synthesis and we have seen that the chloroplasts are largely responsible for this increase in macromolecule synthesis.

Removal of the *polar lobe* at the trefoil stage results, in the mollusk *Ilyanassa,* in developmental abnormalities. Isolated polar lobes synthesize many proteins and thus contain, like sea urchin eggs, a store of maternal mRNAs. Removal of the polar lobe does not affect RNA and protein synthesis before relatively late stages of embryogenesis: no changes can be detected before the first appearance of developmental abnormalities. There is no evidence so far that the polar lobe synthesizes specific proteins and we have therefore no indication for a selective localization of certain translatable mRNAs in this region of the *Ilyanassa* egg.

Curiously, the pattern of protein synthesis changes when isolated polar lobes are allowed to age. These changes are the same as those observed in intact embryos and should result from a selective translation of the maternal mRNAs (which is responsible for 98% of protein synthesis during early embryogenesis in *Ilyanassa)*.

A recent study concludes that the cytoplasm of the polar lobe contains both repressors of translation and activators for unmasking maternal mRNAs: if so, both positive and negative controls of mRNA translation would operate in pure, nonnucleate cytoplasm.

THE IMPORTANCE OF THE NUCLEUS FOR EMBRYONIC DEVELOPMENT

General Background

Overwhelming evidence demonstrates that integrity of the nucleus is required for development beyond the blastula stage: haploidy, aneuploidy (unequal distribution of the genetic material in the blastomeres), in some cases hybridization. Many mutations are sooner or later *lethal* for the embryo.

We shall concentrate on the cases in which lethality is an early event (arrest at gastru-lation or neurulation) because they demonstrate that even primary morphogenesis is under nuclear control.

Haploidy can be obtained, in amphibian eggs, by a variety of means (pricking with a needle, fertilization with UV-irradiated sperm, removal of the maturation spindle just after fertilization). Whether the haploid embryos arise from the egg nucleus *(gynogenesis)* or the sperm nucleus *(androgenesis)*, the result is the same.

After normal cleavage, gastrulation is slowed down and further development leads to the already mentioned *haploid syndrome,* which is characterized by microcephaly, deficiency in blood circulation, reduction of the gills, oedema, ascites and finally death at an early larval stage. Lethality seems to be due to haploidy per se, and not to lethal genes which would not be balanced by their normal alleles in the haploid condition.

If, as may happen accidentally, the number of chromosomes doubles in a partheno-genetic egg before it undergoes cleavage, homozygous diploids are produced. Although they contain two copies of the lethal genes, they can reach the adult stage. This is apparently not true for the mouse in which haploid eggs fail to develop beyond the end of cleavage and parthenogenetic diploids are lethal.

In *polyspermic* eggs, only one of the nuclei of the supernu-merary spermatozoa fuses with the egg pronucleus; the other nuclei remain haploid and may take part in the development.

The resulting embryos are a mosaic of diploid and haploid nuclei; although part of the embryo is diploid, death occurs still earlier than in haploids. In mammals, dispermy, which takes place spontaneously in many species, including man, gives rise to *triploid* embryos. Triploids may also arise from fertilizationn of a diploid egg.

It is lethal during the postimplantation development; only very few triploid embryos have been reported to have developed until term. Spontaneously and cytochalasin B-induced tetraploid mammalian embryos display the same lethality. Interestingly, mosaic embryos constructed by aggregation of diploid and triploid or tetraploid morulae are also lethal.

Aneuploidy (unbalanced chromosome complement) always has serious consequences: if, only one or two chromosomes are lost or added to the normal complement, viability is greatly reduced and morphogenesis is abnormal. If chromosome unbalance is very marked, as a result of multipolar mitoses during early cleavage, for instance, development stops at the blastula stage. Strong aneuploidy is thus as lethal as the complete absence of a nucleus.

A great deal of interest has been recently devoted to aneuploidy in mammals because it is the major cause of prenatal death and abortion or of severe postnatal congenital defects such as the human Down's syndrome. Monosomic mouse embryos (39 chromosomes instead of 40) die before or at implantation, while trisomic (41 chromosomes) embryos survive until at least day 10 of pregnancy.

It is possible to rescue cells from such aneuploids by aggregating aneuploid morulae with normal diploid morulae. This strongly suggests that the chromosome unbalance is not lethal to the cells, but is specifically responsible for the impairment of morphogenetic events.

In the following, we shall deal mainly with lethal *hybrids,* because they have been better studied from the biochemical viewpoint than haploids, aneuploids and polyspermic embryos.

Lethal Hybrids

Biological Observations

Various things can happen when eggs are fertilized with sperm from a different species. The spermatozoon, in many cases, does not even touch the surface of the egg and nothing happens at all. In other cases, the spermatozoon comes in contact with the egg membrane and elicits the activation reaction, but it does not *penetrate* into the egg and does not even act as a parthenogenetic agent.

In the remaining cases, the spermatozoon penetrates into the egg and amphimixy follows. But, in many instances, the paternal chromosomes, which have been introduced into the egg by the spermatozoon, are eliminated into the cytoplasm during cleavage; the eliminated chromosomes soon degenerate and development is of the haploid, gynogenetic type *(pseudohybrids).*

In true hybrids, the paternal chromosomes are not eliminated; development is normal, for a longer or shorter time; then the hybrid embryos develop abnormalities and ultimately die. They are the so-called *lethal hybrids.*

Finally, when the chromosomes of the two parental species are sufficiently similar from the genetic viewpoint, the hybrids are vi*able;* they reach the adult stage, but they are not always able to reproduce. Sterility is frequent, because the homologous chromosomes cannot undergo perfect pairing during meiosis.

Early lethal hybrids will be the main subject of this chapter, because they have a peculiar interest for molecular embryology. When eggs of the common species of sea urchins or frogs are fertilized with sperm from other common species, development is often arrested at the late blastula or *early gastrula* stage.

This is the stage at which, as we have seen, major biochemical changes occur in the embryo; the most conspicuous of them is the resumption of rRNA synthesis. Why the hybrid embryos are unable to go through this critical stage of gastrulation is the question that a number of biochemical studies have tried to answer.

Before we discuss the biochemical studies on early lethal embryos, a last important biological fact must be mentioned. If a piece of an amphibian lethal hybrid, which is arrested in develo-pment at the early gastrula stage, is grafted into a normal host, a "revitalization" usually follows.

If, for instance, the dorsal lip of a blocked hybrid gastrula is grafted into a normal recipient gastrula, the grafted dorsal lip differentiates into chorda and induces, in the host, a secondary nervous system. This revitalization is not species-specific. The same result (chorda differentiation in the graft and neural induction in the host) is obtained when a dorsal lip taken from an hybrid between two species of frogs is grafted in a urodele (triton, axolotl) gastrula.

Revitalization has been explained by supposing that unspecific substances, which cannot be synthesized by the hybrid cells, are given by the host cells to the graft. But it has recently been shown that for certain lethal hybrids at least, it is sufficient to dissociate the cells of the blocked embryo and to cultivate them in salt solution to obtain extensive differentiation of the hybrid cells.

The "conditioned media" prepared from such hybrid cells are very toxic for normal embryos. Therefore, it is possible that lethality is due, at least in part, to the production and retention of toxic substances (repressors) produced by metabolism.

Nuclear transplantation has been used in order to find out whether the block at gastrulation in the combination *R. pipiens* y × *R. catesbeiana d is* due to irreversible changes in the nuclei. Transfer of nuclei taken from arrested hybrid gastrulae into recipient enucleated *R. pipiens* unfertilized eggs showed that nuclei taken from hybrid embryos that have just stopped developing allow normal cleavage; but development stops abruptly at gastrulation (as in the hybrids themselves).

If 'nuclei are taken 10 to 20 h after gastrulation arrest and then transplanted into enucleated unfertilized eggs, cleavage is no longer normal. There are thus no irreversible changes in the nuclei of the lethal hybrids so long as they do not undergo cytologically visible alterations.

Biochemical Studies on Lethal Hybrids

Respiration. Most of the biochemical studies made on lethal hybrids between sea urchins have been done with the combination *Paracentrotus* × *Arbacia.* These hybrids cleave normally, hatch and die during gastrulation.

Paracentrotus (P) eggs have a higher oxygen consumption than those of *Arbacia (A).* The respiration of the P × A hybrid is first similar to that of the maternal species (P). Later on, the respiratory rate decreases in the hybrids, so that intermediary values between those typical for P and A are obtained.

This reduction of the respiratory rate in the P × A hybrids suggests that a negative control, due to the paternal A genes, becomes effective a few hours before the arrest of development.

More data are available for hybrids among amphibians. The best studied combination is one among two American species of frogs: *Rana pipiensy x Rana sylvaticad,* where development stops at gastrulation. Respiration shows a normal increase during cleavage, then remains absolutely constant until the death of the embryo several days later.

This decrease in the respiratory rate affects all parts of the embryo, as shown by measurements of the oxygen consumption made on dissected hybrid gastrulae. The decrease in respiration rate is the same in the ectoblast, the mesoblast and the entoblast. The reduced respiratory rate of the hybrids is linked, as one would expect, to a diminution of carbohydrate metabolism.

Glycogenolysis is slowed down, glycolysis is decreased and the ATP content drops more quickly under anaerobic conditions in the hybrid gastrulae than in the controls after development has stopped.

These results cannot be generalized to all hybrid combinations between amphibians, however. The rule seems to be that as in sea urchin hybrids the oxygen consumption still increases, but at a slower rate than in the controls in arrested embryos.

Nucleic Acid and Protein Synthesis. We shall first consider the lethal hybrids of echinoderms, then those of amphibians.

Protein synthesis has been studied in several combinations of *sea urchins*. The "pool" of soluble amino acids and small peptides remains of the maternal type, even at a time when an effect of the paternal nucleus should become evident.

In the *Paracentrotusy* × *Arbaciaa* combination (P × A), paternal proteins (antigens) can be detected, by immunological methods, in the hybrid blastula. The amount of paternal protein synthesized is about half that which would have been produced by a diploid *Arbacia (AA);* it is identical with the quantity of *Arbacia* antigen which is produced by a haploid, parthenogenetic *Arbacia* embryo at the same stage.

These antigens are not synthesized at all by anucleate fragments of *Arbacia.* It follows that the amount of antigen produced depends on the number of *Arbacia* genes present. It does not matter, in this respect, whether a haploid set of genes is placed in its own *Arbacia* cytoplasm or in the foreign *Paracentrotus* cytoplasm.

However, in other echinoderm combinations *(Strongylocen-trotusy* × *Dendrasterd)* no paternal antigens can be detected in the hybrid embryos, which stop development at a later stage than in the P × A combination. In these hybrids, one enzyme has been characterized: it is the "hatching enzyme" which enables the blastula to dissolve the fertilization membrane and to swim freely in the sea.

This hatching enzyme has been found in the hybrids to be of the purely maternal type. Taken together, these results suggest that in contrast to the P × A combination, the paternal chromosomes are partly or entirely inactivated in the hybrids which show a complete maternal dominance.

This suggestion has been reinforced by recent work on sea urchin hybrids which can reach the pluteus larval stage. Among many proteins synthesized by these hybrids, only a few are specified by the paternal genome.

Analysis of mRNA production in these hybrids has further shown that the paternal mRNAs are strongly underrepresented; however, paternal histones and their mRNAs are detectable already during early cleavage. It has also been found that some of the messages coding for paternal proteins are not translated in the hybrids.

Production of foreign paternal proteins in the hybrids is thus repressed at two levels: transcription of the paternal genes and translation of the mRNAs (except for the histone genes and their mRNAs).

The controls exerted by the egg cytoplasm on the activity of the paternal genes differ whether development of the hybrids stops at gastrulation or reaches the pluteus stage.

A thorough study, with the now available methods, of the early lethal hybrids is needed before one can explain this difference. The present evidence suggests that arrest at gastrulation takes place in eggs in which the foreign paternal genes are not repressed by the egg cytoplasm and are thus expressed.

Development until a more advanced larval stage would be possible when the paternal genes are repressed to a large extent during early development. In this case, development would be almost parthenogenetic as in gynogenetic pseudohybrids. However, in these hybrids the foreign paternal chromatin far from being eliminated, is repeatedly replicated.

Arrest of DNA synthesis is certainly *not* responsible for the block in development. DNA synthesis continues when the P × A embryos are blocked at the gastrula stage, but the rate of synthesis is slower than in the controls.

This probably results, from a discrete elimination of paternal chromosomes during the development of the P × A embryos. Such an elimination also explains why the blocked P × A hybrids contain three times more maternal than paternal DNA, as shown by molecular hybridization.

The situation in *frog hybrids* is not fundamentally different, but a few peculiarities should first be mentioned. Hybridization modifies the properties of the plasma membrane; its permeability to amino acids and phosphate increases. Furthermore, if cells of a lethal hybrid gastrula are dissociated by removal of calcium and magnesium from the medium, they have great difficulties in reaggregating when these ions are added back to the medium.

These differences in reaggregation in hybrid and normal gastrula cells are probably due to differences in the surface glycoproteins; it is likely that changes in cell surface glycoproteins and cell adherence play an important role in the arrest of many hybrid combinations at gastrulation.

Indeed, it is known that treatment of normal *Xenopus* morulae with lectins (which bind to the sugar residues of the surface glycoproteins) prevents their gastrulation.

Another characteristic of the lethal hybrids among frogs is that the *yolk re*serves are utilized at a much slower rate than in the controls. The enzymes present in the yolk platelets also fail to be synthesized or liberated at the normal rate. The reasons for this delay in vitellolysis are not known.

As in the lethal sea urchin hybrids, *paternal antigens* become detectable at the blastula stage and an additional basic protein is found in the hybrids.

The synthesis of several *enzymes* has been studied. Esterases remain of the maternal type in one lethal combination of frogs. In another, which is viable, the enzymes that catalyze the oxidative metabolism of the sugars have values that are intermediary between those found in the two parental species, but if the egg nucleus is removed before fertilization, in order to obtain *a merogone hybrid* (which is lethal at a later stage than gastrulation) the synthesis of these enzymes follows the paternal pattern.

As in the sea urchin hybrids, *DNA synthesis* does not stop, but only slows down after the arrest of development. The same is true for historic synthesis, which remains well coordinated with that of DNA. However, DNA and RNA syntheses stop before protein synthesis.

When the hybrid is close to death and can no longer synthesize nucleic acids, it becomes comparable to the anucleate fragment of an egg, in which protein synthesis can continue for a long time because of the preformed maternal mRNAs and ribosomes.

A situation somewhat different from lethal hybridization has been achieved by Gurdon who injected, into an enucleated unfertilized egg of *Xenopus,* a diploid nucleus taken from a blastula of *Discoglossus.* In this case and in contrast with the lethal hybrids, a diploid foreign nucleus, but no maternal haploid nucleus, is present in the egg.

The eggs that have received such a foreign *Discoglossus* nucleus stop development, as many lethal hybrids do, at the late blastula stage. RNA synthesis has been analyzed in these blocked blastulae. There is no rRNA synthesis and there is a decrease in mRNA and tRNA synthesis.

As one can see, much more work is needed before we can understand the molecular reasons for the lethality in both sea urchin and amphibian hybrids. If one can venture a guess, arrest of development, followed by cell death, might largely be the result of the accumulation of foreign, paternal proteins.

BIOCHEMISTRY OF EARLY DEVELOPMENTAL MUTANTS

There is no doubt that molecular embryology would make tremendous progress if *mutants* deficient in a specific metabolic step were easily available. It is needless to recall that molecular biology would not be what it is today without the extensive study of thousands of mutants in bacteria and phages.

We have a few mutants available in amphibians and their number is increasing every year, but extremely few of them have been submitted to a biochemical analysis. This is unfortunate, since the already discussed *nucleolusless* (nu-o) mutant of *Xenopus* has played an exceedingly useful, if not decisive, role in the understanding of the intervention of the nucleolar organizers in rRNA synthesis and ribosome formation.

The only mutants that have been, to a very limited extent, analyzed from the biochemical viewpoint are the o and *cl* mutants of the axolotl. These two mutations lead to early lethality when they are in the homozygous state.

o denotes *ova deficient* and *cl* denotes *cleavage arrest*. In the heterozygous condition, the o mutants reach the adult stage and are fertile, but many of the individuals die before they become adult. The homozygous o/o mutants die before or during gastrulation.

Interestingly, they can be saved, as Briggs discovered, if one injects into the just fertilized egg the content of a germinal vesicle taken from a normal oocyte. The injection of this nuclear sap into one of the two first blastomeres enables the injected blastomere to develop further than gastrulation, while the other is blocked.

Thus, the nuclear sap of the normal oocytes contains a factor, presumably a protein, that cannot be synthesized by the *o mutant* and that is necessary for development beyond the gastrula stage. Further studies on the chemical nature of this "corrective" factor will be awaited with great interest.

We only know that it is present in the oocyte nuclear sap of a variety of amphibians and, thus, has no species specificity. The corrective factor is absent in the cytoplasm of ovarian oocytes, but is present in the germinal vesicle at all stages of *oogenesis*.

RNA synthesis is very low in o mutants and it is, therefore, possible that the missing gene product is a karyophilic protein involved in RNA synthesis. Nuclear transplantation experiments have shown that nuclei taken from of o-morulae support full development after transfer into enucleated, unfertilized eggs; but nuclei taken from o/o blastulae no longer support it.

Thus, irreversible alterations occur in the nuclei between the morula and the blastula stages in o/o mutants. The *nc* (no cleavage) mutant of the axolotl is also of interest for both molecular cytologists and embryologists: the eggs of this mutant can be fertilized, but they do not cleave because they are unable to assemble their large store of unpolymerized tubulin into spindle and aster microtubules.

The mutation can be corrected, and repeated cleavages will occur after injection of centrioles (basal bodies of flagella) into the cytoplasm of fertilized *nc* eggs. What seems to be missing in these eggs is the capacity to assemble microtubule organizing centers (MTOC).

Comparable results have been found in *Drosophila*, in which a mutation called *deep orange (dor)* causes early death in the homozygous *(dor/dor)* embryos; the defect can be corrected by injecting cytoplasm from unfertilized normal eggs into *dor/dor* blastulae.

It is not yet known whether, as in the o mutant of the axolotl, the repair factor is accumulated in the germinal vesicle during oogenesis. Other developmental mutations of *Drosophila* have been corrected by injection of cytoplasm from normal eggs.

It has been known for many years that the complete absence of the X chromosomes *(nullo-x* condition) results in deficiencies in early *Drosophila* development: gastrulation soon becomes abnormal and then stops. Small deletions in the X chromosomes may also lead to early arrest of development.

It is now well established that the blocks in early development mentioned in this section are due to the fact that mutations, such as o, *cl, dor, nullo-x,* etc. affect the organization of the *cytoplasm* during *oogenesis:* the normal alleles of these genes control the organization (in gradients, for instance) and the chemical composition of the egg cytoplasm. Unfortunately, the products of these very important genes have not yet been identified.

Mention should be made of a few mutations affecting mouse development. The Os (oligosyndactylia) mutation, located on chromosome 8, causes digit fusion in the feet of heterozygous mice. In the homozygous state, it causes lethality between the seventh division and implantation, as a result of a mitotic block in metaphase. However, unlike the *nc* mutation in the axolotl, the Os mutation does not impair tubulin synthesis or microtubule assembly. It seems to be associated with a defect in spindle formation after the sixth division. Another dominant mutation, located on chromosome 2 [*AY* (yellow)] is responsible for the yellow coat of the mutant mice; it also causes early lethality in the homozygous state. Cleavage is arrested before hatching from the zona pellucida and differentiation of the trophectoderm into trophoblast giant cells is totally impaired. Nothing is known about the biochemical target of the mutation.

The *t-complex,* located on chromosome 17, where many different mutant alleles are known, includes the H-2 locus which controls, in the adult mouse, the synthesis of the histocompatibility antigens responsible for graft rejection. Of particular interest, among the many t-mutants, are the t^{12}/t^{12} mutants where development stops already at the morula stage. Neither transcrip-tion nor translation are affected by this mutation until the embryo dies, but energy production must be somehow affected since neutral lipids accumulate in nuclear and cytoplasmic droplets.

"TRANSGENIC" MICE AND TERATOCARCINOMA

An old dream for many biologists has been to create new strains of animals with predetermined genetic changes: in other words, to introduce a selected gene into an egg and to witness its expression not only in the adult, but even in its offspring.

As we have seen, pure genes are now available in relatively large amounts due to progress in molecular biology. Parallel progress in embryology allows skilled experimenters to inject substances into one of the pronuclei of a fertilized mouse egg, to culture in vitro the injected eggs and to implant them into pseudopregnant females. If all goes well, fetuses will develop in their foster mothers and grow into adults, which will give birth to offspring after mating.

Injection of various genes into mouse pronuclei has been successfully achieved in recent years: if the gene coding for rabbit /3-globin, for instance, has been injected into one of the pronuclei of a fertilized mouse egg, adult mice, in which all the cells contain several copies of the foreign (rabbit) gene, can be obtained. Still more important, these mice often transmit the gene to their offspring (or some of their offspring).

These so-called transgenic mice fulfill the old dream of directing mutations, of manipulating Heredity at will. Success requires the integration of the foreign gene in the germ line cells of the recipient eggs and embryos.

Brinster has recently succeeded in performing similar experi-ments with "fusion-genes". A fusion-gene has been made of the promotor/ regulator of the metallothionein I gene from the mouse and the structural human growth hormone gene.

In mammals, metallothionein is normally produced by the liver and growth hormone by the anterior pituitary. When this fusion-gene is microinjected, the liver of the transgenic mice produces large amounts of human growth hormone, whereas their anterior pituitary does not.

As shown in figure elsewhere in this chapter, such mice become giant mice. The same experiments are now feasible with eggs of domestic animals and could possibly lead to the economically important production of giant cows and pigs. From a more fundamental viewpoint, the most important feature of these experiments is that they offer a new way of studying gene regulation during mammalian development.

It is now possible to learn much about the *tissue-specific regulation* of the expression of foreign sequences coding for globin, growth hormone, insulin, transferrin, metallothionein, imminoglobins, etc.

It has been shown that insertion of foreign genes in the germ line cells can ce mutations affecting development. The first case of *insertional mulagenesis* rice was reported by Jaenisch: insertion of viral DNA sequences in mice red in a recessive lethal mutation, which causes, in the homozygous state, the death of the fetus at day 14 of pregnancy.

Using the viral DNA as a probe, it was ible to show that the insertion lies in the 5' end of the al (I) collagen gene; *h* of the embryos could be due to their incapacity to synthesize collagen. Inonal mutations have also been reported in mice carrying human growth hore gene sequences; we have seen that insertion of a mobile genetic element into a *Drosophila* gene induces its mutation.

Similar experiments with human eggs have not yet been reported and it is obvious that very important ethical problems would arise if they were attempted. We all know that many "babies in test tubes" were born and that they grow normally. The methodology of artificial insemination, in vitro culture and replantation in a foster mother works as well for the human as for the mouse egg. "*Transgenic men*" are thus a theoretical possibility. Let us hope that if human eggs are ever manipulated (attempts to introduce new genes or to replace a "bad" gene by a "good" one), it will always be for the good and never for the evil of mankind. In this field, the moral responsibility of the scientist toward mankind is very heavy, not to say frightening.

Exciting material for both embryologists and oncologists is the *mouse teratocarcinoma,* a malignant tumor which originates spontaneously from the ovary or the testis, or can be induced by grafting 7-day-old embryos into the testis of male mice. Some teratocarcinoma cells differentiate into a monstrous embryo; others, called embryonal carcinoma cells, remain undifferentiated and are responsible for the malignancy of the tumor.

If these "*stem cells*" are injected into a mouse, the recipient animal will be killed by a fast-growing tumor made of a variety of tissues as well as undifferentiated embryonal carcinoma cells.

The *pluripotentiality* of the embryonal carcinoma cells makes them attractive to embryologists because they share this property with the ectodermal cells of the inner cell mass. Particularly interesting is the fact that if a small number (three to five) of these cells are implanted into a normal mouse blastocyst, they are cured from their malignancy and can participate in normal development.

Rccently, teratocarcinoma cells have been used as vectors for *gene transfer* in mice: human b-globin genes have been introduced into mouse embryonal carcinoma cells by growing them in the presence of cloned human globin DNA sequences. The transformed cells may be introduced into normal blastocysts which are then grown in the usual way.

The expected result is the production, in the second generation, of transgenic mice, thus of new strains with predetermined genetic changes (production of human b-globin) by another method rather than direct injection of a gene into a pronucleus.

As one can see, embryology is entering into a new era: fascinating, exciting, perhaps even dangerous discoveries lie ahead of us.

CONCLUSIONS

It is now time to draw a few conclusions before we leave the early stages of development and move into the much more complex area of embryonic differentiation. The work that has been summarized in the preceding chapters clearly shows that the chemical machinery for the initial steps of development has already been made during oogenesis.

Not only ribosomes have been produced in such a high number as to allow development until a late larval stage (feeding tadpole) without further synthesis of rRNA or ribosomal proteins, but a great variety of mRNAs, containing a large array of information, have been stored in unfertilized eggs.

New mRNA species are synthesized during cleavage, but they are not actually needed during this period which can normally proceed in the complete absence of RNA synthesis (except in mammals). The result of this economic, almost avaricious policy is that during the early stages of development, the importance of the cytoplasm largely overwhelms that of the nucleus.

The critical point is the end of cleavage; afterward, gastrulation movements, and neural induction become impossible without fresh information, resulting from gene activation and transcription.

But one should not forget that the fundamental, basic processes that control final development occur during these very early stages. Polarity, which will control cephalocaudal differentiation, formation of the grey crescent in amphibian eggs, (an immediate consequence of fertilization, responsible for the dorsoventral organization of the egg), germinal localizations in the "*mosaic*" eggs and embryonic regulation, all occur at stages in which the genes are almost entirely repressed and mRNA synthesis is hardly detectable.

If one returns to the different theories of embryonic differen-tiation which have been presented in other chapter of this book, one cannot escape the conclusion that the ideas expressed by Morgan, almost 50 years ago, remain valid as far as the early stages of development are concerned.

His concept that gene activity is controlled, at these early stages, by the properties of the heterogeneous cytoplasm which surrounds equipotential cleavage nuclei has been substantiated by many experiments (nuclear transplantations, in particular).

It is now clear that the nuclei of adult, differentiated cells and those of zygotes possess the same genes, but cytoplasmic factors present in oocytes and eggs affect their expression. The spectrum of proteins synthesized under the control of an adult nucleus is not the same whether it has been injected into an oocyte or it lies in its own cytoplasm.

The identity of the *cytoplasmic factors* present in oocytes and eggs, which allow the reprogramming of injected nuclei, remains unknown. If one may venture a guess, these factors might well by karyophilic proteins, which have the remarkable capacity to move quickly from the cytoplasm into the nuclei where they soon accumulate and may bind to DNA.

Detailed analysis of the proteins present in the oocyte nucleus and in cleaving egg nuclei is an important and urgent task for the molecular embryologists of tomorrow.

The cytoplasmic factors that have just been discussed certainly continue to play their role in later stages of development, when the cells undergo their typical differentiation from the morphol-ogical, physiological and molecular viewpoints.

But their importance becomes less and less evident as development proceeds. The sequential, selective activation of individual genes or groups of genes now appears in the forefront.

7

INDUCING EFFECTS

As in the chapter on sea urchin development, an attempt will be made here to summarize the various approaches to the problem of differentiation in the amphibian egg. Experimentation has proceeded from the level of tissues and cells to the molecular level.

Development of new experimental techniques and approaches depends on technical advances and also on fundamental discoveries in the field of biochemistry. In recent years, the most intensive effort has been at the molecular level.

The problems are the same as those encountered in the egg of the sea urchin: the nature and origin of egg polarity; the capabilities of various parts for self-differentiation; the interdependence of different regions upon each other.

At the molecular level, the analysis has been mainly concerned with a description of the normal sequence of biochemical events during development of the embryo, most particularly with the synthesis of DNA and RNA and protein during development. The study of transcription and translation of genetic information again provides knowledge of the relationship between geneexpression and differentiation.

Isolation of Blastomers; Evidence of a Regulative System: The first experiment on embryonic material of any kind is generally conceded to be that of Wilhelm Roux who studied the effect of killing one of the first two blastomeres of the frog egg.

The results of this study were published in 1888. The experiment is of considerable historic interest, even though the conclusions reached by Roux turned out to be in error. Roux was the first biologist to work in the area of causal analysis of development. His underlying motive was to resolve problems of development into fundamental processes which he assumed to be the same fundamental processes underlying inorganic phenomena. This was a strong departure from the ideas of the day, which were mainly concerned with explanations of embryological phenomena in terms of comparative anatomy or evolutionary data.

Prior to Roux, an embryonic structure could be considered to appear because it was part of the phylogenetic history of the species to which the embryo belonged. Roux formally put forth the idea that the proper study of the cause and effect during embryonic development was the embryo itself.

He founded a new discipline of causal, analytical embryology which he called developmental mechanics. He is frequently called the "father of experimental embryology."

The experiment reported in 1888 was simple in design and was

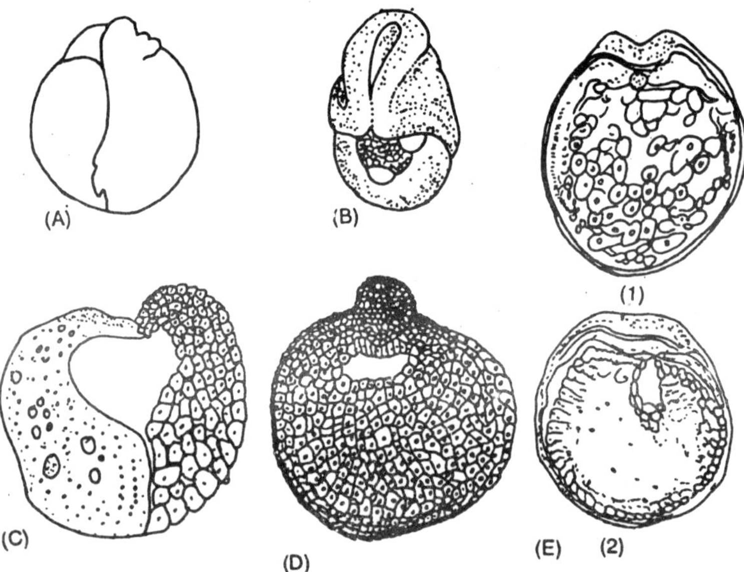

Figure 7.1 : (A) Half embryo of frog, resulting from injury to one of the first two blastomeres; (B) anterior embryo, so-called, after injury to one of the first two blastomeres of an egg in which the first cleavage had been at right angles to median plane. This embryo may also be interpreted as a whole embryo in which the lateral lips of the blastopore have been prevented from coming together by the material of the injured half; (C) section of blastula stage of ½ blastomere; (D) section of ½ embryo like that shown in A. (E) Cross sections through dorsal embryos arising from isolated Triturus blastomeres. (1) Half embryo; (2) quarter embryo, some magnification, (dorsal meaning the gray crescent material was included).

intended to decide the issue of independent versus interdependent (mosaic versus regulative, respectively) development as the underlying principle of differentiation in the frog egg. The experiment was performed at the two-cell stage and consisted of injuring one of the first two blastomeres after the first cleavage.

Since the surviving cell gave rise to half an embryo, Roux felt that each cell developed independently of its neighbour and that development of the whole embryo represented the summation of the development of each part. These conclusions were later, shown to be erroneous for the frog egg.

If the first two blastorneres were isolated from each other (instead of leaving an injured cell in contact with a live one), each blastomere was found to develop into a complete tadpole and frog, on most occasions. Even though Roux's experiment was shown to be of faulty design, the foundations were laid for the new experimental approach to the study of embryology.

Later experiments also showed that two complete embryos resulted from isolation of the first two blastomeres only when the first cleavage occurred in a plane that bisected. the gray crescent area.

In those instances where the first cleavage plane occurred at some angle to the plane of bilateral symmetry instead of parallel to it, only the blastomere containing the gray crescent material developed into a complete embryo as diagrammed in Figure.

The blastomere that lacked the gray crescent developed only into an undifferentiated mass of epidermal cells. These results, along with other studies, served to focus attention on the gray crescent area in later experimental work.

Fusion of Two Embryos; Additional Evidence of Regulative Ability

Two whole amphibian eggs can be merged to form a single embryo of double size. Two 2-cell stages are cross united in the *manner*. The result is regulation to form a single embryo, not a double "monster." Again, this result confirms the theory of interdependence of developing parts rather than theory of independent, self-differentiation of parts.

Experiments on the Gastrula; Interchangeability of Presumptive Neural Plate and Epidermis

The early gastrula, as shown by the fate map, has a roof over the blastocoel composed of cells destined to give rise to neural plate and epidermis. Prior to extensive involution of the presumptive *chordame-*

soderm at the dorsal lip of the blastopore, the roof of the blastocoel can be rotated 180° such that presumptive neural plate and epidermis have exchanged locations. The result of this operation is a completely normal embryo indicating that, at this time, the two tissues are interchangeable (regulative).

Inductive interaction between cells is shown by transplants of the dorsal lip

The developmental fate of the cells turning in at the dorsal lip of the blastopore is largely chordamesoderm. If the chordamesoderm is transplanted to the ventral epidermis of an early gastrula host embryo, the transplant tissue will turn under and come to lie beneath the ventral epidermis of the host embryo.

At this point an entire secondary neural axis will arise on the ventral side of the host. This experiment was first performed by Spemann and Mangold, who went on to demonstrate that the neural tube of the secondary embryo is of host origin and the mesoderm is of graft origin.

The Spemann—Mangold experiment involved the use of two

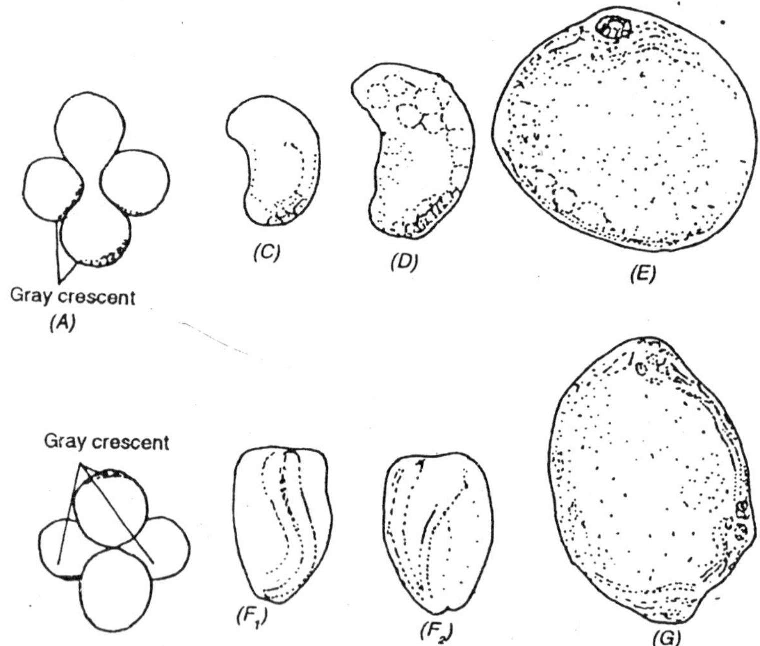

Figure 7.2 : Fusion of two two-cell stages (salamander). Two types of results are found, depending on the first cleavage of the two fused eggs.

salamander species of differing pigmentation as donor and host tissues. Sipenann gave the dorsal lip of the blastopore the functional designation of inmary organizer," reflecting its role in the organization of the neural axis. The presumptive chordamesoderm, upon transplantation or explanation to culture medium, is capable of self-differentiation into those structures assigned to the dorsal lip by fate map studies.

The action of the dorsal lip is apparently, from the experiment above, to organize the overlying presumptive neural plate into the neural axis. This is borne out by the following observations:

1. The neural axis fails to develop in those embryos whose primary organizer regions have been removed.
2. Accessory neural axes are induced in overlying presumptive epidermis in the presence of implanted dorsal blastoporal lip material.
3. The presumptive epidermal and neural areas are interchangeable in the early gastrula, as shown by experiments involving rotation of the roof of the blastocoel through 180°.
4. In embryos induced to form exogastrulae, in which the chordamesoderm self-differentiates but does not come to lie beneath the presumptive neural plate cells, fail to differentiate as neural cells.

The Chordamesoderm is Regionally Specific

Transplant studies show that posterior regions of the chordamesoderm induce posterior neural structures; anterior portions to form. This can be shown by defect experiments, in which defects are made in the chordamesoderm at different times during gastrulation.

Defects are thus caused at different levels in the chordamesoderm and produce corresponding defects in the overlying neural tube. The action of the dorsal lip material (primary organizer) is an example of embryonic induction, in which an inductor tissue evokes a response (differentiation) in tissue competent to respond to the stimulus.

This competence extends to presumptive neura plate areas and to presumptive epidermis. Competence to respond to the primary organizer is lost in later (neurula) stages.

In summary, the. primary organizer induces *the neural* axis of the embryo. The organizer activity is confined to the region of the dorsal lip of the blastopore. The organizer is selfdifferentiating and regionally specific in its action. The location of the dorsal lip and, therefore, the organizer corresponds to the location of the gray crescent material of the uncleaved, fertilized egg.

Cortical Transplants Provide Evidence for Regionalization of Egg Cytoplasms

As described earlier, the gray crescent is a cortical area characterized by a retreat of pigment from the equator at fertilization. It lies at the boundary between the black and white hemispheres of the frog egg and marks the future dorsoposterior side of the embryo.

It will be re-called that the gray crescent region of the fertilized frog egg corresponds to the region of the primary organizer and the chordamesoderm. Development of isolated blastomeres is complete is both isolated cells only if both contain gray crescent material in 'nearly equal amounts.

These observations focus attention on the gray crescent cortex of the egg as the initial organizer center. Recently it has appeared possible to transplant or remove the cortical gray crescent material. The results of these operations are summarized.

The gray crescent material behaves in much the same way as the dorsal lip material with regard to neural induction. Removal of the gray

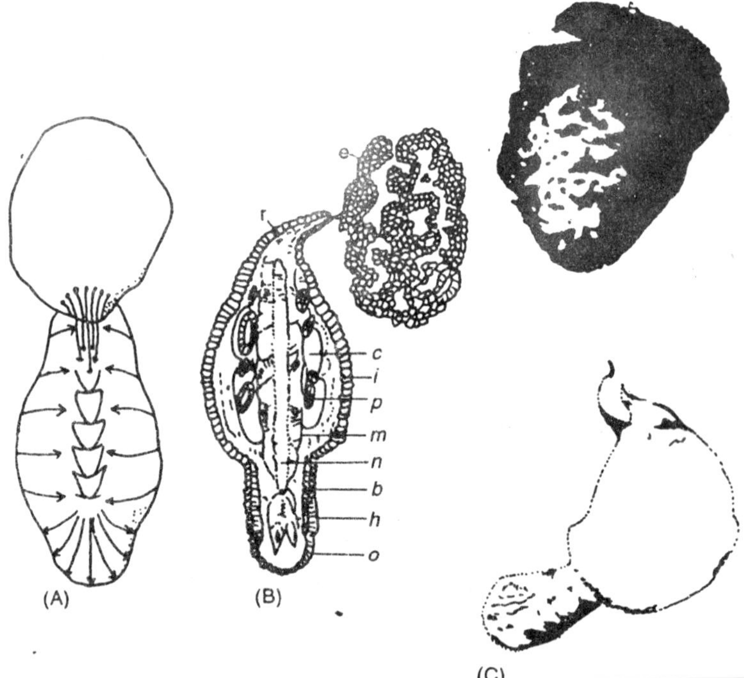

Figure 7.3 : Complete exogastrulation of a salamander germ.

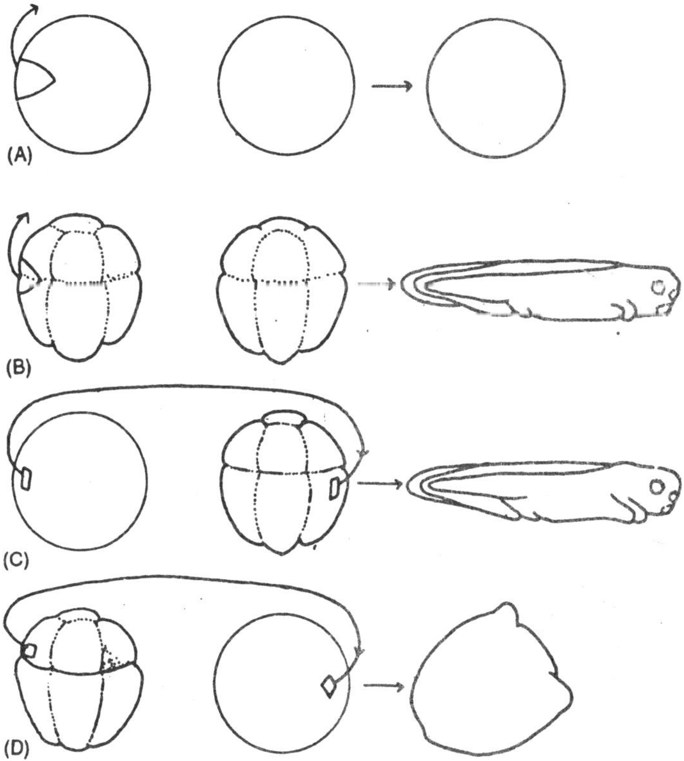

Figure 7.4 : Cortical transplant in Xenopus. (A) Excision of the cortical gray crescent area the one-cell stage; no gastrulation; (B) same experiment at the eight-cell stage; normal embryo; (C) graft of the gray crescent cortex of the one-cell stage to the ventral part of the eight-cell stage does not result in the induction of a secondary embryonic axis; (D) graft of the gray crescent cortex from the eight-cell stage to the ventral margin of one cell stage a secondary embryonic axis.

crescent (up to the eight-cell stage) terminates neural development. Prior to the eight-cell stage, implanting an extra gray crescent produces an extra neural axis.

The gray crescent cortex has apparently discharged its function by the eight-cell stage, after which removal or mplantation of extra material does not affect development. The cells of the dorsal lip will possess organizer activity even hough the gray crescent (at eight cells) has been largely or entirely removed.

In many respects, the gray crescent material is analogous to the micromere cortex of the sea urchin egg, in that it seems to be the center of egg polarity at early stages and possesses inductive potential.

Changes in Composition of the Gray Crescent Cortex

The cortex of fertilized eggs in the one-cell and eight-cell stages has been tested for organizer activity in other independent studies. These more recent studies involved implantation of the cortical material in the host blastocoel. The implant comes to lie beneath the ventral epidermis. The gray crescent cortex from one-to eight-cell stages was able to induce neural structures under these conditions.

The inductive potency of the cortex increases from one-cell to eight-cell stages. In these studies, the blastula stage was obviously able to respond to the cortical stimulus, perhaps because of the position of the implant in the blastocoel (other studies above found embryos beyond eightcells refactory to the action of implants in the cortex of cortical material).

The controls in these experi-ments, involving implants of cortical material derived from regions other than the gray crescent cortex, were uniformly negative. This indicates the activity is localized in the gray crescent cortex. Further examination, using micro-electrophoresis techniques, allowed study of the protein components of the cortical areas.

The cortex was found to be regionally different in composition (the gray crescent differs from other regions) and it was found that the composition within the gray crescent cortical are changes between the one-celled and eight-celled stages.

It is tempting to assign a causal role to the proteins found in association with the gray crescent, since differences in localized protein composition are correlated with differences in morphogenetic activity. Unfortunately, there is no evidence yet to support or negate such speculation.

It has been confirmed, from two sources, that the gray crescent cortex contains information for axiation and neuralization of the embryonic ectoderm.

Chemical Nature of the Primary Organizer

That living cells are not necessary for primary induction is shown by the ability of killed (boiled) dorsal lip material to induce neural structures when transplanted beneath the ventral (competent) epidermis of an early gastrula host.

Induction may also take place across a non-living membrane which allows the passage of macromolecules but not cells. However, it has been difficult to establish that induction across membranes of known porosity does not involve the intercellular contact of fine cell processes which are able to insert themselves through the membrane pores. Induction will not take place across cellophane, which allows neither macromolecules nor cell extensions to penetrate.

It is therefore uncertain whether or not normal induction involves the release and uptake of materials from inducer tissue to competent tissue, and whether or not cell contacts are required for induction under normal circumstances.

With the above restrictions in mind, the task of identifying the chemical naturally responsible for neural induction is a Viable one. Many extracts and substances have been tested and many were active.

The problem is still to find a means of examining the natural inductor (the dorsal lip) for the molecules it may release during induction. Because of the small quantities of material available, conventional approaches are difficult, especially when it is realized that the active material may be present in low concentration and hidden by the diverse organic molecules normally encountered in cells.

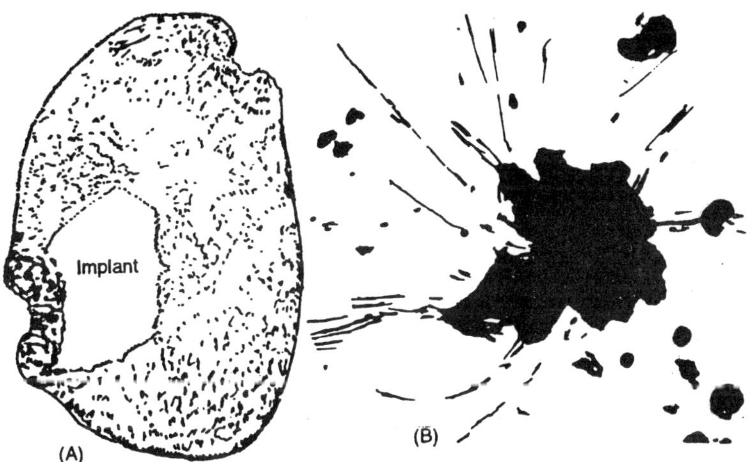

Figure 7.5 : (A) Induction of secondary neural plate in an axolotl embryo by means of an implanted piece of agar containing substances extracted from muscle by boiling the muscle in water. (B) Neural tissue with radiating nerve fibers and melanophores developed from a piece of young Triturus ricularis ectoderm cultivated in conditional medium.

A technique has been devised whereby isolated dorsal lip tissue can be cultured in defined medium, in drop cultures. Under these conditions, the explant will selfdifferentiate into a number of mesod-ermal structures. Since self-differentiate of the tissue occurs, it is reasonable to assume that other events associated with chordame-soderm development will also occur, including the release of molecules concerned with inductive changes in presumption neural plate.

These "inductor molecules" would presumably be located in the culture medium, since the dorsal lip cells were cultured without responsive cells. Using this system and attendant assumptions about the release of active molecules, the following biological assay was developed. Medium in which presumptive chordamesoderm had been growth for a number of days was recovered, free of cells.

This medium contained the original inorganic components, plus any molecules, contributed by the differentiating chordamesoderm cells. In new cultures the chordamesoderm medium (termed conditioned medium) was used to culture pieces of presumptive epidermis.

After some days of culture, the tissue was examined for evidence of neural development (presence of cells with nerve fiber outgrowths; chromatophores derived from neural crest cells. The presence of neuralization in conditioned medium and the absence of neutralization in control cultures prepared using medium conditioned not by chordamesoderm but by presumptive epidermal cells,' were taken as evidence that the inductor molecules were present in the test-conditioned medium.

In practice, conditioned medium consistently produced cultures showing evidence of neutralization, while control cultures were negative. The conditioned medium was found to contain many biological molecules, among them nucleic acids and proteins, as revealed by the absorption spectrum of the conditional medium in ultraviolet light.

The system can be used for finer analysis of the chemical nature of the inductor molecules: Specific enzymes can be employed to specifically inactivate different kinds of biological molecules. Proteolytic enzymes (trypsin) would be expected to remove inductor activity associated with proteins. Nucleases (RNase and DNase) would be expected to remove activity associated with nucleic acids.

Actual experiment shows that only trypsin completely abolishes the inductor activity of conditioned medium, though RNase diminishes the activity without abolishing it completely. These experiments suggest that the material involved in the induction of the neural axis by the underlying chordame-soderm is protein in nature, possibly a ribonucleoprotein.

Inductions Using Foreign Materials

Obviously, the amount of tissue available for analysis in the dorsal lip of the blastopore is extremely limited. In the course of experiments on killed inductors (cell extracts, or heated prep-arations), it was found that mammalian tissues yielded preparations having inductive activity when tested on amphibian embryos. Thus, boiled guinea pig bone marrow caused the competent, early gastrula, presumptive epidermis to develop into spinocaudal structures (tail musculature and other mesodermal derivatives).

Other tissues, such as guinea pig liver, yielded material causing the induction of anterior neural structures. The active materials all seem to be protein in nature. Later, it was discovered that active material could also be extracted from whole amphibian gastrulae.

These abnormal inductors have been highly purified and preliminary word has been done on the mode of action of the proteins. The spinocaudal factor (mesodermal or vegetalizing factor was inhibited by nucleic acids.

The inhibition was due, at least in part to the binding of the factor to the nucleic acid (DNA or RNA). The source of the nucleic acid was not of consequence, chick DNA or RNA serving as well as amphibian DNA or RNA. This has been interpreted to mean that the inductors might be acting by processing RNA some way, facilitating its transport to cytoplasmic regions.

These speculations have little to espouse them. There is little evidence for the activity of these factors in the normal inductive process. The situation is reminiscent of the situation encountered in artificial parthenogenesis. Many different agents can act to activate the egg. None is necessarily involved in normal fertilization processes.

The ability of an "*inductor*" to alter the fate of a cell is of interest as a system of cell differentiation, but does little to elucidate what actually happens during the normal determination of the presumptive neural plate of the amphibian gastrulae.

Interestingly, these kinds of results have led some to theorize that induction involves changes in intracellular concentration of various ions or changes in *intracellular* pH. Lithium, sodium, or calcium can all induce differentiation of neurons from presumptive epidermis.

Later treatment with the same ions also induces differentiation of these same neurons into pigment cells. The observations are supported by the occurrence of great changes in concentrations and proportions of cations *in vivo* during the induction of the nervous system.

The effects of the conditioned medium and other preparations are apparently not medicated through damage to the competent ectoderm cells, but through the interaction of the molecules with the cells. There is reason to believe that inductor molecules act indirectly on competent tissue to cause a response. The extent to which the abnormal inductors are identical to the natural substances from the dorsal lip is not known.

Inductive Action is Indirect and Mediated Through the Responsive Genome

These analyses involve the use of two species; one, the source of the inductor; and the other, the competent responding tissue. The response to the inductor is always in accord with the genetic information contained in the responding tissue. For example, salamander larvae possess a structure on either side of the head just behind the jaw angle and beneath the eye. This structure is called a balancer.

Anuran embryos possess homologous structures beneath the mouth variously called mucous glands, adhesive organs, or oral suckers. These structures (balancers and mucous glands) 'are induced in overlying epidermis by inductive action of the anterior roof of the archenteron. Speaking first of balancers, there exist some species of salamander that lack these organs.

If epidermis from any part of the body of a species normally possessing a balancer is transplanted to replace the balancer epidermis of a species not possessing a balancer, a balancer will develop. In the reverse experiment, epidermis from a species not possessing balancer fails to develop a balancer when placed over the balancer inductor in the species normally possessing balancers.

The balancer inductor is obviously operative in both species. The ectoderm of the species lacking balancers is genetically not competent to respond to either inductor, whereas the epidermis from the species having balancer is competent in either location. The main point is that the genetic lack of competence is not overridden by the inductive stimulus which is present in both species. Inductors cannot alter the genetic nature of the responding tissues.

Similar experiments involve the exchange of epidermis from the flank of frog and salamander embryos to the balancer or mucous gland region. One then has a salamander embryo (neurula) with frog belly epidermis overlying the balancer inductor and a frog embryo with salamander belly epidermis overlying the mucous gland inductor.

The result is that the salamander epidermis forms a balancer on the frog in response to the mucous gland inductor of the frog, and the frog epidermis forms a mucous gland on the salamander in response to the balancer inductor of the salamander.

Again, it is seen that the inductors evoke a response in the competent tissue, but the response of the competent tissue is strictly in accordance with the genomic information it possesses. The inductor itself is not a direct instructor.

CELL AND TISSUE INTERACTION IN MORPHOGENESIS

Movements of cells within the embryo make possible later inductive relationships, such as exist between chordamesoderm and overlying neural plate. Although differential growth rates, as discussed in other chapter of this book, may account for some of the cell movements observed during amphibian morphogenesis, some happenings, such as gastrulation and germ layer formation, are not adequately described by models based *solely on* differential growth.

In addition to growth patterns, some cells appear to exhibit a "preference" for certain other cells as neighbors during gastrulation and the later segregation of germ layers. Numerous experiments, including some very early ones, have been aimed at evaluating the possible elective behavior of cells during gastrulation. The most definitive experiments along these lines using gastrula material will be summarized here.

Isolation and Self-differentiation of the Endoderm

Pure endoderm isolated from the vegetal pole of an early gastr-ula first rounds up into spherical aggregates of cells. Neighbouring aggregates will fuse to form larger aggregates. Each sphere then flattens and the cells spread out over the substratum. Mitosis is infrequent, and the cells glide over the glass surface by producing pseudopodia.

A few individual cells may separate from the mass and migrate independently. The tendency to spread on the glass subtratum corresponds to the time (in a normal embryo) when the edges of the floor of the-trunk endoderm move dorsally, eventually fusing on the midventral line to form the intestinal tube.

Though the migration in culture is only in the horizontal plane, the topog-raphy of tissues within the whole embryo may guide the endoderm cells into the proper three-dimensional movements during normal development. It was. found, for example, that the migrating endodermal cells will spread over the outside of a tissue substratum and form an inverted intestinal epithelium.

If no substratum is present and the aggregates of endoderm cells are made to float freely in the medium, no epithelial sheet is formed and the aggregated cells give rise to a cytologically differentiated epithelium containing spaces filled with secretions. The ability of the endoderm to so differentiate in culture indicates that the medium used as without

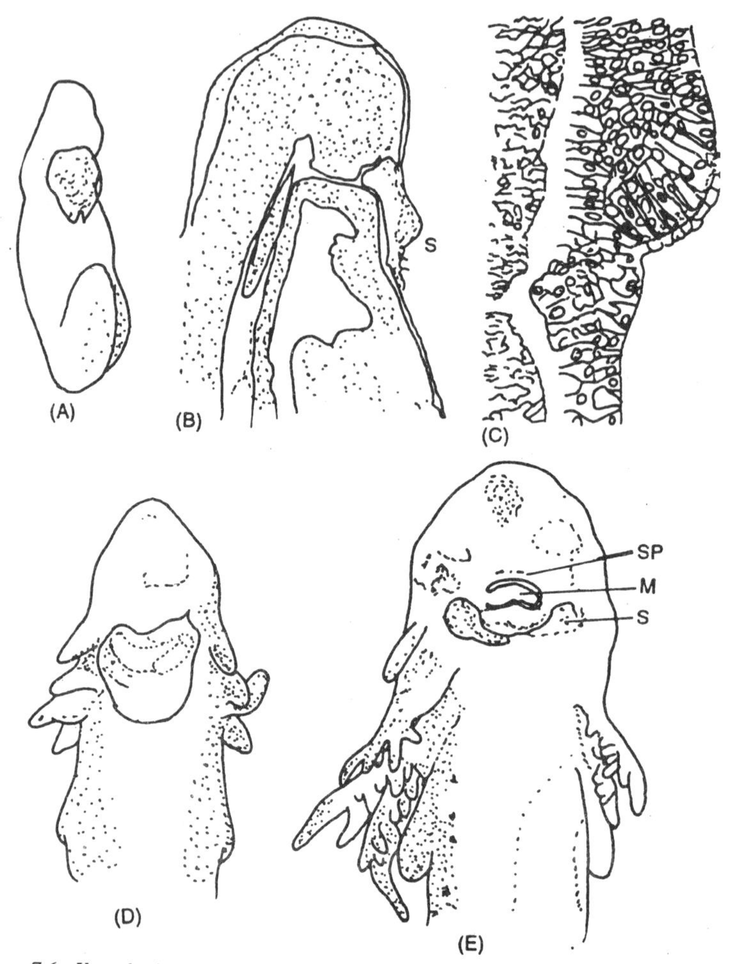

Figure 7.6 : Xenoplastic transplants; ventral skin of Rana esculenta implanted into Triturus gastrula. A, tail bud stage; B, later stage, longitudinal section; C, adhesive disc regions- of the embryo of B greatly magnified; D, E urodele larva with an implant of anuran epidermis in the mouth region; F, showing sucker discs, tadpole mouth with horny jaws (M) and horny spines (SP) on the lips. S, sucker discs.

adverse effect on the cells and that the endoderm his self-differentiating capacity.

Changing affinition of ectoderm and Endoderm During Gastrulation

Ectoderm and underlying endoderm can be excised together from an early gastrula from the mediocaudal region opposite the blastopore.

Accident inclusion of ventral mesoderm is to be avoided. The two cell types are loosely connected at first. The white endoderm cells form a pile within the pigmented ectoderm cells, which tend to curl closely around the endoderm.

After one to two days the two cell types separate from one another, forming discrete spheres of ectoderm and endoderm. If a suitable substratum is present, the endoderm cells will then migrate over it is with pure endoderm isolates, and differentiate as intestinal epithelium. The ectoderm remains as a ball of epidermal cells.

The separation of the two types of cells was achieved by directed autonomous cell activities and not by differential growth or budding. These kinds of separations occur between various cell types during normal development and car be thought of as arising from a negative affinity between ectoderm and endoderm.

This affinity changes with time, as evidenced by the initial covering of endoderm by ectoderm, followed by the separation of the two. Isolated ectoderm three to four days old will not adhere to endoderm even when the endoderm is freshly isolated from the early gastrula.

Role of Negative Affinity in Gastrulation

The temporary incompatibility of ectoderm and endoderm seems to play an important role in steering the involution process during gastrulation. The problem of why involution rather than exogastrulation normally takes place has not as yet been clarified. In any event normal gastrulation seems to operate initially by way of an active moving apart of heterologous tissues.

Reassortment of Tissues in Explants Involving all Three Germs Layers

Endoderm and mesoderm combinations are obtained by isolating a laterocaudal piece of an early gastrula containing elements of the lateral plate mesoderm, which normally gives rise to connective tissue. Ectoderm is also included in the isolate. The isolated mass first rounds up into a sphere.

The endoderm remains incompletely covered by the ectoderm. Again, after one to two days, the endoderm protrudes from the sphere and a constriction develops at the junction of the ectoderm and endoderm. Instead of progressing to complete isolation, the constriction will later relax.

A vesicle is then formed with an interior of mesenchyme cells which form a loose network in a secreted fluid. The aggregates of ectoderm and endoderm cells do not remain solid, but form the walls of the vesicle. The point of juncture of the ectoderm and endoderm is marked with a slight fissure and the two are held by the internal mesenchyme. If a larger piece of ectoderm is provided, the structure resembles that diagrammed.

The vesicle is completely covered by ectoderm, in two layers. The mesoderm lies between the inner ectoderm and the endoderm, which have separated a provide a cleft into which the mesenchyme can move. The inner endoderm forms a lumen lined with intestinal cells and containing a secreted fluid.

The arrangement of tissues and the inward orientation of the endodermal cells in producing the intestinal epithelium are typical of the spatial arrangements found in normal embryos.

Combinations of Ectoderm and Mesoderm and Induction

When small amounts of ectoderm are isolated with dorsal lip material normally forming somites and notochord, a variety of forms results. Endoderm is excluded from these cultures. These mesodermal-ectodermal isolates first round up into spheres; then the mesoderm elongates, separates from the ectoderm, and self-differentiates in contact with the substratum into somites and notochord.

The epidermis then reassociates with the mesoderm over a widening area and, subsequently, segregates into epidermis and neural tissue (through induction by the mesoderm). Thus, the affirnity of ectoderm for mesoderm progresses from a negative to a positive affinity.

The degree of specificity of cell affinities (or cell adhesion) is of the order encountered in the interaction of antigen and antibody molecules. The phenomenon is encountered in many development systems, in both plants and animals.

In simple terms, it means that cells are able to "recognize" neighboring cells or neighbouring cell-bound molecules and "elect" to adhere to these or not. The phenomenon is, undoubtedly, a surface phenomenon, dependent on surface molecules present in or protruding through the surface membrane.

SELF-DIFFERENTIATION OF FIELDS AND STEPWISE DETERMINATION

An explant of presumptive mesoderm of the early gastrula is capable of self-differentiation into a variety of mesodermal tissues. The endoderm is capable, at this same time, of cytodifferentiation into intestinal epithelium. Ectoderm, destined to form either epidermis or neural tissue, is dependent upon induction by the chordamesoderm for differentiation into neural tissue. The presumptive mesodermal areas of the early gastrula are capable of-self-differentiation and are referred to as fields, such as the chordamesoderm field and the heart field.

If an amphibian embryo is induced to exogastrulate (using Li), the chordamesoderm will fail to underlie the presumptive neural plate area, which then develops as epidermal tissue while the chordamesoderm self-differentiates into notochord, scmites, nephric structures, and so forth.

The failure of the presumptive neural plate to differentiate is due to the failure of the chordamesoderm to provide the necessary inductive stimulus, as described above. If a piece of presumptive neural plate is isolated prior to gastrulation, the result is the same.

If the isolation is made after gastrulation, the explant will develop into nervous tissue. It is said to have become determined. This determination does not take place in isolated, early ectoderm.

In the embryo the entire neural axis (neural field) is determined by inductive actions along the length of the underlying chordam-esoderm. In a corresponding manner, the ectoderm developing as epidermis loses its ability or competence to respond to neural inductors.

As time progresses the neural axis becomes divided into subfields (forebrain, midbrain, and hindbrain; spinal cord). The regions of the neural field then act as in ductors, causing the lens to develop opposite the optic vesicle and the otocyst to form toward the posterior end of the hindbrain.

The lens then becomes deter-mined, for example, and will form lens tissue in isolation. In the intact embryo, the eye lens exerts an influence (induction) on the overlying ectoderm, causing the development of the cornea. The process of differentiation when seen in this context becomes a series of stepwise determinations and inductions, beginning with the initial action of the chordamesodermal field (determined during oogenesis) on the overlying ectoderm.

LIMB DEVELOPMENT: AN EXAMPLE OF A FIELD

The beginnings of forelimb development are to be seen in the somatic mesoderm ventrolateral to the pronephros and behind the gill region. The mesoderm becomes thickened, as does the overl-ying epidermis. A similar development takes place just anterior to the anus. These thickenings are the forelimb and hindlimb buds, respectively.

The mesoderm element in the anterior limb bud is determined at an early stage, shortly after the closure of the neural folds. Pieces of somatic mesoderm removed from the forelimb region can be transplanted to other areas of the body beneath the epidermis, where they will cause the development of heterotopic limbs (limbs forming in areas not normally giving rise to limbs).

That any portion of the epidermis can, at this time, participate in limb formation is shown by experiments placing epidermis from various parts of the neurula over the limb mesoderm. All such pieces, regardless

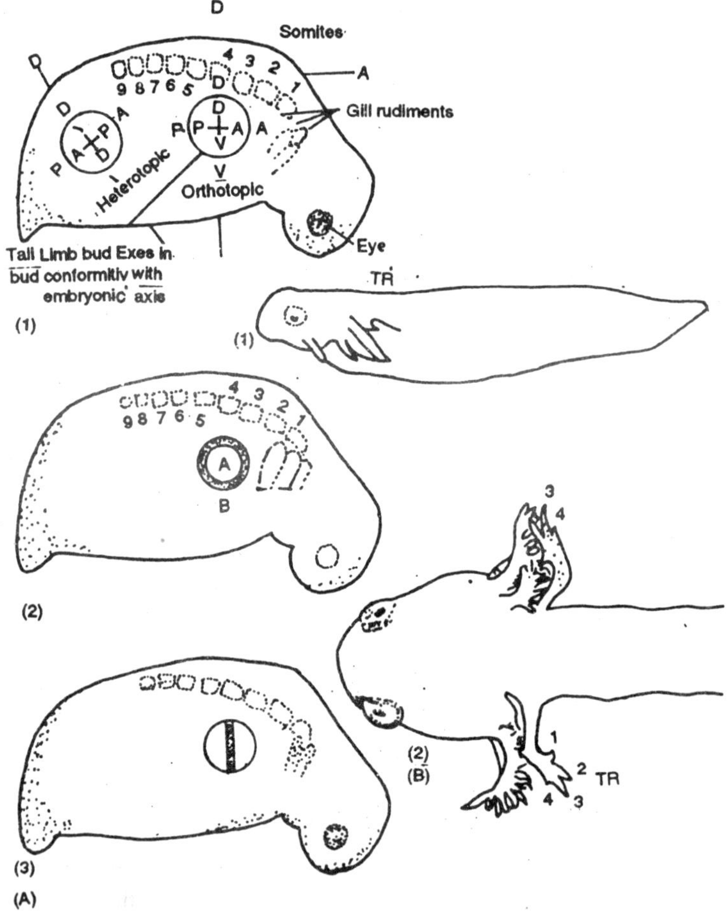

Figure 7.7 : (A) Experiments on the limb-bud differentiated in salamander (Ambystoma) during early tail-bud stage.

of source, are able to cooperate with the meso-dermal element to produce a normal limb.

This does not mean that the ectoderm of the amphibian limb bud is passive in limb development. A limb structure will not develop from isolated mesoderm; a covering of ectoderm is required even though ectoderm from a variety of sources will suffice for this purpose. The role of the ectoderm is more pronounced in amniote embryos.

The Early Limb Bud is a Regulative System

The early limb bud can be divided into two areas, or two limb buds can be placed side by side, slightly overlapping in normal orientation, or one-half of the limb bud can be removed. In all cases, a complete single limb will develop. The system is a regulative one.

If the area of the limb bud is removed entirely, a peripheral area not normally involved in limb development, will replace the lost limb bud. This larger area of limb potential is the limb field. It includes the limb bud proper and a narrow zone of cells surrounding the bud which are capable of limb development, though they are not normally involved in limb formation.

Later Limb Development Involves the Stepwise Determination of the Limb Axes

The limb field is determined as limb at the time of closure of neural folds. The extent to which the limb itself is determined has been studied primarily using salamander embryos. If the early forelimb bud is moved from the right to the left side of the embryo by moving the bud around the anterior end of the embryo, the anteroposterior axis will be reversed, though the dorsoventral axis will remain in the normal orientation with respect of the embryo

This results in a limb with reversed anteroposterior polarity, indicating that this axis is determined in the embryo immediately after neurulation. At this time, the dorsoventral axis may be inverted by *transplanting'* a limb bud from right to left by moving it over the dorsal side of the embryo, thus leaving the anteroposterior axis undisturbed.

This results in a normal limb, thus indicating the ability of the dorsoventral axis to right itself, or to regulate to the pattern imposed on the limb by neighbouring tissues.

In later stages, the dorsoventral axis loses this regulative ability, and inversion of this axis will result in a limb with the palmar surface facing upward instead of downward. The third axis is the proximodistal axis (in-out axis), as determined by reversal of the mesodermal element of the limb bud beneath the epidermis. It is still able to regulate at a, time when the dorsoventral axis has been determined. Later stages show abnormalities of limb development if this axis is inverted, indicating a determination of this axis.

These experiments are difficult to interpret precisely, since a limb cannot actually reverse itself and grow into the body; to do so would mean a loss of contact with the ectoderm essential for normal limb development, in addition to encountering structural barriers to limb growth from other Body tissues.

The three major axes of the limbs are seen to be determined at separate times, and the limb bud as a whole is determined in a stepwise fashion. The later axial sequence of limb determination is in a proximodistal direction. The girdle is determined first, the phalanges last.

OTHER FIELDS

Stepwise Determination is Evident in the Development of the Central Nervous System

The normal forelimb movements of urodele larvae depend on innervation by three pairs of spinal nerves (3, 4 and 5) comprising the branchial plexus. If this area of the neural plate is exercised and replaced with a more posterior segment, the graft will morphologically and functionally fill the role of controlling the limb movements.

A similar transplant at the tail-bud stage results in limb innervation, but leads to abnormal limb movements. The posterior segment can no longer participate in the development of the branchial plexus.

At the tail bud stage, the section of neural tube which is to form the mudulla may be excised and rotated 180°. The replaced graft develops into a morphologically normal medulla, and all nervous responses of the operated larvae may be perfectly normal. Experiments on later tail bid larvae lead to abnormalities of medulla morphology and function.

Consideration of symmetry determination in other fields (e.g., heart field) would yield similar conclusions; the field is initially regulative after being specified as a field and later becomes a determined mosaic of smaller parts, which again regulate within the subfields until they also become specified, presumably by induction from neighbouring cells.

The beginning of this sequence of dependent differentiation is, again, the initial induction of the nervous system by the chordamesoderm and the continued self-differentiation of the endomesoderm.

SPECIFICATION OF SYNAPSES AND THE DEVELOPMENT OF INTEGRATED BEHAVIOUR

Background Considerations

In some invertebrates, where the total number of neurons in the animal is relatively small, the neural connections (between nerve fiber and muscle, sense organs, or other interconnecting fiber) can be mapped to some degree.

Neural connections are reproduced among these animals with great fidelity, presumably on instructions from the genome. In vertebrates, there are far too many neurons for one to know with certainty the degree of order among neural connections specified by the genome.

Two extremes exist. On the one hand, neural connections might be made in random fashion. The useful connections could then be sorted out by some learning procedure.

On the other hand, the entire circuitry might be specified in the genome and manifest in specific surface moleccules which make each nerve cell and its connections unique. The truth most probably lies somewhere between the two extremes.

Those against genetic specification sometimes argue that there are 'more neurons and synapses in a vertebrate brain than there are genes in the genome. However, in a number of cell recognition systems, the recognition molecules are protein-polysaccharide complexes. Specificity has been associated with the branching arrangement of the polysaccharide moiety, a property that could manifest many configurations within the instructional frame-work of a single gene.

The possibility exists for complete genetic specification, though this is perhaps not the most likely situation. Alternatively, there is considerable evidence against the hypothesis of random connections later sorted out by learning activities.

The development of behaviour is obviously dependent on the development of the nervous system of receptors and effectors and interconnecting pathways (neurogenesis). Motility and bioelectric activity are essential to behavioural patterns.

There is a close correlation between structural and functional maturation. Young neurons are capable of impulse transmission, and primitive synaptic connections serve to mediate early motility. Each behavioural advance follows closely on the completion of new synaptic connections. As will be seen below, sensory input and practice, or trial and error, have little to do with the development of integrated behaviour in vertebrates. In salamanders, behaviour is integrated as meaningful swimming movements from the time of hatching.

In higher vertebrates such as birds and mammals, initial movements are unorganized. In the bird, these random movements are superceded by quite different, intricate, and coordinated movements that precede hatching. The latter movements do not grow out of the former, since the uncoordinated movements continue to take place even after hatching.

They appear to be the result of indiscriminate and massive electrical discharges that excite many motor neurons and cause movement, which may be important for the normal differentiation of joints. The intricate hatching movements appear to be integrated from their beginning without benefit of practice movements in earlier stages.

These observations tend to deny any role of sensory input in development of early integrated behaviour. It is now widely held that the neural machinery for integrated behaviour differentiated autonomously, as a "*printed circuit.*"

Clâssic Experiments with Specification of Sensory Connections

These experiments deal with the regeneration of the optic nerve fibers of larval and adult salamanders and frogs. The eyes originate as a pair of lateral out-pocketings from the walls of the dienc-ephalon. The optic vesicles reflect back on themselves, forming the two-layered optic cups.

The inner layer of the cup is the retinal layer and contains the sensory elements (rods and cones), the interconnecting neurons, and the optic nerve cell bodies which send their axons to the optic lobes (optic tectum) in the midbrain region. Experiments with the optic tectum in the midbrain have established that the retinal field is projected, point by point, onto the optic lobes.

Specific stimulation of the retinal cells (one or two cells at a time) gives signals to specific cells of the lobe corresponding to the portion of the retinal field being stimulated. The optic fibers are predetermined as to their point of termination in the optic lobe even before they have grown back to the midbrain.

This can be shown by rotation of the optic cup prior to optic nerve growth. This operation results in inverted vision, leading to the conclusion that the optic nerve fibers deviated from their normal path to journey from their position in the inverted retinal layer to their normal terminals in the tectum.

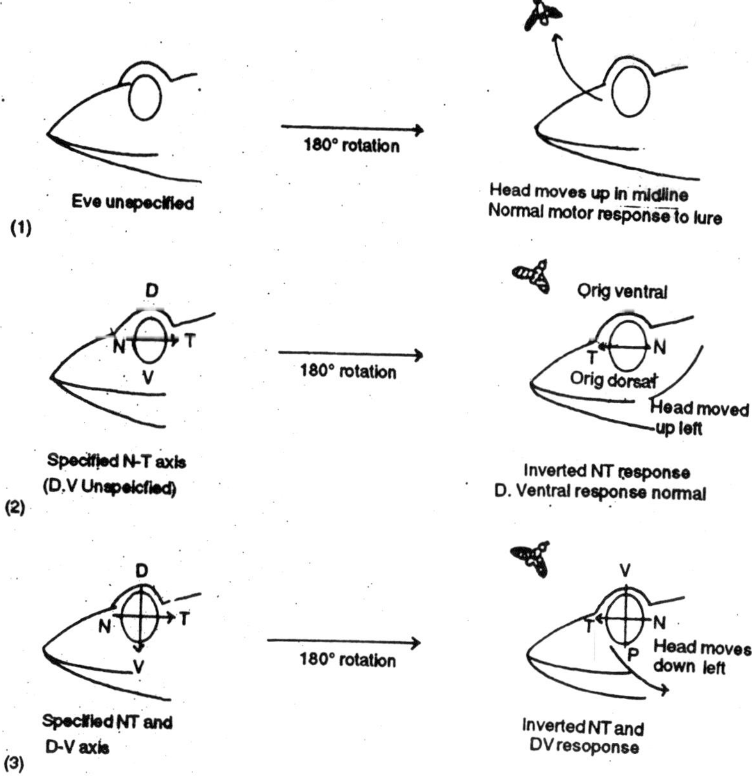

Figure 7.8 : Diagram of development of retinal polarity in eyes of amphibians rotated at various embryonic stages. Embryonic state of determination at rotation in left column; resulting adult behaviour in right column. DV, dorsoventral; NT, nasotemporal. The fly acts as a lure.

Crushing of the optic nerve in salamander larvae or adult frogs can result in optic nerve regeneration and restoration of normal vision and normal responses to visual cues in the environment. Rotation of the eye in the orbit after crushing the optic nerve also results in the restoration of vision in a significant number of operated animals.

The vision of the inverted eye is, however, inverted, as are the responses to visual cues in the environment. The implication of these studies is that the optic nerve fibers returned to their original connections in the optic lobes through a thick mat of apparently randomly regenerating nerve fibers.

Now the environmental stimulations impinge on inappropriate portions of the retina (a dorsal stimulation now focuses on the ventral retinal field, for example). Responses to the stimuli are correspondingly inverted. The inappropriate responses continue throughout the life of the adult frog. Most animals continue to snap at lures or food (flies) in a reverse manner; a fly dangled dorsal to the nose will be lunged for between the front legs.

These responses in some animals were replaced by indifferent reaction, as though the frog had simply given up on the impossible chase. Thus, there may be a tendency in some to ignore the lure or to inhibit the responses. However, in no animal was a positive attempt at correction of the defect noted.

Evidence from Lesions in the Optic Tectum of the Optic Lobes

Specific lesions (scotomas, or blind spots) produced in the tectum result in production of expected visual deficiencies in animals with regenerated optic nerves. An anterior lesion in the tectum abolished responses in the anterior region of the visual field, and so with lateral and posterior lesions.

In those animals with inverted eyes, the lesion in the posterior region produced deficient responses in the anterior visual field. In summary, it can be concluded that the ingrowing optic fibers reestablish functional connections in the same topological areas of the optic lobe where they originally terminated during normal neurogenesis.

Implications of These Studies

The foregoing data support the contention that each retinal locus possesses functional connections with brain centers differing from all other loci. The data also support the contention that the orderly restoration of central reflex connections, which occur regardless of the orientation of the retina, must be independent of functional adaption or learning.

It is also implied that the retinal fibers must possess specific properties by which they are recognized in the optic lobe according to their respective retinal origins. In studies on regeneration of the optic nerve in fish, these conclusions have been borne out by direct anatomical observation, not only with respect to the terminal connections established in the tectum, but also with respect to the pathways taken by the optic fibers to reach their central, terminal points.

Experimental Evidence; Motor Activity

The concept of the "*printed circuit*" of neural activity as a part of the genetic endowment of a species was supported by early narcotization experiments on salamander embryos. In this study a clutch of salamander eggs was divided into two groups. One group was kept as a control. A second group was anesthetized with chloretone from the preswimming stages to beyond the swimming stage.

When the anesthetized tadpoles were removed from the anesthetic, they were found to swim in a fashion equalling the coordination of the control group.

Additional Information from Inverted Limbs

The patterning of the central neural pathways would thus seem to depend on specific neuronal connections. This type of selective connection (or affinity) is presumably based on molecular recognition of one nerve cell by another, or of recognition of particular muscles by nerves. Evidence for a chemoaffinity (as opposed to trial and error) mechanism operating between nerve fibers and between nerve and muscle is found in studies of inverted salamander limb buds.

These limb buds were inverted at a time when their anteroposterior axis had been determined, and thus grew out in a reverse position. It was found that from the beginning of motility the limbs also moved in reverse. This implies that the spinal center for locomotion develops independently of the inverted limb and produces a "*malfunction*" that is never corrected.

In another experiment, the legs were deprived of sensory nerves before leg movement had started. Coordinated locomotor function was not impaired in the treated limbs. It has also been found that the nerves of the lumbosacral plexus (8, 9, 10) will grow around a mica barrier, taking an abnormal path to the developing hindlimb.

The normal innervation of the hindlimb is possible if the barrier is not too extensive. This points to a selective affinity of these nerves for muscle fibers of the hindlimb.

It has been found that a forelimb transplanted to the head region will be innervated by cranial nerve fibers, sensory and motor. The limb will not function in coordinated fashion with other limbs, but rather tends to move in concert with anterior structures (e.g., movements of the gill chamber) also innervated by the nerves supplying the abnormally placed limb.

The abnormally placed limb is innervated non-specifically in an abnormal situation. Under these conditions it functions as though it were a part of the motor apparatus in the abnormal position and not in co-ordination with other normally placed limbs. The basis of the apparently non-specific attraction of cranial nerves to the abnormally placed muscle is puzzling and represents a question to be resolved.

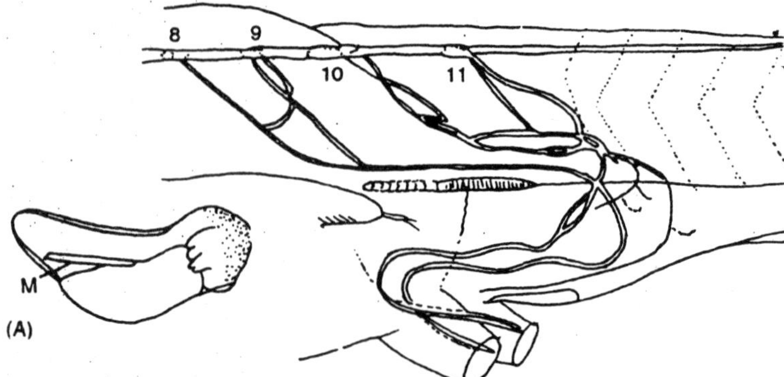

Figure 7.9 : Response of outgrowing nerve fibers to physical barrier in normal pathway. (A) A mica plate (M) inserted in the path of outgrowth of the spinal nerves to the hindlimbs in a frog embryo: (B) the nerves have grown around the obstacle and have reached the hindlimbs.

Gradual Specification of the Nervous System

Beginning with the induction of the central nervous system by the dorsal lib of the blastopore, the development of the functional nervous system proceeds through a period of plasticity to an extremely high degree of specification.

Nerves maintain the same specific connections made during development after regeneration, so that motor nerves innervate the appropriate muscle, and sensory nerves innervate the proper end organ in establishing reconnections.

Clinical attempts in humans to supply facial muscles with cervical nerves (as when the facial nerve , is paralyzed) may result in innervation of facial muscles by the cervical nerve. However, the facial muscles. function in concert with cervical voluntary muscle rather than with other facial muscles.

This situation is not corrected centrally and persists throughout life. In the section on human development additional information is available on plasticity during development and on the essentially inflexible situation found in the adult. The basic mechanism governing nerve cell outgrowth and synaptic connections is essentially unknown and represents one of the most fascinating and challenging areas of experimental biology.

GENOMIC CONTROL OF DEVELOPMENT

In the introductory chapter, it was stated that during develop-ment genetic information is neither gained nor lost in the various cell lines. This concept implies that each cell of the adult contains the same

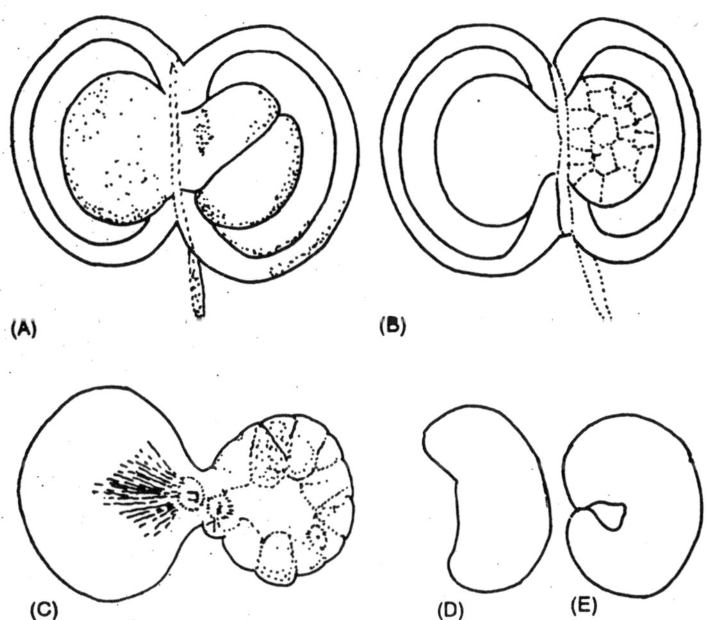

Figure 7.10 : Nuclear equivalence during cleavage. A restrictive noose is placed around the egg at the first cleavage (A, B). The nucleate half cleaves. Later (C) a cleavage nucleus passes into a nucleate half. resulting in older and younger twins.

genetic complement, and that differentiation of this clone of cells requires control a gene activity at either the transcriptional or translation level. In this section, we will examine some of the evidence of nuclear control of differentiation process as it involves the nuclei of embryonic and adult cells.

Cleavage Nuclei are Equivalent

Cleavage nuclei were first tested by a simple but ingenious. experiment. The fertilized salamander egg was constricted by a hair loop into a dumbbell shape. The fusion nucleus was located in one portion that proceeded to cleave, while the portion without the nucleus remained unsegmented.

The constriction is made parallel to the anterior-posterior axis of the embryo and bisects the gray crescent. After cleavage to perhaps 32-64 nuclei, the constriction allows one of the cleavage nuclei to pass to the enucleated egg. The selection of the nucleus which passes to the undivided side is completely random.

The newly nucleated portion of the egg then begins to segment, and the loop is pulled tight to completely separate the two halves. The result is the formation of two identical twins, one of which will be somewhat younger as regards the stage of development. Since the selection among the cleavage nuclei was completely at random, it follows that all cleavage nuclei are equivalent and totipotent at the stage at which the nucleation of the uncleaved portion of the egg took place.

Equivalence is Extended to the Blastula Through Nuclear Transplantation

In the 1950s the technique of nuclear transplantation was developed using the frog egg. This procedure allows for the direct testing of embryonic nuclei for equivalence and totipotence. In essence, the technique involves the enucleation by microsurgery of an artificially activated egg by removal of the second polar spindle and associated chromosomes.

This is followed by the introduction of a test nucleus into the egg via a micropipet. The test nucleus is taken from any source by pulling a cell into a small-bore pipet with gentle suction. This procedure usually breaks the cell membrane. Transplantation of isolated nuclei without transfer of cytoplasm has not been accomplished, since nuclei isolated in artificial media are inactivated as far as future nuclear division are concerned.

The transplanted cytoplasm is diluted by the egg cytoplasm by about 1: 40,000 and is usually neglected in interpretation of results. By the late blastula stage, the original nucleus has proceeded through 13 or 14 generation producing about 5,00016,000 cells. About 80 per cent of the eggs receiving transplants of a blastula nucleus cleave, and a smaller proportion proceed through all stage of embryonic development and metamorphosis.

This experiment directly demonstrates the equivalence and totipotence of nuclei of the late blastula. The technique has been extended to the gastrula and later stages, nuclei of cells of an adenocarcinoma of the adult frog kidney.

Transplants from Later Stages Prove Difficult to Interpret

Using *Rana pipiens*, the percentage of transplant embryos showing normal development is markedly reduced as the age of the donor nuclei increases. However, in *Xenopus*, gastrula nuclei, tadpole intestinal nuclei, and nuclei from the germ line and adult tissues (excluding brain nuclei) reportedly give rise to normal tadpoles, as do the nuclei of the frog adenocarcinoma. Part of the difficulty with nuclei from embryos of later

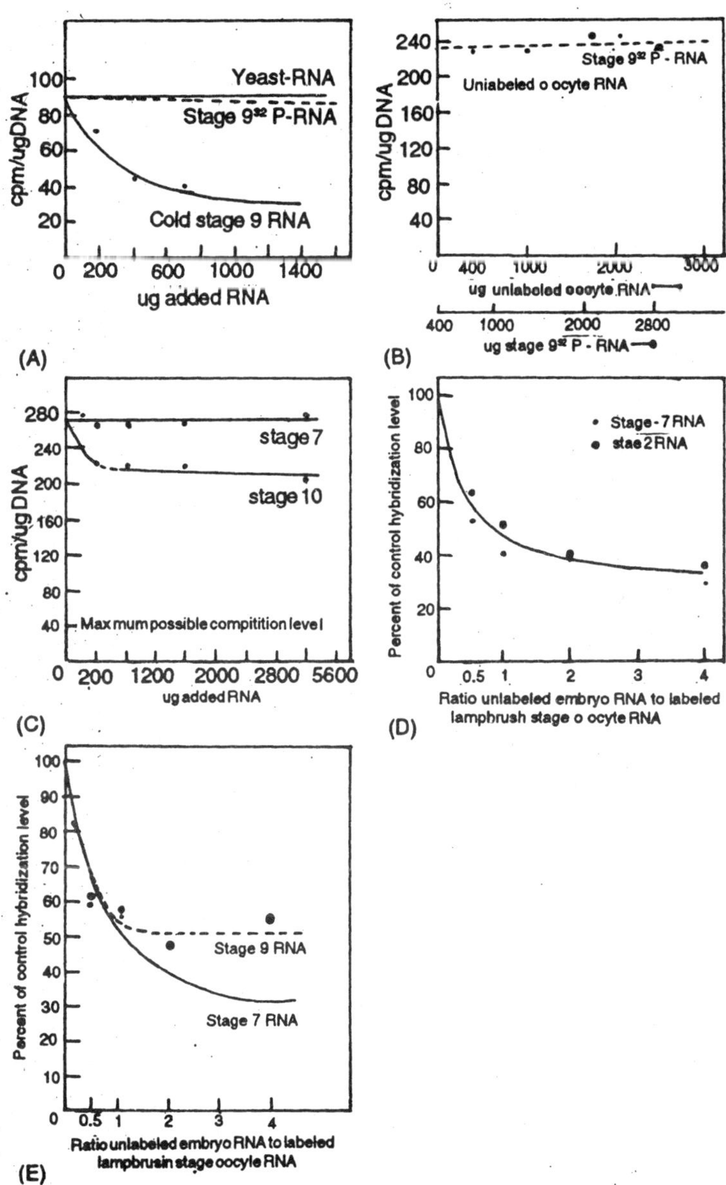

Figure 7.11 : (A) Competitive NA: RNA binding experiments indicating degree of homology between RNAs of various embryonic stages and RNA of lampbrush oocytes of Xenopus.

stages may be technical and involve difficulties in handling the small cells without injury to the nuclei and chromosomes. There is also a biological difficulty resulting from the faster generation times during cleavage as compared to later stages (about once every hour or two versus once every two or more days).

The replication of chromosomes may fail to keep up in transplanted nuclei from older embryos or adults, leading to partial replication of chromosomes may fail to keep up in transplanted nuclei from older embryos or adults, leading to partial replication of chromosomes and loss of chromosome material. Fragmentation of chromosomes is observed in *Rana* nuclei transplanted from later stages.

Success obtained with older material might be due to parthenogenesis, resulting from failure of the experimenter to remove the egg nucleus. Such operational errors have been recognized and eliminated by using markers in the transplanted nuclei.

Both triploidy and number of nucleoli have been used as cytological markers for transplanted nuclei. Genetic markers could be used as well. The marker appears in all the cells of the embryo resulting from nuclear transplantation if the transplant nucleus has indeed participated in cleavage in the absence of the egg nucleus.

The successful nuclear transplants in the *Xenopus* system, as well as those using adenocarcinoma nuclei of frog kidney, utilized marked nuclei. There is no mistake about these positive results. It has been argued, however, that the percentage is so low as to raise the question of the involvement of exceptional, undifferentiated cells in the successful experiments.

The Basic Question Involves two Concepts

The first is the question of the totipotence or pluripotence of the somatic cell nucleus. The second is the question of the reversibility of nuclear changes (differentiation) which occur during development. Somatic nuclei may be possessed of all their nuclear genome and may thus potentially function as zygote nuclei.

However, the process of differentiation, involving selective inactivation of most of the genome, may require special circumstances for reversal which are not met by the activated egg cytoplasm. Indeed, the nuclear changes may not be reversible at all. This is certainly the case in somatic tissues of some insects which develop giant polytene chromosomes.

In these cells the entire genome is represented, but each chromosome has been replicated throughout its length perhaps 1000-fold, leading to structural specializations which preclude its function as zygote nucleus, perhaps on purely mechanical grounds. Systems other than nuclear transplantation (such as the development of tumors of primordial germ cells) also indicate totipotence of nuclei from differentiated tissues.

It is necessary to consider all the evidence at hand and not just nuclear transplantation data alone, in arriving at the conclusion that the genome is conserved during development.

The Cytoplasm Controls Nuclear Activity

At the molecular level, the transplanted nuclei undergo changes similar to those noted in the male pronucleus after normal fertilization. The sperm nucleus first swells and then begins synthesis of DNA. A transplanted nucleus from the midblastula stage, normally synthesizing DNA, will increase in volume and continue DNA synthesis at the times specified by the egg cytoplasm.

Neurula nuclei, normally synthesizing ribosomal RNA but infrequently DNA, will, upon transplant to an activated enucleated egg, cease ribosomal RNA synthesis and begin DNA synthesis after the nuclear-volume increase. During cleavage, these nuclei synthesize some messenger RNA.

The transplanted neurula nuclei will again synthesize ribosomal RNA upon reaching the gastrula and neurual stages, following the transplant. Nuclear transplants may also be made to primary oocytes (actively synthesizing RNA, not DNA) and to unfertilized, unactivated eggs (making neither DNA no RNA).

In growing primary oocytes, transplanted nuclei from blastulae are induced to cease DNA synthesis and commence RNA synthesis. In unactivated eggs, nuclei are quiescent with respect to RNA synthesis. The transplant chromosomes condense to resemble the egg meiotic chromosomes.

Foreign nuclei (mouse liver) transplanted to enucleate, activated frog eggs are induced to synthesize DNA and to stop RNA synthesis. Clearly, changes in nuclear activity are dictated by the cytoplasm, and in a universal language as well. The several responses of nuclei to different cytoplasms argues for pluripotence of these nuclei.

Preprogramming of the Oocyte and RNA Synthesis During Development

Transcription of the genome occurs during oogenesis, as well as during embryogenesis. The question arises as to the qualitative nature of the mRNA formed at various times between oogenesis and the completion of differentiation. Some mRNA is transcribed during oogenesis for later use by early embryos (through blastula). To this extent, the egg may be said to be preprogrammed with differentiation controlled at the translational level.

The paragraphs below examine gene activity during oogenesis through late tadpole stages, with respect to synthesis of the various classes of RNA (ribosomal, transfer and messenger).

During oogenesis lampbrush chromosomes are present, as are numerous nucleoli. The nucleoli have been found to represent amplified DNA sequences engaged in the production of ribosomal subunits. This gene amplification represents a specialization for the accumulation of the large amount of ribosomal RNA required during early development through to the hatching tadpole stage.

The rate of ribosomal RNA synthesis is maximal at the lampbrush stage and predominates to the point of obscuring synthesis of messenger RNA. In mature oocytes, 95 per cent of the total RNA is ribosomal. Though measurement of the template mRNA is difficult in the presence of large amounts of ribosomal RNA, it has been estimated that about 2.5 per cent of the total RNA of *Xenopus* oocytes is template active.

This template or messenger RNA seems to have been synthesized during the diplotene lampbrush stages, particularly stage 4, and constitutes the bulk of the messenger RNA inherited by the embryo as maternal template. In terns of unique-sequence DNA transcripts oocyte RNA contains about 9 per cent of the information content of uniquesequence DNA.

This is remarkable close to the value for *ilrechis* where a similar complexity is found. At ovulation, a burst of messenger RNA synthesis occurs under the influence of pituitary hormones. It is of some interest that the egg cytoplasm does not acquire the ability to induce DNA synthesis in transplanted nuclei until after break-down of germinal vesicle and ovulation occur.

Preovulation oocytes do not support DNA synthesis in transplanted nuclei. This property appears after injection of pituitary hormones and may be dependent upon the burst of messenger RNA synthesis at ovulation. Maturation itself is not dependent on the hormone-induced RNA synthesis.

In the sea urchin, it was seen that early development was dependent upon template RNA *synthesized* during oogenesis. Later development required new message transcription, beginning at the late blastula stage. A summary of RNA synthesis in amphibian embryos *(Xenopus)* shows similar patterns. An early synthesis of messenger (template) RNA exists; which increases in intensity during cleavage to midblastula. During late portions of the blastula stage an abrupt increase in gene transcription occurs over the whole embryo.

The newly synthesized messenger RNA contains a new program of genetic information. Transfer RNA synthesis is activated during this same period. During cleavage, the synthesis of ribosomal RNA is at a low level, if present at all.

New ribosomal RNA is synthesized at gastrulation and becomes essential at the hatching tadpole stage. Ribosomal RNA in *Xenopus* has been studied using anucleolate mutants. These mutants are actually deficient in ribosomal DNA though a deletion of area of the genome responsible for organizing the nucleous.

The heterozygous mothers of homozygous anucleolate embryos synthesize a normal complement of ribosomes. The embryos homozygous for the deficiency, however, do not make new ribosomal RNA and die as early tadpoles even though messenger RNA is synthesized in normal amounts.

The effects of transcription inhibition by actinomycin D has also been studied in amphibian embryos, including *Xenopus*. Actinomycin does not affect cleavage, though gastrulation and neurulation—are completely blocked. Inhibition of translation using puromycin blocks cleavage in *Xenopus*, as it does in the sea urchine embryo.

Organization of the Xenopus Genome

Studies involving reannealing of sheared *Xenopus* DNA fragments indicate interspersion of longer section of single copy DNA with shorter, repetitive segments. The sheared fragments of various lengths were incubated to a Cot value to include only repetitive segments, using 450 nucleotide long DNA fragments in excess to drive the reaction.

As the length of the fragments increased, these longer fragments contained more and more unique-sequence DNA, a result expected if the single copy sequences were interspersed with repetitive sequences.

The use of Xenopus Oocytes in mRNA Assay; the Question of Translational Control

Oocytes contain mRNA which is transcribed during oogenesis and "masked" for controlled use during early development. The nature of the mask is not known. However, mRNAs ("unmasked") from various diverse sources can be microinjected into oocytes and will be translated along with the messages normally available for protein synthesis in these cells.

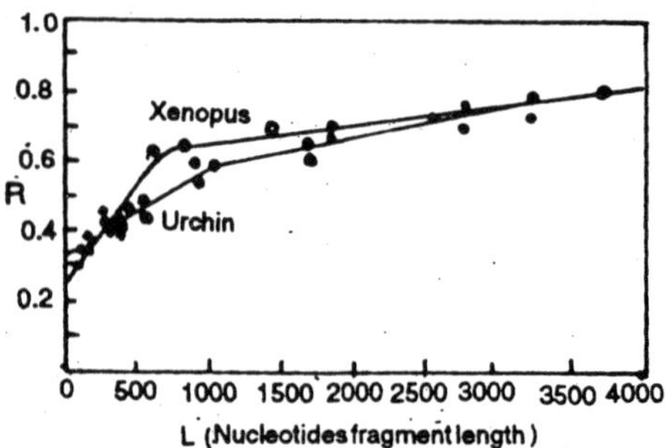

Figure 7.12 : Interspersion of single-copy and repetitive DNA in Xenopus. Relation between fragment length and hydroxyapatite binding in the DNA of the sea urchin (Strongylocentrotus purpuratus), compared to similar curve for Xenopus DNA.

The oocytes are readily permeable to amino acids, and the newly synthesized proteins can be labeled and separated from one another by various chromatographic or immunological procedures. A number of species of eucaryotic mRNA have been used, in purified form, to test the oocyte system.

Included are mRNAs for calf lens protein, duck, rabbit and mouse globin chains, trout protamine and rabbit hemoglobin. To reemphasize, the protein products of translation of mRNAs are stable and have been characterized as belonging to the species providing the mRNAs and not to the frog. Since oocyte is not engaged in synthesis of any of the proteins coded by the exogenous mRNAs, recognition of the translation product is possible.

The fidelity of translation has been examined and found to be of the same order as when the mRNA is translated in the donor cells. This reflects a marked lack of species specificity in the process of translation mRNA.

The oocyte system is very efficient, and messages are translated with the same rapidity encountered in normal donor cells. Small amounts of mRNA in a crude RNA preparation can be detected in this way, and the system has proved to be most useful in this regard.

Quantitatively, using single injected messages, a linear relationship between mRNA injected and protein synthesized exists up to a saturation level. Even at saturation the amount of endogenous oocyte protein synthesized is not altered.

This would indicate that the oocyte has a definite reserve translation capacity beyond normal endogenous levels, which may not be the same for all mRNAs injected. Some have argued that the presence of this spare translation capacity rules against controls at the translational level. To put it another way, it appears that any mRNA present in the cytoplasm will be translated, regardless of origin. The translation activity of exogenous message continues for days in injected oocytes. Normal translation activity is controlled by "masking" endogenous mRNA.

The proper masking control for exogenous message seems not to exist. Translation can proceed in embryos in the absence of transcription (actinomycin studies) in controlled fashion, and one must conclude that the "foreign" message is somehow treated differently from that produced by the oocyte nucleus. A number of controls ("masks") have been proposed. It is possible that oocyte messages are transcribed, but are retained in the nucleus until time for translation.

It is also conceivable that messages are produced by the oocyte nucleus and immediately combined with a cytoplasmic repressor that is removed at the proper time, such as after fertilization. These controlling events may be inoperative in the injection of foreign messages into the oocyte cytoplasm.

Both transcriptional and post-transcriptional controls are utilized by cells. A major problem for developmental biologists is evaluation of the relative importance of these levels of control in bringing ' about and maintaining a differentiated state. Repressing a gene at the level of transcription of DNA might seem more economical than synthesis of message to be followed by the masking or inactivation of that message.

Transcriptional control is seen in terminal stages of development, in which all genes save one sequence may be inactive. In early development some control seems to be at the level of controlling translation of preexisting maternal message, as occurs in sea urchins and amphibians.

Between the two extremes, the cell probably utilizes both methods. Under certain conditions, transcription may be speeded and translation of mRNA accelerated, either by increased translational rate, or by decreased degradation of the message.

Controls are therefore complex and involve levels of controls not only at the DNA and at the ribosome, but also in the length of functional life of the mRNA molecules. Most differentiated tissues do develop very stable mRNA molecules for production of tissue-specific proteins. 'The cells of the eye lens, as epithelial cells, synthesize the crystalline protein of the lens fibers.

Actinomycin inhibits this synthesis in the lens epithelial cells, but not in the fiber cells into which the epithelial cell develop. The fiber cells possess an extremely long-lived mRN that operates in the absence of any transcription for long periods of time, obviating the need for nuclear activity. A similar situation exists in cells synthesizing hemoglobin. In this respect, translation control and long-lived messages seem to be characteristic of the beginning and the end of development.

Transcriptional control seems to operate during differentiation, though information in support of such generalizations awaits a thorough analysis of at least one example of cell differentiation.

Sources of Information and Raw Materials for Early Protein Synthesis

Most of the evidence concerning the function and fate of early messenger RNA tends to support the suggestion that these molecules are largely stored for later use in protein synthesis. The observation that both enucleate and actinomycintreated embryos synthesize as much protein as control embryos during cleavage suggests that the amount of protein synthesis dependent upon newly synthesized messenger RNA is, at least quantitatively, insignificant.

Protein synthesis, as measured by incorporation of labeled amino acids, takes place in the unfertilized egg. No increase in protein synthesis occurs after fertilization. What is actually being measured by following incorporation of radioactive amino acids into protein is the synthesis of new protein molecules from a pool of amino acid precursors derived from the breakdown of yolk platelets.

The yolk protein is thus broken down and resynthesized into different protein molecules according to the needs of the embryo. Some protein molecules may be qualitatively the same as those made in the oocyte. New types of protein appear at gastrulation, as evide-nced by serological experiments. These experiments demonstrate that antigens are present in the gastrula and neurula which are absent from the earlier stages.

These proteins may be translated from the messages synthesized during the burst of messenger RNA synthesis that occurs just prior to gastrulation. They may also be the resultd of the unmasking and translation of oocyte mRNA already present. In any event, the rate of protein synthesis does increases during gastrulation.

8

DIFFERENTIATION

The basic problems involved in embryonic differentiation have already been discussed at the beginning of this book. A typical question, which we have already asked before, is the following: How is it possible that red blood cells, which have synthesized large amounts of a very specific protein, hemoglobin, appear only in a very localized area of the embryo, at a very precise stage of its development?

A good example of embryonic cell differentiation; it is a section through a young tadpole and shows, side by side, two types of highly specialized cells, which have just undergone embryonic differentiation. *Cartilage* cells and *muscle* cells lie side by side and are easily recognizable by their morphology.

The cartilage cells are surrounded by an amorphous matrix made of special glycoproteins, called *chondroproteins;* the muscle cells contain fibrils (the myofibrils), which are made of entirely different proteins, in particular, the contractile proteins *myosin* and *actin*. The structure, the function and the biochemical composition of the cartilage and muscle cells are thus very different and specific for each type of cell.

Yet, if the section had been made a couple of days earlier, it would not have shown such highly differentiated cells; histological studies of the embryonic development, during this short 2-day period, would have shown that cartilage and muscle cells have the same mesodermal origin. In a group of mesodermal cells, which all look alike, if not identical, some cells will form glycoproteins in large amounts and become cartilage; others will synthesize contractile proteins and differentiate into muscle cells.

Since we know, due to the nuclear transplantation experiments of

Gurdon, that all the nuclei of the tadpole and even the adult contain the same genes as the fertilized egg (the zygote), differen-tiation can only be explained by assuming that in muscle cells, certain genes that direct the synthesis of the contractile proteins are active, while those that direct the hemoglobin and chondro-proteins are silent.

In the same way, the hemoglobin genes must be active in the differentiating red blood cells and the chondro-protein genes must be functioning in the young cartilage cells. Thus, the genes involved in the synthesis of large amounts of the other specific proteins (which Holtzer called "luxury" proteins) must be inactive in these cells.

On the other hand, the genes that control the synthesis of proteins which are needed by all cells must necessarily always be active. Any cell, whether it is a muscle cell, a red blood cell, a cartilage cell, etc.; requires energy and must possess all the enzymes needed for oxidative phosphorylations (cytochrome oxidase, dehydrogenases, etc.).

Differentiated cells contain variable amounts of these enzymes; Holtzer called them the "housekeeping" proteins, and they must be present if the cell is going to survive. Thus, the problem of cell differentiation is largely a problem of *gene regulation.* We have seen that many different mechanisms may operate in the control of gene activity.

During cell differentiation, where one or a few genes coding for specific luxury proteins are selectively expressed, there is little doubt that the main control takes place at the *transcriptional* level. As we have seen, "active chromatin" displays several peculiarities: high sensitivity to DNase digestion, frequent undermethylation of the DNA molecule and quantitative or qualitative changes in the DNA-binding proteins are well documented to day.

But this is still a crude approach and we do not yet know how specific genes are turned on or off in a very precise and delicate way. We do not even know by which mechanisms the hemoglobin genes are undermethylated in the chromatin of red blood cell precursors and fully methylated elsewhere. How apparently undifferentiated cells become "*committed*" to differentiate along a given developmental pathway (erythropoiesis, myogenesis) remains, despite many efforts, an unsolved problem.

One of the hypotheses proposed to explain cell differentiation is that of "quantal mitoses", cell divisions which would give rise to two *non-identical* cells: one of them would remain an undiffere-ntiated *stem cell,* the other would be committed to differentiate according to a given program. By repeated cell divisions, this committed cell would ultimately give rise to *a clone* of nearly identical differentiated cells. This clonal theory of embryonic differentiation is now widely accepted.

However, it should be pointed out that the analysis of allophenic (tetraparental) mice has shown that all the muscles of a mouse derive from a small number of cells committed to become myocytes (muscle cells) in the somites, i.e. they do not derive from a single clone of cells, but from a few independent clones of cells.

One usually speaks today of *terminal* differentiation, a term which seems to imply that cell differentiation is an irreversible process. As we shall see, there is at least one case in which differ-entiated cells may *dedifferentiate* and then redifferentiate along another pathway. A shift in the differentiation program, called *transdifferentiation,* is thus a possibility, but it remains a rare event in the animal kingdom.

It is often believed that there is an antagonism between *cell division* and *cell differentiation;* indeed, mitotic activity, in general, slows down or even stops completely when cells fully differentiate. But one should not be too dogmatic on this point. There are several examples in which dividing cells have been found to be fully differentiated, as evidenced by the production of specific luxury proteins.

Two different biological systems can be used for the study of cell differentiation: developing embryos and in vitro cultures of embryonic cells committed to differentiate along a particular developmental pathway. The two systems are adequate for the analysis of cell differentiation, but both have their advantages and disadvantages.

Whole embryos represent a more "natural" system than cultured cells; although the latter are easier to work with, the results obtained with them do not necessarily apply to whole embryos. This is not surprising since, in embryos, in contrast to cultured cells, cellular movements, cell-to-cell interactions, morphogenetic gradients and inductions play a decisive role.

When embryos are dissociated into individual cells, these cells lose the precise localization they had in the morphogenetic gradient where they interacted with their neighbors. The result is the loss of what has been called *positional information.* This, as well as the poor permeability to drugs of whole embryos from certain species (amphibians, in particular), explain why some agents which, as we shall see, suppress cell differentiation in cultured embryonic cells have very little effect on intact embryos.

It is impossible to give a full account, in a few pages, of all the work that has been done and is being done on cell differentiation. We shall concentrate on *embryonic* differentiation and speak more superficially of the studies made by the workers who use tissue cultures as their main method.

Before we come to the presentation and discussion of the known facts, a last word of introduction should be said. The study of cell differentiation has more than an academic interest. It is a problem which is of fundamental importance for mankind, because cancer is, to a large extent, a disease in which the cells divide, but fail to differentiate.

The solution of the cell differentiation problem might lead sooner or later to a solution of the cancer problem. If malignant cells could be made to differentiate, they would stop dividing and would perhaps no longer be malignant.

SPECIFIC PROPERTIES OF CELL MEMBRANES

Already at the gastrula stage,-as-shown-by Holtfretex,, cells of the different parts of the embryo differ in their *tissue affinities.* If cells of the ectoblast and the entoblast are joined together in the same explant, they separate from each other after a few hours of culture. They have *negative affinities,* since they reject each other.

On the other hand, the mesoblast cells stick to either ectoblast or entoblast when they are placed in contact; they have *positive affinities* for the cells of the two other cell layers. Interestingly, treatments with actinomycin or cycloheximide modify the ectoblast tissue affinities: their acquisition or maintenance thus requires the synthesis of macromolecules.

Comparable findings have been made with cells isolated from much older embryos. For instance, Moscona dissociated cells from the retina of chick and mouse embryos and mixed them with other dissociated cells, coming from the kidney of the same embryos.

He found that retina cells recognize each other and form *a chimaeric* retina, made of mouse and chicken cells; in the same way, the dissociated kidney cells form *chimaeric* kidney tubules, made of cells originating from the two species of embryos. In this case, the cells recognize the other cells which belong to the same tissue and are members of the same family.

Many attempts are being made today to identify the molecules responsible for specific cell-to-cell adherence in embryonic tissues and in lower invertebrates, such as the sponges (where reaggreg-ation of dissociated cells is easy to obtain).

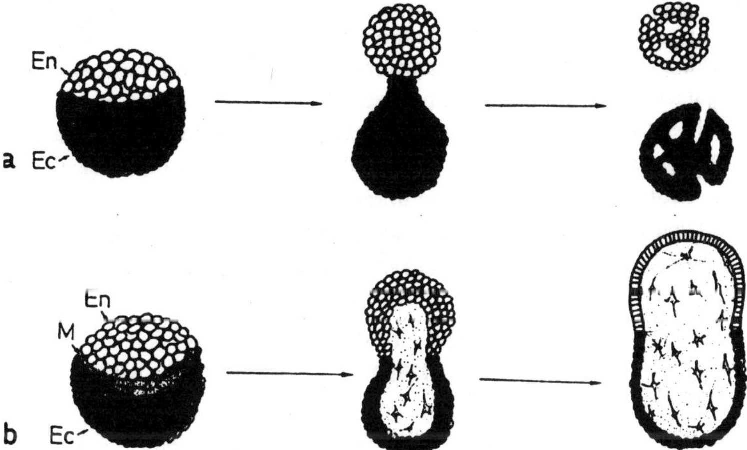

Figure 8.1 : a,b. Tissue affinities. a Ectoderm (Ec) and endoderm (En) show "negative" affinity; they quickly separate from each other and form unorganized gut and epidermic cells. b If a piece of mesoderm (M) is inserted between the two fragments, ectoderm and endoderm no longer separate; they both have a "positive" affinity for mesoderm; a vesicle containing mesenchymatous cells is obtained.

These molecules (called *CAM* for cell adherence molecules or Cognins) display tissue specificity: L-CAM from liver favors the reaggregation of liver, but not of nerve cells. In contract, N-CAM from the nerve cells acts specifically on the reaggregation of neurones and has no effects on liver cells.

All these molecules are large acidic glycoproteins, with a molecular weight of more than 100,000. Schematically the relationships between N-CAM, a nerve cell and the outer medium. N-CAM is formed of two closely related polypeptide chains with molecular weights of respectively, 140,000 and 170,000.

At the end of the amino terminal region is a binding site for another N-CAM molecule, while the other end (carboxyl terminal region) penetrates into the cell through the cell membrane. Specific cell adhesion molecules are already present at early stages of development and are responsible for the reaggregation of dissociated sea urchin blastula cells and for compaction of 8-cell-stage mouse embryos.

It has been recently proposed by Edelman that CAM genes control the morphogenetic movements required for induction. An additional word should be said about the immunity reactions which follow the grafts made in amphibian embryos. These grafts are very well tolerated when a piece of tissue (an organizer, for instance) is grafted in a host of another species.

A normal head, with eyes, nose, etc., can be induced in the host, but sooner or later, this secondary head is rejected and destroyed. Incidentally, such experiments show that the inducing capacity of an organizer does not display any species specificity, but the response of the host to the inductor is species-specific and thus gene-determined. A frog organizer grafted into a newt embryo induces newt nervous tissue.

From the viewpoint of the immuno-response, there is a difference between amphibian and mammalian embryos. If a foreign tissue is grafted in the latter, it is not rejected even when the immuno-system has been built up after birth, (acquired tolerance).

EFFECTS OF TISSUE EXTRACTS ON CELL DIFFERENTIATION

Many efforts have been made in the past with the hope of isolating substances from differentiated tissues, which would induce the tissue-specific differentiation of undifferentiated cells. These efforts have so far been frustrating and the limited effects obtained should be ascribed to the presence of many factors in tissue extracts, which stimulate or retard cell proliferation and, as a result, may affect cell differentiation in an indirect way.

Substances present in adult tissues which inhibit mitotic activity of homologous embryonic cells in a tissue-specific way are called *chalones*. For instance, an extract of adult kidney reduces the number of mitoses to a greater extent in embryonic kidney (pronephros) than in all other tissues of an amphibian larva.

Many different chalones have been described; they seem to be very heterogeneous molecules, since their molecular weights may vary from less than 10,000 to more than 100,000.

It is beyond the scope of this chapter to discuss the numerous substances *(growth factors)* which stimulate cell division and may indirectly slow down differentiation. Many of them bear some similarities with the pancreatic hormone *insulin* and act by binding to specific receptors present on the cell membrane.

The growth factor-receptor complex is internalized by endocytosis and partly broken down; the growth factors ultimately stimulate DNA replication in the cell nucleus. The best known of these growth factors is the *ectodermic growth factor* (EGF); this small polypeptide (6000 daltons, 53 amino acids) exerts many early effects on fibroblasts (increases in Na' uptake, intracellular pH, amino acid and glucose transport, etc.) which finally lead to the stimulation of nuclear DNA synthesis.

IMMUNOLOGICAL STUDIES ON EMBRYONIC DIFFERENTIATION

Immunological (serological) methods are exceedingly valuable for the detection *of specific proteins*. If a protein (an *antigen)* is injected into an animal, a rabbit for instance, this animal will produce in its serum an *antibody*. This is a protein which is called an *immunoglobulin* and which reacts in a highly specific way with the antigen.

The antigen-antibody complex is less soluble than the antigen and the antibody. The presence of the antigen can thus be detected by its immunological precipitation after addition of the antibody. If one binds to the antibody, by chemical means, a fluorescent dye or a radioactive marker, it becomes possible to detect, under the microscope, the cells which contain the antigen.

The present technology of *monoclonal antibodies* production makes antibodies of exquisite selectivity available; they are produced by *hybridomas,* which result from the fusion of a malignant myeloma cell and a cell taken from the spleen of an animal which has been immunized against the substance of interest.

Early work in which embryos crushed at various stages of their development were used as antigens showed that new antigens appear at amphibian gastrulation and that others can be detected at later stages of development.

This merely confirms the data that physical methods for the separation of newly synthesized proteins had shown. It is probable that the use of monoclonal antibodies would lead further, but the most important application so far of immunological methods to embryology has been the analysis of the formation of the eye *lens* proteins (the α, β and γ-crystallins).

The *lens* is induced by the optic cup, at the expense of the ectoderm, at a late neurula stage. Immunological analysis shows that the lens antigens can be detected in the head of the young neurula. They are present in all cells, but in low concentrations.

When the lens is induced, the specific lens antigens become detectable in the still undifferentiated lens in higher concentrations than elsewhere. Their concentration considerably increases at the time of cell differentiation, characterized by the appearance of fibers, made of crystallins, in the center of the embryonic lens. This is due to a selective activation of the crystallin genes under the influence of the inducing retina cells. There is thus a progressive restriction in the synthesis of the lens protein from the whole head to the eye lens only during development.

Another system of interest is the so-called *Wolffian regener-ation* of the lens in adult amphibians or advanced tadpoles. When the fully differentiated lens is removed by surgical operation, a new lens forms at the expense of the superior rim of the iris.

The various biochemical phases of lens regeneration have been very accurately studied by Yamada. The first event is *a dedifferentiation* of the iris cells, which lose their black pigment (melanin). Depigmentation is followed by DNA synthesis and cell multiplication. When mitotic activity stops, a period of intensive RNA synthesis sets in.

Large nucleoli make their appearance and rRNA synthesis becomes very conspicuous. This phase of RNA synthesis is very important for the success of the regeneration. The latter is inhibited by actinomycin, which blocks RNA synthesis.

The mRNAs for the *crystallins* are also synthesized during this period and these proteins become detectable, by immunological methods, immediately after the phase of extensive RNA synthesis. The various crystallins are synthesized one after the other. They are found, at first, in both the cytoplasm and the nucleus; the latter degenerate afterward and completely disappear.

These experiments show that inducing agents originating from the retina produce a derepression of the genes present in the iris. As a result, synthesis of the crystallins is induced in the regenerating lens. Cell differentiation (formation of lens fibers) and synthesis of the crystallins are closely linked and go hand in hand.

Wolffian regeneration of the lens is an interesting case of *transdetermination,* the phenomenon first discovered by Hadorn in transplanted imaginal disks of *Drosophila* larvae. A similar phenomenon, which has been called *transdifferentiation,* can be obtained with cultured neural retina embryonic cells: some of the pigmented cells lose their pigment and differentiate into lens cells.

Small lenses, called lentoids make their appearance in the cultures. As in Wolffian regeneration, inactive crystallin genes are activated: crystallin mRNAs are produced in large amounts and the corresponding proteins accumulate in the lentoids.

ENZYME SYNTHESIS AND EMBRYONIC DIFFERENTIATION

It has often been suggested that differentiation might be due to the induction of enzyme synthesis resulting from the appearance, in a region of the embryo, of the substrate for the enzymatic reaction, or to the repression of enzyme synthesis by the localized accumulation of the end product of the enzyme reaction. Such control mechanisms of enzyme synthesis operate efficiently in bacteria and the elucidation of their genetic and molecular bases is at the root of the famous JacobMonod model of gene regulation in bacteria.

But eggs are not bacteria, and all efforts made to demonstrate substrate-induced synthesis in embryos have been unsuccessful. The presence of many enzymes in unfertilized eggs makes an enzymological approach to the study of cell differentiation difficult: some of the enzymes synthesized during oogenesis are kept ready for use at much later stages of development. However, this approach has been valuable in a few cases, as we shall now see.

Several *hydrolytic enzymes* are known to increase in activity in the digestive tract when it becomes functional. This is true for amylase, proteases, phosphatases, etc. in sea urchin plutei as well as in amphibian tadpoles and chicken embryos. In such cases, enzyme synthesis is clearly linked to the acquisition of a physiological function.

The activity of *cholinesterase,* an enzyme which plays an ess-ential role in the nervous system, is already important in the unfer-tilized amphibian eggs. It increases considerably when, after neural induction, the nerve cells begin to differentiate into typical neurones which are physiologically active. Similar changes in cholinesterase activity have been recorded in the cells of a tumor of the nervous system, *neuroblastoma.* The tumor cells have a low cholinesterase activity.

But, if they change into more typical neurones (provided with nerve fiber expansions) as the result of a change in the in vitro culture conditions, cholinesterase activity very markedly increases as the result of the synthesis of new enzyme molecules.

Changes in enzyme synthesis, correlated with modifications of the outer medium, have been recorded in other systems as well. For instance, *glutamine synthetase,* an enzyme which is also involved in nervous system metabolism, is synthesized, at a given time of development, by the retina of the chick embryo.

If the retina cells are dissociated, and then allowed to reaggregate, glutamine synthetase synthesis is accelerated by several hours. Still earlier synthesis of glutamine synthetase is obtained when the retina cells are treated by corticosteroid hormones. The use of macrom-olecule synthesis inhibitors suggests that glutamine synthetase synthesis is regulated at both the transcriptional and translational levels.

It is, of course, impossible to give here a complete summary of all the work that has been done on enzyme synthesis during development and we shall close the discussion with a last example, that is, the synthesis of the *pancreatic enzymes* (trypsin, chymo-trypsin, amylase, etc.) which has been the subject of very interesting studies by Rutter.

The pancreatic "acini," forming the glandular site of digestive enzyme synthesis, are induced by the surrounding *mesenchyme* (the still undifferentiated mesodermal tissue). Their induction is of the "transfilter" type: it is successful even if the reacting and the inducing tissues are separated by a millipore membrane.

Rutter succeeded in isolating a protein from the filters, which certainly plays a great role in the induction. This protein induces in the reacting cells the synthesis of a particular kind of low molecular weight DNA. This synthesis is followed by the replication of the main DNA, which leads to sustained mitotic activity.

Only when a sufficient number of cells becomes available does RNA and then protein synthesis occur. The various pancr-eatic enzymes are synthesized in a constant order, each following the other. There are no indications that enzymes which serve in an identical metabolic pathway, protein or sugar metabolism, for instance, are synthesized together.

This speaks against the exist-ence of an "*operon*," similar to those of the bacteria. It is clear that the structural genes coding for the different pancreatic enzy-mes are activated in a sequential and individual manner, since there is a concomitant increase in their respective mRNAs: for instance, the content of the pancreatic cells in amylase mRNA increases 400 times during their differentiation.

In this case, the main control of enzyme production is clearly at the transcriptional level. Transcriptional controls are particularly clear when organs respond to *hormonal* stimulation by the production of large amounts of a specific protein. For instance, steroid hormones induce the secretion of very large quantities of ovalbumin in the chick oviduct, and of vitellogenin in the amphibian liver.

The hormone binds to a receptor in the target cell; in the nucleus, the hormone-receptor complex activates selectively the ovalbumin or the vitellogenin genes, probably by binding to them. As a result, chromatin in the structural gene and its flanking regions takes the loose conformation of "active" chromatin.

The gene is now exposed to RNA polym-erase activity and the final result is an enormous production of the corresponding mRNA, which is immediately translated into protein: the latter is finally excreted. While there is less than one molecule of ovalbumin mRNA in each cell of the unstimulated chick oviduct, thousands of molecules per cell are found after steroid hormone administration. As we have seen, the response of the amphibian oocyte to progesterone is fundamentally different since gene transcription has very little, if any, importance for protein synthesis. This process is controlled at the translational level.

DIFFERENTIATION OF CULTURED EMBRYONIC CELLS

A Brief Description of a Few Biological Systems

Numerous studies on the in vitro differentiation of undittere-ntıated, but already "committed" embryonic "blast" cells have led to the conclusion that the key event is the selective activation and expression of a small number of specific genes.

Myogenesis (muscle differentiation) is one of the favorite syst-ems for the in vitro study of cell differentiation. Undifferentiated *myoblasts* fuse together in culture and form *myotubes,* whereby the latter finally differentiate into contractile fibers, and simult-aneously, several muscle-specific proteins (in particular, muscle actin and myosin) make their appearance.

DNA synthesis stops when myoblasts fuse together. Experiments with actinomycin have shown that myoblasts synthesize unstable mRNAs, while differen-tiated muscle cells contain stable mRNA molecules. Large amounts of the messengers coding for the muscle-specific proteins are produced when the myotubes differentiate into muscle cells. In contrast, the bulk of the mRNA population (about 17,000 different mRNAs) does not vary greatly during myogenesis.

Another interesting system is *chondrogenesis* (cartilage formation), in which embryonic *chondroblasts* differentiate into *chondrocytes* which incorporate inorganic sulfate and build up an extracellular matrix of sulfated proteoglycans.

Chondroitin sulfate and chondroproteins are typical markers of cartilage differentiation. Soluble factors released by in vitro growing chondrocytes stimulate differentiation of chondroblasts into carti-lage.

There is growing interest in the differentiation of some fibroblast strains (preadipocytes) into *adipocytes* (fat cells). Treatment with insulin triggers or accelerates the accumulation of triglycerides in the cytoplasm which becomes filled with a large lipid droplet; simultaneously, the cytoskeleton breaks down.

Many papers have been devoted to *erythropoiesis,* the forma-tion of red blood cells, which is characterized by the synthesis of an easily detectable protein, hemoglobin. Erythropoiesis is stim-ulated by the addition of *hematopoietin,* a protein hormone made mainly by the kidney. The molecular organization of the α- and β-globin genes coding for the two globin chains of hemoglobin is very well known: we can follow, at the gene level, the shift from embryonic to fetal and finally adult hemoglobin.

We know that the α- and β-genes lie on different chromosomes, and that there is no amplification of these genes when large quantities of hemoglobin are synthesized during erythropoiesis. We also have a growing knowledge of the multiple molecular causes of the *thalassemias,* hereditary diseases in which there is an unbalance in the synthesis of the α- and β-globin chains.

It is the dream of many molecular biologists to cure these diseases by transfer of the correct gene into the deficient genome of the patient (gene therapy); success is not very far ahead. Finally, a large amount of work is being done on Friend's murine *erythroleukemic* cells (MEL cells): these cells are malignant because they are infected by Friend's leukemia virus.

They grow continuously in culture, but do not differentiate. However, differentiation, evidenced by the synthesis of hemoglobin and a few other marker molecules, can he induced by treating MEL cells with simple chemical agents (dimethylsulfoxide, butyrate, etc.). In all these cases, whether hemoglobin synthesis occurs sponta-neously or is induced experimentally, there is a close parallelism between the number of globin mRNA molecules present in a cell and the amount of hemoglobin it synthesizes.

Thus, during erythropoiesis, a small number of genes are activated sequentially. Among the other systems in which cell differentiation can be followed in vitro, mention should be made of the formation of *melanocytes* (pigment cells) and *neurones* (nerve cells). Melano-cytes contain a brown pigment, called melanin, which makes these cells easily recognizable under the microscope.

Neurones possess *neurites* (axones and dendrites), have typical electrophysiological responses when they are excited and contain a number of specific proteins. The outgrowth of neurites is stimulated by *a nerve growth factor* (NGF) which has been well characterized.

Cell lines derived from brain tumors (neuroblastomas, neurogliomas) grow actively in culture; these cells are undifferentiated, but they can be induced to differentiate and to form neurites by treatments with a number of physical and chemical agents which tend to slow down their growth.

It has been possible to isolate, from teratocarcinomas, embr-yonal carcinoma cells (ECC) of different origins and to produce a large number of established cell lines. Some of them remain pluripotent after prolonged cultures, while others became restricted in their differentiation potentialities. Sometimes they gave rise to a single committed cell type, which is the case for the so-called neuro-teratocarcinoma stem cells which differentiate in vitro only in neurones.

Another ECC line, F9, differentiates into visceral endoderm if treated with retinoic acid (a vitamin A derivative) and into parietal extraembryonic endoderm if treated with a retinoic acid and cAMP mixture. Differentiation of F9 cells into visceral endoderm cells is 'accompanied by the activation of the a-fetoprotein gene. This gene remains silent in F9-derived parietal endoderm cells, which synthesize instead, plasminogen activator, cytokeratin filaments, laminin and type-IV procollagen.

Experimental Analysis of Cell Differentiation in Culture

Cell Fusion

A very interesting approach has been developed for the study of cell differentiation: it is the technique of *cell fusion,* also called *somatic hybridization*, which was first devised by Ephrussi. The properties of the cell membrane can be altered by treatment with a hemolytic virus, called the Sendai virus, even when this virus has been inactivated and is unable to multiply in the infected cells, or with polyethylene glycol.

These changes in membrane properties allow the easy fusion of two (or more) cells of the same or different types. As Harris has done. when a nucleate hen red blood cell is fused with a human cancer cell (HeLa cell), very remarkable changes occur.

The nucleus of the red blood cell (erythrocyte) was, before fusion, in a completely inactive, repressed form. It synthesized neither RNA nor DNA; the HeLa cell nucleus, on the other hand, is extremely active in both DNA and RNA synthesis. When the two cells are fused together, a *heterokaryon is* prod-uced.

There are two nuclei, of different origins (hen and man) in a common cytoplasm. Analysis of this somatic hen-man hybrid shows that the erythrocyte nucleus is quickly *reactivated.* It becomes capable of synthesizing DNA and RNA. The hen erythrocyte nucleus contained no

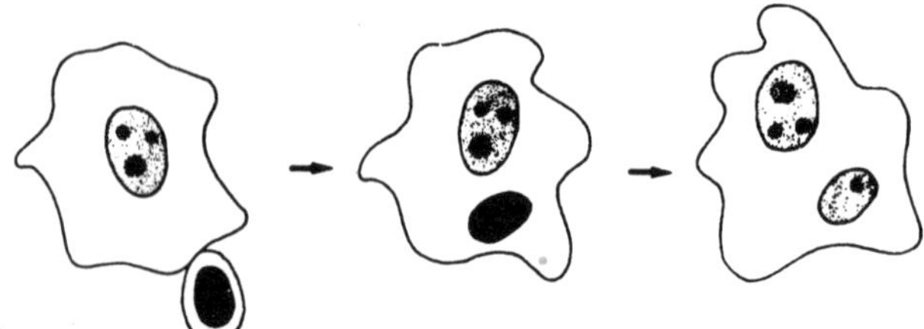

Figure 8.2 : Fusion of a chicken red blood cell (small, with a condensed nucleus) with a mammalian cell: the reactivated red blood cell nucleus forms a nucleolus.

visible nucleoli before the cell fusion; after cell fusion, they become very conspicuous and RNA-rich.

The nucleolar organizers are now active again. At that same time, it becomes possible to detect, by immunological methods, the formation of chickspecific antigens in the cell membrane of the somatic hybrid. The HeLa cell *cytoplasm* has thus exerted a positive. derepressing effect on the previously wholly inactive chick nucleus.

This effect is entirely similar to the one we met before when the experiments of Gurdon on unfertilized *Xenopus* eggs were described. Injection of a nucleus taken from an adult organ (brain, liver), where there is no longer any DNA synthesis, into young cytoplasm is followed by a resumption of DNA synthesis.

Clearly, the same kinds of positive control that the egg cytoplasm exerts on the injected adult nuclei still operate in active adult cells. This elegant technique of cell fusion has been used by Ephrussi in the study of cell differentiation. He fused together *melanocytes* and still undifferentiated embryonic cells *(fibroblasts)* and cultured these hybrids.

The result, in this case, was that the undifferentiated cell exerted *a negative* control on the differentiated one. The hybrid between the two cells soon lost not only melanin, but even the capacity to synthesize tyrosinase. Similar results were obtained with liver and nerve cells. The former are no longer capable of synthesizing a number of enzymes and blood proteins after fusion with an undifferentiated cell.

Nerve cells no longer synthesize a specific protein (called S 100) of the nervous system after fusion with embryonic fibroblasts. The control exerted by the undifferentiated cell on differentiated functions is not *always* negative but this is an exception to the general rule.

Studies on the fusion of malignant and normal cells have shown that the situation prevailing in somatic hybrids can be very complex. Malignancy usually behaves like a recessive character, but the interpretation of the results is complicated by the fact that cell fusion followed by culture is often followed by the elimination of part of the chromosomes (this also often happens, as we have seen, in the ordinary sexual hybrids).

Experiments, in Ephrussi's laboratory, have shown that if the melanocytes contain twice as many chromosomes as the fibroblasts, the ability to produce melanin is no longer lost in the hybrid between the two cells. This clearly shows the importance of "gene dosage" for the expression of the phenotype (melanin production).

Recent studies indicate that the rule, after cell fusion and culture, is the following: the differentiated phenotype (formation of pigment, liver enzymes, S 100 protein, etc.) is lost soon after cell fusion; but it reappears when a certain number of chromosomes have been lost during cell divisions in the hybrid line.

The differentiated phenotype is first extinguished, then re-expressed. In order to gain some insight in the relative roles of the nucleus and the cytoplasm in the phenotypic expression of differentiation, cytoplasm from enucleated fibroblasts (called *cytoplasts)* has been fused with intact cells.

In such *"Cybrids"*, extinction of the differentiated character does not last more than 12 to 20 h; the differentiated phenotype is restored 48 h after fusion. This experiment shows that a cytoplasmic factor exerts a negative control on cell differentiation, but that this factor has a short life and that it is not renewed in the absence of the nucleus.

Since enucleation induces the differe-ntiation of neuroblastoma cells (neurite formation), it is likely that the nucleus contains or produces factors which inhibit cell differentiation. This is another case of a negative control exerted by the nucleus on the cytoplasm. It is likely that extinction of the differentiated phenotype is due to changes in the molecular organization of chromatin.

But it is too early to draw strong and general conclusions in a fast moving and complex field. We still know very little about the molecular interactions which take place between nucleus and cytoplasm in somatic hybrids and cybrids.

Effects of Bromodeoxyuridine (BrdUr) on Cell Differentiation

There is another interesting approach to the analysis of cell differentiation: Holtzer and colleagues found that an analogue of thymidine, bromodeoxyuridine (BrdUr) specifically inhibits cell differentiation in most cultured cells. At low concentrations, which do not stop DNA replication and cell division, BrdUr inhibits the in vitro differentiation of melanocytes, cartilage cells, muscle cells, etc.

It has no remarkable effects on already differentiated cells and on cell viability, but it specifically inhibits the expression of the *luxury* proteins: melanin, chondroproteins, contractile proteins in respectively, melanoblasts, chondroblasts and myoblasts.

It has also been reported that treatment of very young chick embryos with BrdUr suppresses the formation of red blood cells and the synthesis of hemoglobin. However, treatment with BrdUr of fertilized or cleaving sea urchin, ascidian, amphibian, hen and mouse eggs has led to disappointing results: all one can see is a slowing down of development, due to cell injury and death.

There is no specific inhibition of tissue differentiation when the drug is added to whole eggs or embryos. Why BrdUr exerts highly specific effects on the expression of the luxury proteins remains obscure at the molecular level. Incorporation of the drug into replicating DNA obviously takes place, but it occurs randomly and we have only hypotheses for explaining the specific effects of the analogue on the genes which control the synthesis of the luxury protein.

One possibility is that incorporation of BrdUr into DNA affects regulatory sequences controlling the activity of the structural genes coding for the luxury proteins. Another possibility is that certain chromosomal proteins bind specifically to the substituted DNA sequences: transcription of these sequences would then be shut off. Indirect effects on the cell membrane have also been suggested, but it is unlikely that they play a major role in the still mysterious BrdUr effect.

Effects of Phorbol Esters and Retinoids

The cancer problem lies outside the scope of the present book, but we cannot avoid mentioning it when cell differentiation is under discussion. Cancer is believed to be a multistep process (like embryogenesis!) leading to "*mad*" cells in which the regulatory mechanisms which control proliferation and differentiation in normal cells no longer function properly.

All carcinogenic agents (physical, like X-rays; chemical, like the polycyclic hydrocarbons; viral, etc.) affect DNA. The cells in which DNA has been injured by a genotoxic agent will not necessarily become malignant: after *initiation* by a carcinogen, *promotion* by a co-carcinogen (prom-oter) is required as a second step. If. as shown by I.

Berenblum almost 30 years ago, one applies locally a single dose of a carcinogen to the ear of a mouse, cancer will seldom develop, but if one rubs the ear of this mouse with croton oil, fast-growing tumors will form. Croton oil thus contains *a co-carcinogen* (promoter); the active principles of croton oil are *phorbol esters,* the more potent being 12-O-tetradecanoyl-phorbol-13-acetate (TPA).

At very low concentrations, TPA stimulates cell proliferation and, like BrdUr suppresses differentiation in cultures of myoblasts and chondroblasts. It is even able to induce the dedifferentiation of cultured myotubes and to disrupt selectively their myofibrils. Differentiation of cultured epithelial cells is also suppressed by TPA: they fail to synthesize their marker protein, keratin. In induced MEL cells, TPA inhibits hemoglobin synthesis.

So far, the effects of TPA on whole eggs and embryos have been seldom studied: the most striking, in both sea urchin and amphibian eggs is a dissociation of the superficial ectodermal cells. This suggests that the cell membrane is the initial site of action for TPA.

There is now strong evidence for the view that TPA indeed acts on the *cell membrane:* in particular, cell-to-cell communication (i.e. the transfer of small molecular weight substances from a cell to its neighbors) is strongly decreased in TPA-treated cell cultures. Recently, it has been demonstrated that the cell membrane possesses specific receptors for phorbol esters.

After binding to these receptors TPA activates a newly discovered membrane protein kinase (which requires phospholipids and Ca" for activity). It is now believed that this *protein kinase C* is the phorbol ester receptor. Its activation by TPA results in the phosphorylation of cytoplasmic proteins involved in the organization of the cytoskeleton and in protein synthesis.

It should be added that, as in the cases of cell hybridization and BrdUr treatment, the effects of TPA on cell differentiation are not always negative. For instance, TPA promotes the differentiation of a few human leukemic cell lines. We do not know why they make exceptions to the general rule; all one can say is that drugs like TPA and BrdUr exert "*pleiotropic*" effects, but this word masks our present ignorance.

It is interesting to note that the sites of action of BrdUr (DNA) and TPA (the cell membrane) are entirely different; yet the two drugs exert very similar effects on cell differentiation. Only more work will tell us whether TPA and BrdUr are ultimately acting on a single, still unknown, key step which would be absolutely required for cell differentiation.

This step might perhaps turn out to be as simple as a change in the free calcium ion concentration or in the intracellular pH, since we know that such changes may have far-reaching consequences for the cell.

Derivatives of vitamin A (the *retinoids* related to the retina pigment retinene) often antagonize the promoting effects of phorbol esters on cancer cells. In certain malignant cell lines, *retinoic acid* (the most active of the retinoids) can induce the differentiation of undifferentiated cancer cells.

These favorable effects on cell differentiation are linked to the fact that retinoic acid slows down the in vitro multiplication of malignant cells. The most striking effect of retinoic acid on cell differentiation has already been mentioned in this chapter: "nullipotent" (thus incapable of any kind of differentiation) embryonal carcinoma cells differentiate into endoderm if they are treated with retinoic acid.

It is very likely that the action of the retinoids (which are liposoluble compounds) is mediated by a cytosolic retinoic acid binding protein (cRABP).

We do not know how the signals received by the cell, after treatment with either TPA or retinoic acid, are transmitted to chromatin where a few genes involved in cell differentiation must, in a selective way, be activated or shut down.

Other Inducers and Inhibitors of Cell Differentiation

It is likely that these signals are transmitted from the cell membrane to the nucleus, e.g. through *second messengers, cyclic nucleotides* or *calcium ions*. Increasing the cAMP content of the cells often slows down proliferation and favors differentiation.

However, there are many exceptions to this rule. Similarly, there are numerous exceptions to the rule which states that a high cGMP content favors cell proliferation and prevents cell differentiation. As we have seen, a release of membrane-bound calcium ions triggers amphibian oocyte maturation and sea urchin egg fertilization.

It is likely that changes in free Ca^+ and in internal pH play a role when cells stop dividing and begin differentiating. It is probable that cyclic nucleotides and calcium ions, when they trigger cell differentiation, activate protein kinases which can phosphorylate serine, threonine and tyrosine residues in proteins which are important for differentiation.

A few attempts to modify cell differentiation by the addition of so-called *conditioned media* to differentiating cells have been made. Conditioned media are culture media in which cells have been grown for a few hours or days.

They contain substances which have been released or excreted in the surrounding medium by the growing cells. Conditioned media are valuable for the demonstration and isolation of substances which affect (positively or negatively) cell proliferation and of molecules involved in cell aggregation.

However, no differentiation inducing substance has been isolated so far from conditioned media with the exception of a factor, present in the conditioned media of chondrocytes which stimulates the differentiation of these cells into cartilage. This factor, of unidentified chemical nature, stimulates the synthesis, by chondrocytes, of chondroproteins and collagen.

Popular today among the students of differentiation is *5-azacytidine,* which like BrdUr, is incorporated into replicating DNA molecules. The interesting property of this cytidine analogue is that it inhibits *DNA methylation. As* already mentioned, undermet-hylation of the cytidine residues in DNA molecules is a frequent—but not a constant—marker of "*active*" chromatin.

In contrast to BrdUr, 5-azacytidine exerts *positive* effects on cell differentiation: for instance, it induces the differentiation of pre-adipocytes into adipocytes, of myoblasts into myotubes and of chondroblasts into cartilage cells. On the other hand, it has no effects (except sheer toxicity) on sea urchin and mouse eggs; unlike retinoic acid, it does not induce differentiation in nullipotent embryonal carcinoma cell lines. The most curious effect to date of 5-azacytidine is the *reactivation of the X chromosome* in females of the mammals.

Culture of a man-mouse hybrid cell line in the presence of 5-azacytidine is followed by reactivation of the X chromosome as shown by genetic analysis. However, only negative results have been obtained when normal diploid fibroblasts were treated with 5-azacytidine.

Evidently, this substance is a promising tool for the analysis of cell differentiation; but more work is required before a close link between this process and DNA undermethylation can be accepted as a general fact.

EMBRYONIC DIFFERENTIATION AND CANCER

Although carcinogenesis and embryogenesis are two different topics, the two processes have so much in common that a short comparison is not out of order here. Embryonic differentiation is the last step prior to senescence and death; embryogenesis is a multistep process which starts at gametogenesis. Malignant transformation, as mentioned before, is also a multistep process.

It leads to the production of "transformed" cells, which can give rise to permanent cell lines. These cells are *immortal* and, in contrast to normal cells, never show signs of *senescence* (large size, low rate and finally arrest of DNA synthesis, modifications of the cytoskeleton, etc.). Instead, transformed cells go on dividing continuously. Although the initial lesion in the multistep carcinogenic process affects nuclear DNA, other cell constituents are also affected.

In particular, the cell membrane properties are deeply modified in transformed cells. These cell surface changes allow transformed cells, in contrast to normal cells, to grow without attachment to a solid substratum (anchorage-independent growth). This, and the production by transformed cells of proteolytic enzymes, which can destroy the walls of the vessels, are responsible for the formation of the *metastases* which ultimately kill the host.

We have seen that our differentiated tissues derive from the proliferation of one or a few clones of cells; cancers also result from repeated division of a single malignant cell (clonal origin of cancers). All the cells which form a clone of normal, differentiated cells (for instance, all muscle or cartilage cells) look very much alike.

This is not true for cancer cells: cells present in different metastases originating from the same tumor cell and even cells present in the same metastase display a striking *diversity. This* can be shown with immunological methods, which demonstrate a greater heterogeneity in malignant than in normal cells belonging to the same tissue.

This heterogeneity might be due to genetic instability, leading to a high frequency of somatic mutations, in cancer cells. It is indeed well known that chromosomal abnormalities are frequent in cancer cells. However, the reasons why clones—in contrast to those of normal cells - of cancer cells undergo an increasing diversity when they expand by repeated cell divisions, remain mysterious.

It has often been suggested that cancer cells are more similar to embryonic cells than to adult cells. Cancer cells would result from the dedifferentiation of adult cells and this would allow them to proliferate repeatedly. This view is certainly an oversimplification of a more complex problem, but it cannot be disregarded.

A classical example is the synthesis of two related proteins, a-fetoprotein and albumin, by liver cells. Embryonic liver synthesizes a-fetoprotein; around birth, the a-fetoprotein gene becomes silent; the albumin gene is switched on and albumin is synthesized during the whole lifetime.

However, in liver tumors (hepatomas), the a-fetoprotein gene is reactivated and the albumin gene becomes inactive: thus, hepatomas synthesize an embryonic protein, a-fetoprotein. Another well-documented example is the frequent occurrence of *carcinoembryonic antigens* in tumors: these surf-ace antigens are present in mammalian embryos and are no longer detectable in adults, but they are found in many tumors, in particular, in those of the colon.

A particularly interesting case is that of the so-called F9 antigen of nullipotent malignant mouse embryocarcinoma (teratocarcinoma) cells. This antigen is shared with mouse sperm and morulae; it quickly disappears during development and it is absent from differentiated teratocarcinoma cells.

The fact that teratocarcinomas may arise when early mouse embryos are grafted into an ectopic site and that malignant embryocar-cinoma cells can undergo embryonic regulation, i.e. participate in normal development, after their introduction into a normal mouse blastocyst, show further that embryos and cancers must have much in common. The frequent reexpression of fetal genes in cancer cells has been called *retrodifferentiation.*

However, there is an obvious difference between cancers and embryos: cancer cells proliferate and invade the host tissues in an anarchic way: embryos develop in a harmonious way because they are under the control of gradients, fields of differentiation and organizers. Positional information plays an essential role in normal morphogenesis; although it seems to be lost in neoplastic growth.

The origin of cancers has been the subject of extensive research in recent years. Since cancer is a multistep process *(initiation* by a genotoxic agent, *promotion* by agents which, like TPA induce a new phenotype, *proliferation* of the modified cells leading to a tumor by clonal expansion, formation and growth of *metastases),* it is impossible to pinpoint a single cause for all cancers.

However, one of the factors which certainly plays an important role in malignant transformation, is the activation of cancer genes *(oncog-enes),* which have been recently identified by molecular biologists. It has been known, for many years, that a number of viruses induce tumors (leukemias, sarcomas) in birds and mammals (including man).

Some of these viruses contain DNA, others RNA; the latter are called *retroviruses* because they contain an enzyme, *reverse transcriptase,* which copies the viral RNA into the complementary DNA which is integrated into the host genome as *a provirus.* Proviral DNA contains an oncogene responsible for malignant transformation.

When these oncogenes are expressed in virus-infected cells, transforming factors are produced. In many cases, the viral oncogene product is a protein kinase, which phosphorylates several cellular proteins and alters their conformation, leading to the transformed phenotype. Recently, it has been found that certain viral oncogenes increase the production of growth factors or of their receptors on the cell surface.

One of the major surprises in this fast-moving field has been the demonstration that *normal cells* contain oncogenes *(cellular oncogenes)* very similar to the viral oncogenes. Activation of normal cellular oncogenes could thus lead to malignant transformation.

Today we know two possible mechanisms for *oncogene activation.* One of them is *point mutation:* replacement of a single base by another in a cellular oncogene will lead to the production of the homologous transforming viral oncogene. The products of the two oncogenes will differ by a single amino acid substitution and this might have farreaching consequences for the biological properties of the encoded proteins.

The other mechanism for cellular oncogene activation is *translocation* of the gene from its normal site to a different site located on another chromosome. We have seen that chromosomal aberrations are frequent in tumors; among them are translocations in which a fragment of a chromosome is integrated in another chromosome.

If a cellular oncogene is inserted in a chromosome region, which is favorable for transcription, it will be activated and the protein it is coding for will be expressed. Several examples in which cellular oncogenes are activated by translocation into the very active immunoglobulin locus (where antibodies are synthesized) are now known.

What is not known is the possible role of the cellular oncogenes in the normal cells which harbor them. These genes are silent in normal adult cells; but it is not excluded that they are active in the control of growth and differentiation in embryos.

The fact that cellular oncogenes have been highly conserved during evolution and that they are present in *Drosophila* as well as in mice and men, indicate that they must have important functions. It will be an important task for the molecular embryologists of tomorrow to identify the products of the cellular oncogenes during development and to determine whether they play a role in proliferation and differentiation. If so, molecular embryology will benefit greatly from the recent progress made in cancer research.

THE FUTURE OF MOLECULAR EMBRYOLOGY

It would be absurd, for someone who has witnessed the stupendous progress made by molecular biology during the last 40 years to make predictions about the future development of molecular embryology.

The discovery of a favorable material, which could be analyzed genetically, show good cell differentiation and lend itself to the preparation of in vitro systems, would be enough to make a big jump ahead. We need, among the eukaryotes, an organism which can be compared, in simplicity, with the bacteria and the viruses.

Whether the ideal material for the study of cell differentiation really exists seems doubtful. All we can do now is to select systems that have advantages for the study of a particular problem and compare them.

This is what we did, for example, when we compared *Acetabularia* and sea urchin eggs from the viewpoint of the morphogenetic and biochemical potentialities of nonnu-cleate cytoplasm. There is no doubt that the developing embryo, despite its complexity, remains the best material for the study of cell differentiation.

In the first edition of this book, we discussed a number of problems which seemed particularly important. We thought that their solution would be a turning point for molecular embryology, but this solution seemed so remote that they looked more like dreams than realities.

Due to the astounding progress made by molecular biology, within little more than 10 years, these dreams have indeed become realities; molecular embryology has greatly profited - and will profit more and more in the future from the availability of pure genes due to recombinant DNA technology.

Cloned genes (or the cDNAs of specific mRNAs) can be used for the detection, by in situ hybridization, of individual gene localization on metaphase chromosomes. We can now see, under the microscope, where specific genes (the hemoglobin genes, for instance) are located.

Purified genes can be injected into the nucleus of a *Xenopus* oocyte where they will be accurately transcribed; these genes may be modified by chemical or enzymatic manipulation prior to their injection and this will tell us more and more about the control of transcription.

Injection of mRNAs into the cytoplasm of *Xenopus* oocytes is now a standard method for the analysis of message translation. These oocytes are often a more convenient and efficient material than ribosomes isolated from bacteria, wheat germs or reticulocytes for those interested in the mechanisms of protein synthesis.

It is now possible, as we have seen, to inject pure genes into one of the pronuclei of a fertilized mouse egg and to obtain *transgenic* mice which express the injected gene. This seemed, 10 years ago, a very remote dream; the obtainment of transgenic mice is a major victory for molecular and experimental embryology.

There is no doubt that we shall learn a good deal more about gene control in different tissues from this kind of experiment. It should be added, although the topic is outside the scope of this book, that the introduction of cloned, pure genes into somatic cells (fibroblasts, for instance) by addition of DNA to these cells is now a routine experiment: hundreds of genes have been introduced into cells by such *transfection* experiments and their fate (integration into the genome followed by replication, or degradation of the DNA molecules which have not been integrated) has been followed.

In addition, somatic hybrids, in particular, those between mouse and human cells, have provided invaluable information about the localization of many genes on human and mouse chromosomes. Those who have witnessed the slowness of our progress in gene "*mapping*" on human chromosomes by classical genetic analysis of pedigrees are amazed to see that the mouse and human genomes might soon be known in as much detail as that of *Drosophila.*

We wondered, 12 years ago, whether the progress made in mammalian embryology would lead to the production of "babies in test tubes". We know that they now exist; artificial insemination of human eggs is now routinely performed in many hospitals; culture of the fertilized eggs and implantation until birth in their own or a foster mother has been successfully achieved without ill effects so far.

We wish to point out again the ethical problems raised by experimental work on human eggs and sperm: by properly selecting eggs and sperm, one could try to obtain a higher proportion of Nobel Prize winners or football players.

By separating blastomeres during cleavage prior to implantation, one could obtain, at will, identical twins. When one is able to separate the spermatozoa which possess the X or Y chromosome, it will be possible to shift, at will, the proportion of men and women in the world. Injection of genes into a pronucleus of a fertilized human egg might lead to transgenic men, a rather frightening prospect. All this would be Huxley's *Brave New World,* a novel which would be a perfect anticipation of a future society, if the key word was not missing: DNA.

The day will come when, thanks to the development of more and more sophisticated machines for DNA sequencing, we shall know the whole base sequence of the human, mouse, *Drosophila,* sea urchins, etc.

genomes. This will not explain embryonic develo-pment, but we shall know completely the information received by the fertilized egg, which will finally lead to the formation of an adult belonging to the same species.

DNA is a unidimensional molecule and living things are three-dimensional beings: to understand morphogenesis, we must know much more about the tridime-nsional structure of proteins and other large molecules. From the DNA base sequence, the primary structure of the encoded proteins may be deduced, but not the interactions between the tridimensional structuresof these proteins with lipids or with other proteins.

We doubt that, as was once humorously suggested by Sidney Brenner, if one fed all these data to a computer, a living mouse would come out of the machine (unless it was living in the computer).

Falling now from dreams to reality, we think that research in molecular embryology will go on very much as it does now: the experimental, reductionist approach, which has been so successful, will be followed for many years to come.

Science progresses slowly: an experiment destroys a wrong idea, and this is progress; a new technology appears, often in a very different field, and this is another source of progress. All this leads to a better understa-nding of embryogenesis; perhaps the day will come when theoretical biology will be able to explain, in a mathematical form-ula, the unity and diversity paradox of the living world: but this is another dream and there are more urgent tasks ahead of us.

We should know more about gene regulation before we can understand embryonic differentiation; we should know much more about the molecular nature of the germinal localizations, of embryonic regulation and of inductions before we can indulge in theories.

These are tasks for today and we do not doubt that outsta-nding progress will be made in the years to come. How long will it take before we understand how, at the molecular level, a hen egg becomes a chicken: a few decades, a few centuries? Perhaps we shall never know.

9

HATCHING

Terminating the embryonic existence, a process usually described by the deceptively simple word '*hatching*', is in fact an extremely complex procedure and it is not surprising, therefore, that there is a peak of mortality at this time.

Unlike the mammal, essentially all the stimuli are generated by the embryo itself, though it is able to respond to stimuli from both the parents and the other members of the clutch.

Superficially there are three major events in the sequence of hatching: the onset of pulmonary respiration, pipping (that is the single point fracture of the shell, usually in that overlying the air space), and the emergence of the hatchling.

However, underlying these more obvious events are many others, e.g. maturation of muscles, changes in the circulation, withdrawal of the yolk sac, all following a sequence which will ensure a successful outcome of incubation.

PULMONARY RESPIRATION

Introduction

Birds and reptiles pass through a stage of development unknown to the mammal—a period when the embryonic and adult respiratory surfaces, the chorio-allantois and the lungs, function side by side. This period is conveniently described as the parafetal period and the animal, the parafetus. These terms are due to Romijn and were originally used to describe the period from the onset of breathing to pipping. Here we shall use the term to cover the period when the two respiratory systems function simultaneously.

The Respiratory System

The embryology of the system is fully described by Hamilton and Romanoff. In brief the system develops from two distinct sites: the larynx and trachea are derived from the median laryngotracheal grooves; the lungs, bronchi and air sacs develop from paired endodermal diverticula of the embryonic foregut.

The air sacs, which appear on the sixth day, are initially found as six paired structures but two pairs fuse to form the median clavicular sac and, in some species including the fowl, another pair fuse to form the median cervical sac. Thus there are normally eight or nine sacs : the median (or paired) cervical sac(s), the median clavicular and the paired cranial thoracic, caudal thoracic and abdominal sacs.

These sacs are connected, via the ostia, to the bronchi and are concerned, in the hatched bird, with the movement of air within the system. Their innervation, in the fowl and mute swan, is described by Groth.

The architecture of the bronchi is complex. Each lung is supplied with one primary bronchus (the mesobronchus) which gives rise to the secondary bronchi (ento-, ecto-, latero- and dorsobronchi) and these in turn give rise to the tertiary bronchi (parabronchi). No tertiary bronchus ends blindly and manyy join to form long, curved circuits.

It is these tertiary bronchi that are concerned with gaseous exchange. Each is pierced by numerous openings which lead in turn to the atria and then the infundibula. The infundibulum leads to a network of fine anastomosing air capillaries which are in intimate contact with the blood capillaries. Further details of the structure of the lung of the hatched bird may be found in King & Molony.

The flow of air within the respiratory system is complex and the air sacs, far from being evolutionary relics or buoyancy bags, have an important function in maintaining the correct pattern of air flow. They are not concerned with gaseous exchange *per se,* however.

There is no true diaphragm. The pattern of air flow is fully discussed by Brackenbury, Bouvert & Dejours, Bretz & Schmidt-Nielsen and Scheid & Piiper. It will be sufficient to point out here that both inspiration and expiration are active processes and, because of the flow pattern, fresh air is probably passing unidirectionally through the tertiary bronchi continuously.

The avian respiratory system is, then, an extremely specialized system, designed to meet the heavy energy requirements of flight, and, as a result, is able to supply the bird with more oxygen per unit time than any other system.

The Onset of Breathing: the Pulmonary Stimulus

A prerequisite for breathing is the removal of the fluid invading the whole of the respiratory tract. Undoubtedly a major component of this fluid is of amniotic origin, but within the lung itself the fluid may contain a large proportion of an ultrafiltrate of the embryo's blood.

The uptake of amniotic fluid, by active imbibition, is begun, in all birds examined, when 70% of the incubation period is complete and is virtually complete just prior to the initiation of breathing.

Thus in the fowl imbibition is completed about a day before the chick hatches. While some of the fluid remaining in the respiratory tract proper may flow into the pharynx and then be swallowed, some remains and presumably is taken up by the lung tissue itself.

Early attempts to identify the pulmonary stimulus, though not completely successful, served to indicate that it is probably gaseous and that it is possibly a high partial pressure of carbon dioxide in the arterial blood $\left(Pa_{CO_2}\right)$.

Thus Windle & Barcroft with the fowl, and Windle & Nelson with the duck, found that clamping the umbilical vessels of the mature embryo stimulated deep rhythmic respiratory movements almost immediately. That carbon dioxide was the more likely stimulus was suggested by the observation that respiratory movements could be initiated by raising the P_{co2} in the atmosphere surrounding the egg by only 12 mmHg while a reduction of 68 mmHg in P_{O2} was required.

It is only comparatively recently that there has been further interest shown in the nature of the pulmonary stimulus. Visschedijk found that increasing or decreasing the permeability of the shell over the air space did not lead to any change in the timing of pulmonary respiration.

However, Freeman using paraffin wax instead of liquid paraffin for reducing permeability found a significant advance of five hours. He confirmed that perforating the shell was without effect as was ventilating the air space.

At first sight it would seem that either the enhanced P_{CO2}, reduced P_{O2} or both in the air space (Visschedijk, 1968c) was responsible for stimulating breathing. There is an important objection to this interpretation, however: perforating or ventilating the air space failed to delay breathing.

Another approach to the problem has been made possible by the recent development of micro-techniques suitable for analysis of gases in the small volumes of blood available from the avian embryo. Towards

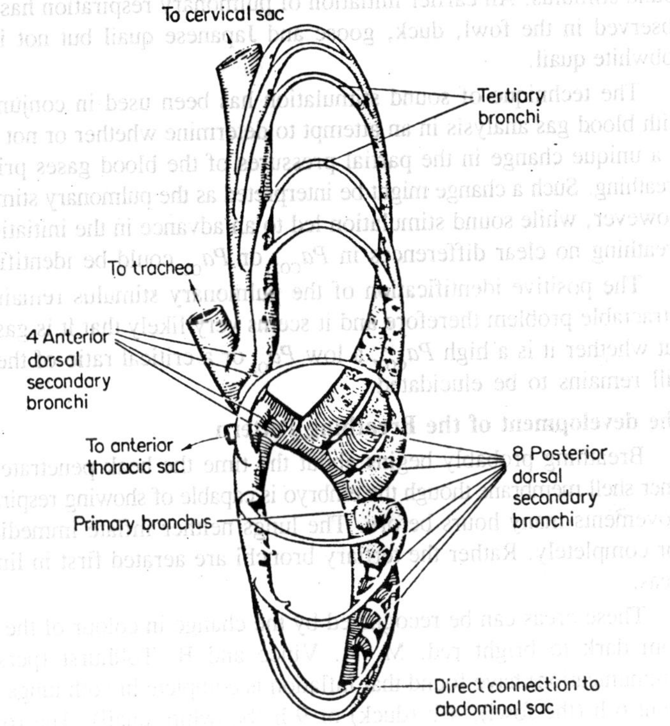

Figure 9.1 : Bronchi of the goose lung. Dorsal view of the right lung; the ventral series of posterior secondary bronchi has been omitted.

the end of the embryonic period the Pa_{CO2} may reach the very high level of 60 mmHg (Dawes & Simkiss, 1969; Freeman & Misson, 1970). Pa_{O2} is relatively low - about 20 mmHg - but not exceptionally so.

The likelihood of a high Pa_{CO2} being the stimulus is strengthened therefore by these direct observations, though the further observation that there is a significant fall in Pa_{CO2} before breathing commences weakens the hypothesis.

In an attempt to resolve this paradox we have examined a modification of the hypothesis suggested by the above results - that the stimulus is a particular ratio of Pa_{CO2} to Pa_{O2} as it is in the sheep - but without success.

A different approach has been suggested by the demonstration that all aspects of the hatching process can be substantially advanced by sound - 'clicking'. This suggests that under certain circumstances the gaseous stimulus can either be initiated earlier or overriden by the sound stimulus. An earlier initiation of pulmonary respiration has been observed in the fowl, duck, goose and Japanese quail but not in the bobwhite quail.

The technique of sound stimulation has been used in conjunction with blood gas analysis in an attempt to determine whether or not there is a unique change in the partial pressures of the blood gases prior to breathing. Such a change might be interpreted as the pulmonary stimulus. However, while sound stimulation led to an advance in the initiation of breathing no clear differences in Pa_{CO2} or Pa_{O2} could be identified.

The positive identification of the pulmonary stimulus remains an intractable problem therefore and it seems very likely that it is gaseous but whether it is a high Pa_{CO2}, a low Pa_{O2} or a critical ratio of the two still remains to be elucidated.

The development of the Breathing Pattern

Breathing probably begins about the time the beak penetrates the inner shell membrane though the embryo is capable of showing respiratory movements many hours before. The lungs neither inflate immediately nor completely. Rather the tertiary bronchi are aerated first in limited areas.

These areas can be recognized by the change in colour of the lung from dark to bright red. M. A. Vince and B. Tollhurst (personal communication) have found that inflation is complete in both lungs after about 6 h (the fowl), 4 h (duck) or 9 h (bobwhite quail). The role of surfactant in facilitating breathing has not been completely elucidated.

Initially breathing is both irregular and intermittent. Gradually regular patterns of medium amplitude breathing emerge and come to dominate the pattern. In the third phase both frequency and amplitude increase and each breath is often accompanied by a '*click sound*'. This pattern has been observed in the several species examined though the length of any one phase may vary from species to species.

The rate of breathing in the hatching fowl reaches about 90 or 100 breaths per minute while in the duck it is a little higher at 110 breaths per minute.

Recent work by Dawes (1973) indicates that the control mechanisms of respiration are at least relatively mature at hatching. Thus during experimental anoxia he found that breathing amplitude rose quickly indicating some peripheral stimulation.

Thereafter the parafetus began to gasp, breathing rate slowed and finally ceased illustrating the excitatory and depressant effects respectively of anoxia on the respiratory centres of the brain.

Circulatory and Associated Changes

The circulation of the embryo is characterized by two short circuits—the ductus arteriosus between the pulmonary and aortic arches and the intraatrial foramina, the avian equivalent of the foramen ovale. The ductus venosus, unlike its mammalian counterpart, is lost on the seventh day of incubation.

When the bird commences breathing these short circuits, together with the chorio-allantoic circulation, have to be closed off in order that a double circulation can be formed.

It has been found that the walls of the ductus arteriosus have well developed muscle layers, particularly at the proximal end. As breathing begins these muscles contract to restrict and finally prevent blood flow.

This is a gradual process, and is complete in 91% of the chicks at hatching. The physiological mechanism is uncertain, neither acetycholine nor catecholamines have marked stimulatory properties on the muscle cells.

The intra-atrial foramina are so constructed that they normally act as at their great valves. During diastole of the right atrium the blood flows into the chamber Blood press and pushes the flaccid inter-atrial septum towards the left thereby opening and continue the formina and allowing a proportion of the blood to flow directly into the left atrium.

During systole the pressures in the atria become equalized, thus allowing the septum to take a medial position and in so doing the

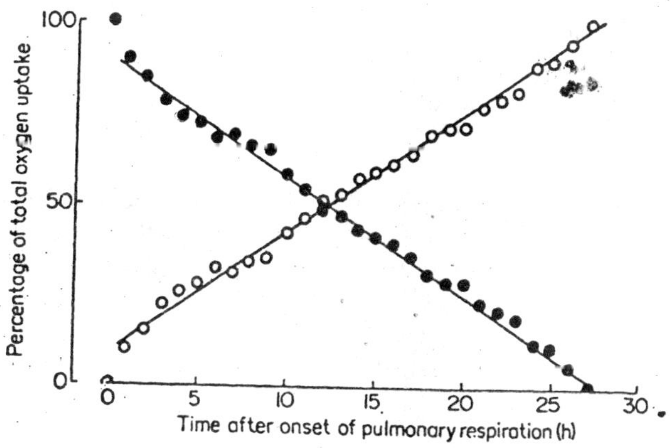

Figure 9.2 : The percentage contributions of the chorio-allantois (I) and the lungs (O) to the oxygen requirements of the hatching chick. The data for this figure were calculated from Visschedijk (1962). The embryo pipped the shell after 10 hours and hatched 27 hours after the commencement of breathing.

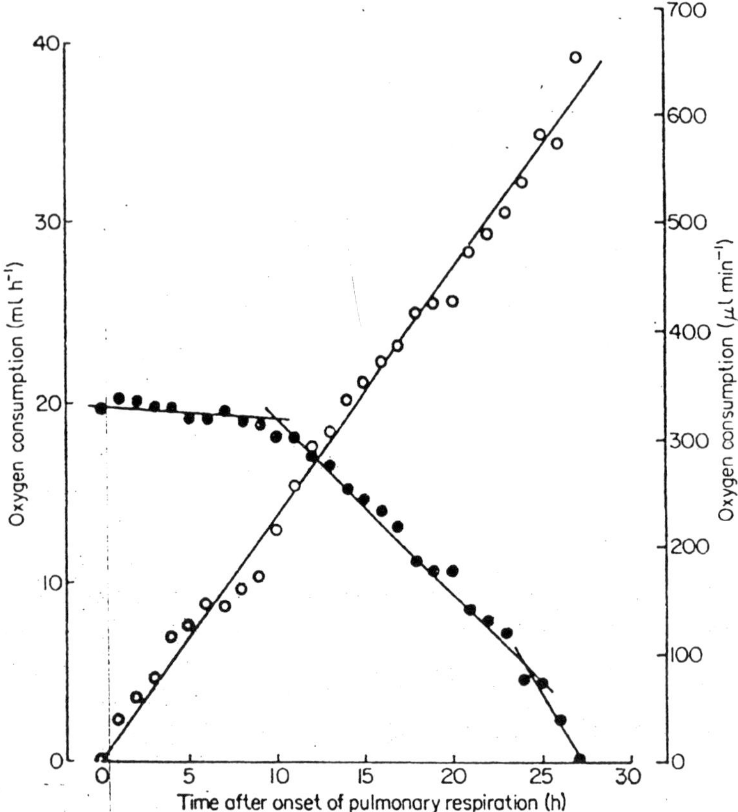

Figure 9.3 : The contributions of the chorio-allantois (I) and the lungs (O) to meeting the oxygen requirements of the hatching fowl. The degeneration of the chorio-allantois as a site for gaseous exchange does not begin until some 10 hours after breathing commences and thus coincides with the time of pipping.

foramina are closed. As breathing becomes established more blood flows into the left atrium from the now-functioning pulmonary vein and as a result the pressures within the atria tend to be equal. This closes the foramina and their permanent closure follows, usually within four or five days.

The mechanism by which the blood flow, via the umbilical vessels, to and from the chorio-allantois, is progressively restricted is uncertain though the closing of the umbilicus itself may be involved.

Clearly this is a gradual process, for gaseous exchange continues for more than 20 h. But it should be noted that the volume of blood flowing through the chorio-allantois begins to decline some four days or so earlier. Why this is so is uncertain and in some ways paradoxical. It is precisely at this period of development that the problems of oxygenating the embryo are probably at their greatest.

Blood pressure rises markedly during hatching to at least 43/23 mmHg and continues to increase in the immediate post-hatching period and the heart rate increases from about 260 beats per minute to 295 beats per minute or more.

Pipping

Some 8 or 9 h after breathing is established in the fowl the shell over the air space is usually fractured at one point—'pipped'. The timing of this event can be advanced by waxing or oiling this part of the shell, delayed by perforating it or virtually abolished by ventilating the air space with atmos pheric air.

Thus pipping, like breathing, would seem to result from a gaseous stimulus. In a series of elegant experiments Visschedijk (1968c) has established that either a reduction in the P_{02} or a rise in the P_{CO2} in the air space advances the time of pipping and that for a given change, carbon dioxide is twice as effective as oxygen.

Just prior to pipping the P_{02} in the air space of the domestic fowl may fall to as little at 60 mmHg while the P_{CO2} may reach 60 mmHg. At this point in the hatching process the lungs are meeting about 40% of the total oxygen requirements of the parafetus and it might be expected that the parafetus will be faced with increasingly hypoxic conditions.

However, there is little evidence that this occurs : cardiac stores of glycogen are normally low, suggesting few problems during hatching though a transient depletion can be detected at pipping. There is, furthermore, no significant accumulation of lactate and the Pa_{02} and Pa_{CO2} do not rcflcct hypoxia.

General activity, including that of the 'hatching' muscle *(muscu-lus complexus),* increases as the parafetus moves further into the air space. Eventually the beak comes into contact with the calcareous shell and the enhanced muscular activity leads to the egg tooth—the reinforced tip of the upper beak—penetrating it.

Once the shell is pipped the parafetus enters a more quiescent period, at least in terms of physical activity, which persists until 3 or 4 h before the chick emerges from the shell.

ACTIVE HATCHING

The final activity of the parafetus is directed towards cutting round the shell and the emergence of the chick. That this period is independently controlled was first suggested by Freeman. He found that the oxygen consumption of the parafetus began to rise an hcur or two before active hatching began. It was suggested that this increase in metabolism resulted from the release of a hormonal stimulus which leads to the active cutting round of the shell.

It is now generally agreed that the hatching stimulus is not gaseous. Visschedijk and Freeman found that altering the permeability of the air space shell did not have any effect on the time of hatching. The suggestion that it is hormonal in nature was strengthened by the demonstration that the thyroid hormones accelerate hatching and that they stimulate active hatching specifically.

However, it has been suggested that progesterone is the hormone responsible, though Oppenheim has been unable to repeat their experiments. An important role has been ascribed to the muscle, *musculus complexus,* in the hatching processes.

This muscle shows marked development just before hatching begins and acts to raise the head of the parafetus and is presumed to provide most of the force necessary for breaking down the shell. The timing of its maturation seems to be unique but attempts to explain this have largely failed.

WITHDRAWAL AND FATE OF THE YOLK SAC

As incubation draws to a close there remains outside the embryo a variable amount of yolk in the yolk sac. This is progressively withdrawn into the abdominal cavity, from the nineteenth day in the fowl, probably as a result of the activity of the abdominal musculature, the process in the fowl being completed some 14 h before emerging from the shell.

The control of the withdrawal activity is uncertain, though there is limited evidence that. the thyroid or adrenal or both may be involved. Although the yolk sac is connected directly with the duodenum there is little or no movement of material via this route even after hatching. Instead it is absorbed through the yolk sac membrane and transported to the chick by the omphalomesenteric vessels.

There are approximately 5 g of yolk in the yolk sac of the neonate fowl but this is rapidly utilized by the bird and the yolk sac is usually vestigial by the fifth day.

OXYGENATION AND ENERGY METABOLISM DURING HATCHING

With the onset of breathing the oxygen requirements of the parafetus

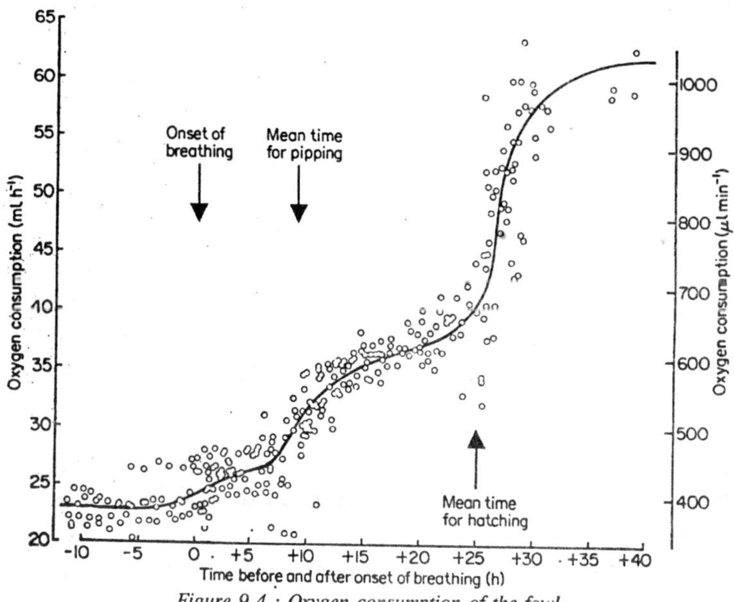

Figure 9.4 : Oxygen consumption of the fowl during the hatching period.

are met by the combined activities of the chorio-allantois and lungs. It is possible to determine their relative contributions during the parafetal period from the data of Visschedijk.

Given that the reduction in the chorio-allantoic component is similar over its whole surface, we can see that after the initial 3 h or so the change-over from chorioallantoic to pulmonary respiration is linear and is not completed, at least in the fowl, until the very end of the hatching period.

However, the degeneration of the chorio-allantois would not seem to be initiated until 10 h after the initiation of breathing for it continues to provide the same amount of oxygen up to that point. Thereafter its importance as a respiratory surface declines rapidly.

The oxygen requirements of the parafetus, not unexpectedly, rise during hatching. Typically the fowl consumes 25 ml h^{-1} prior to breathing, 35 ml h^{-1} prior to active hatching and about 40 ml h^{-1} at the moment of escape from the shell. After this, at least in the homeothermic species, a further substantial rise occurs.

It would seem that the bird's requirement for carbohydrate increases markedly during hatching for there is a dramatic mobilization of glycogen stores a little before breathing commences. It is likely that the glucose that is made available is necessary for the proper functioning of the central nervous system. Indeed a marked rise in the glycogen stores of the brain have been noted at this time. The overall respiratory quotient does not show any increase during this period, however, remaining at about 0.7. A transient period of mild hypoxia is indicated by the fall in cardiac glycogen stores at the time of pipping.

NEWLY HATCHED BIRD

Previous chapters have shown that, as Hinde points out, hatching is not a zero-point for the development of behaviour'. Rather, the chances of successful hatching are maximized by the embryo's activity during the last few days of incubation.

As lung ventilation is established before hatching, and as, for many species conditions within the nest are very similar in the pre- and post-hatching periods, it seems possible that there is more continuity during the trans-hatching period for the avian embryo and chick, than there is in mammals during the corresponding period (for continuity between the human fetus and neonate).

Areas where continuity has been observed, and areas where new behavioural features have been demonstrated will be considered very briefly below. First, the distinction between altricial and precocial species should be made clear.

The early development of species which are, in varying degrees, altricial or precocial has been described by Nice. Young birds which (i) feed themselves, are covered with down, have their eyes open and leave the nest during the day or two following hatching, are classed as precocial but to different degrees : (a) Megapodes are totally independent of their parents, (b) ducks and many shorebirds follow their parents but find their own food and (c) species like quail and fowl follow their parents but are shown food. Young birds which (ii) are fed by their parents include a special precocial group which leave the nest and follow their parents (grebes and rails) then, more dependent (a) semi-precocials which stay at the nest after hatching, although they are able to walk (gulls and terns), *(b)* semi-altricials which, although covered by down are unable to leave the nest (herons and hawks, eyes open after hatching) and those with eyes closed such as owls.

Finally, (c) most dependent, are the true altricials, which have closed eyes, very little or no down, and are unable to leave the nest for some time like, for example, passerines.

Differences in the timing of developmental features in altricial and precocial species have been mentioned. These differences are very much in evidence after hatching. There are descriptions of altricial and precocial development in a few species. For example Nice (1983) has given a full description of the altricial song sparrow, while the altricial blackbird has been considered by Messmer & Messmer (1986) and the great tit by Gibb (1994). Kruijt (1994) has given a full description of posthatching development in the precocial Burmese red junglefowl.

REQUIREMENTS OF THE NEONATE

Warming and Cooling

In most species parental brooding behaviour continues without much change during this transition stage between the incubation and the nestling periods (the blackheaded gull). In this way conditions suitable for warming or cooling the newly hatched bird are maintained at least until it dries out.

In altricial species parental brooding behaviour continues, if only intermittently, until the nestlings gain control over their body temperature. Precocial chicks call loudly and in a distinctive fashion (the 'distress' call) when they are cold, or find themselves isolated and such calls stimulate parental brooding behaviour.

This call is the same as that which may be heard from an egg which is cooling and in fact the alternation between this call and the 'pleasure' call, given by the embryo in response to warmth and gentle turning of the egg, also occurs in newly hatched chicks in response to warmth and contact.

Megapodes provide an important exception to this general picture of care which continues into the post-hatching period. According to Frith (1982) the Mallee fowl's parental behaviour ends, in the case of the female with the laying of the egg, and in the male with his work of tending the incubation mound.

The newly hatched Mallee fowl chick makes its way to the surface unaided, except indirectly by a loosening of the sand resulting from the male's building activities. (If a chick appears on the surface while the male is working it is scratched out of the way with the building material.) Once it has reached the surface it may rest in the sun for about an hour, and then run off into the scrub.

Righting and Standing

Kovach has considered a possible developmental continuity between the co-ordinated and stereotyped movements which bring the embryo into the position for hatching, and contribute to cutting round the shell before emergence, and postnatal righting reflexes. In newly hatched domestic fowl he observed extension in both legs when the chick lies on its back with the head and neck stretched straight out; when the head is bent to the left, the left leg remains extended and the right leg is flexed, while the opposite occurs when the head is bent to the right.

These postural reflexes are fully developed in the 19-day-old embryo and form a part of the co-ordinated and stereotyped movements from that stage onwards.

After righting itself the chick first creeps on its tarsi and later stands by extending its legs. The time for a chick to get up on to its feet and walk can be affected by its rate of development before hatching; artificially accelerated chicks took a little longer to stand up than unstimulated chicks.

Similarly, abnormal prehatching conditions can have slightly damaging effects after hatching. In the bobwhite quail, lengthening the time spent in the shell before hatching lengthens also the time required for uncurling the toes after emergence (equivalent effects in the human neonate).

As in righting and standing, other reflex activities have become functional during the incubation period, although the extent to which they function during that time is not yet understood, in all cases. Detailed trans-hatching studies are needed here.

Social Attachments in Young Birds

In precocial species the parents and young leave the nest within the first few days (or even within a few hours after hatching) and move round together, with intermittent bouts of feeding and brooding. Here there is an obvious need for a mechanism which will keep the group together.

The concept of 'imprinting' (although it arose in the context of species recognition) has led to the analysis of factors involved in the 'following' response in young precocial birds. Many years ago it was found that a newly hatched fowl, duckling or gosling will follow almost any moving object in much the same way as it will follow its parent.

Spalding considered this behaviour to be an instinctive response to visual and auditory stimulation : 'chickens as soon as they are able to walk will follow any moving object. And, when guided by sight alone, they seem to have no more disposition to follow a hen, than to follow a duck, or a human being ... there is the instinct to follow and ... their ear, prior to experience, attaches them to the right object'.

In addition to parameters related to the formation and maintenance of this parent-chick bond recent work has considered the part played in its establishment by pre-hatching experience and activity. Tschanz and Impekoven have demonstrated that *embryos* of the guillemot and gull respond to parental calls in such a way as to query the last part of Spalding's statement: in fact neonates may well have become attached to their parents' calls while still in the egg.

Further, Gottlieb has shown that the duckling's preference for the calls of its own species has been facilitated by hearing calls from neighbouring eggs during the last few days of incubation, and also, by hearing its own calls after it begins to breathe about 3 days before hatching.

Feeding Behaviour

Given the right temperature conditions, a full complement of reflex responses and the necessary social adjustments, the newly hatched bird's more basic requirements are completed by food. Much of the feeding situation is obviously a new development, but continuity with pre-hatching activities is still discernible even apart from swallowing movements which began half-way through incubation.

In the smaller altricial species food is an immediate and urgent need if the nestling is to grow quickly enough to be viable and here food is supplied by the parent. Food is brought to the nest either fresh or partially digested.

In response to mechanical stimuli (the parent landing on the nest) or parental vocalizations, the nestling 'begs' for the food, by stretching up its neck vertically, with the beak wide open and with typical 'begging' calls. This situation has been analysed in nestling thrushes by Tinbergen & Kuenen.

The 'begging' posture in the great tit nestling is shown in Figure; in this species the vertical feeding position is exaggerated, presumably owing to the exceptionally deep nest cup. Although it is weak in the first day or two, the begging pattern can consist of the extension of the limbs, neck and body, raising the possibility that it could be derived from hatching movements if, as has been suggested, these are the same in altricial as precocial species.

Pecking for food in the domestic fowl has been much studied. Spalding observed chicks with no previous visual experience and noted that their pecking was accurate, although they were less successful in seizing the object between the mandibles at the moment of striking.

Subsequently many attempts have been made to separate this skill into innate and acquired factors. In the meantime Kuo observed head, beak and swallowing movements in the domestic fowl embryo and came to the conclusion that pecking in the chick arises from elements which have already matured and been exercised in a co-ordinated fashion in the normal course of embryonic development.

This view, however, does not take into account the perceptual problem where new and important factors are introduced and which will be discussed below.

The transition from incubation to feeding of the young in the ring dove has been analysed by Lehrman. The squabs are fed 'crop milk' regurgitated by the parent. Lehrman found that one of the factors leading to regurgitation is tactile stimulation of the crop; this occurs when the newly hatched squab first lifts its head unsteadily and moves it against the parent (i.e. a 'begging' movement from the position in the nest in which the squab has hatched).

Thus he found that a successful first feeding results partly from the parent's hormonal state and partly from the presence and activities of the young, both arising in the normal course of incubation.

In the semi-precocial herring gull, Tinbergen has given a detailed account of the stimulation eliciting feeding behaviour. Gull chicks beg for food by pecking at the red spot on the underside of the parent's bill and their activities result in the regurgitation of food by the parent.

In a very detailed study of pecking in the laughing gull Hailman has analysed the type of stimulation eliciting pecking in this species and the form and accuracy of their pecking movements. This brings us again to problems beyond the scope of embryonic life; the perception of objects, colour, distance and visually guided movements in space.

Oiling and Preening

Care of the plumage is. important for both precocial and altricial birds. In precocial species preening begins very early; Kruijt (1984) mentions preening movements on the first day after hatching in the Burmese red junglefowl, in particular pecking and nibbling movements, although restricted to a small area of the plumage, occur at that time.

The possibility that these movements could be derived from beak clapping in the embryo, needs to be considered. In the altricial song sparrow and redstart Nice reports that the first preening movements occur before there are any feathers to be preened, and they begin about 5 days after hatching.

In ducklings, which may be led to water on the first or second day after hatching, oiling of the down is an essential factor in preventing their drowning. The way in which this occurs stresses the importance of continuity between the conditions of incubation and those in the post-hatching period. Many species do not have functional preen glands at hatching, and the ducklings' down becomes oiled by the mother's plumage as she broods them.

WHAT IS NEW IN THE LIFE OF THE NEONATE?

The precocial embryo may well have had some experience of light and shade, may well have responded to auditory, tactile and proprioceptive stimulation and will have exercised its muscles for a considerable time, during the last few days of incubation.

Nevertheless it will be faced with new conditions and new problems from the time of hatching. Of these, movement in space, patterned vision and presumably also a wide range of tactile stimuli seem likely to be among the more important.

In precocial and semi-precocial species new co-ordinations appear after hatching; pecking (although it may have pre-hatching antecedents) at the red spot on the parent's bill is a clear example of this. Perception of distance appears to be good, but not perfect; for example Hailman finds that the laughing gull chick, although it will haltingly approach a specific stimulus such as food, may not always position itself at the optimum distance for striking in the first trials.

Response to depth, as in the 'visual cliff' experiment varies from one species to another, the difference appearing to depend on the nesting habits of the species. For example, Kear has demonstrated a difference between newly hatched ground-nesting and holenesting ducks in that ground-nesting ducklings avoid moving over the 'precipice' whereas the hole-nesting ducklings (which have to jump to the ground soon after hatching) do not.

Wehrlin & Tschanz have demonstrated that the ledge-reared guillemot chicks are able to perceive a precipice optically and avoid it, and the same response occurs in the cliff nesting kittiwake. However, comparing results from two surface- and one cliff nesting species Hailman suggests that chicks of all gull species are hatched with an ability to perceive and avoid a cliff-edge.

In such cases centres of co-ordination must have been already formed in the nervous system, although they could not have functioned before hatching. Colour, again, must be a new experience; it has been found, however, in certain species (such as ducks) that the neonate pecks more frequently and more vigorously at certain colours, such as green, while domestic chicks respond more readily to orange and blue.

Taylor *et al.* and Sluckin *et al.* have been able to demonstrate the existence of a tactile discrimination in the domestic chick. This, associated with warmth, must play a part in making brooding effective. Although tactile sensitivity occurs early there would seem to be a possibility of change in the type of stimulation received cutaneously after the chick has dried. Sluckin *et al.* and Taylor *et al.* found chicks to prefer a smooth to a rough surface.

Movement in the chick has been mentioned already in areas where it could rather obviously be derived from pre-hatching patterns of behaviour; walking, running, hopping on different types of substrate (or indeed on the ground *at all)* involves the development of new, albeit very rapidly developed skills.

An example of this is given by Kruijt in the prococial junglefowl a day or two after hatching. Here, again, the muscular co-ordinations will probably have been exercised before hatching, but the relationship with the ground, balance, and the stimuli which call forth this exquisitely polished performance must be new.

EFFECTS OF NEW ENVIRONMENTAL STIMULI ON NEURAL MECHANISMS

In dark-incubated domestic fowl Bateson & Wainwright have shown that a 30-min previous exposure to light has a dramatic effect on the rate at which a chick acquires a preference for a visual stimulus, in the imprinting situation. More recently, Bateson & Seaburne-May have found that the first approach to a flashing light occurs more rapidly the longer the previous exposure to the light.

These findings are consistent with those of Dimond and Adam & Dimond, showing a con tinuity across hatching. Bateson and Seaburne-May suggest that activation of the visual pathways simply by use (exposure to light) may mean that visual stimuli become more effective in evoking behaviour thereafter.

This effect is specific, as after previous exposure to sound chicks were found to be less responsive to the flashing light. (Possibly this specificity could explain Hunt's work on prenatal conditioning to sound; this was found to be effective postnatally only until the chick was given experience of light, food, water, etc.)

Thus, although there is obvious continuity across hatching, there is also change. There is evidence that the functioning of higher levels of the nervous system matures rather rapidly after hatching, and there is evidence also that this follows from the enormous increase in stimulation which occurs at this time. Growing evidence that exposure to stimulation

Figure 9.5 : Hopping in the chick of the Burmese red junglefowl.

can affect the connectivity of neurons in the central nervous system should help to elucidate the nature of the change which comes after hatching.

CONCLUSIONS

Considering the hatched chick in the light of previous chapters there is evidence of behavioural, as there is of physiological continuity. This aspect of development has been stressed by Kuo and Schnierla, who have both been interested in embryonic antecedents of chick behaviour.

Although there is obvious continuity some of Kuo's examples of it are now known to be unlikely (for example, that active movements can be Stimulated by heart beats in the first few days of incubation, and also that activity can be conditioned at an early age.

Even so, the idea that pecking in the chick arises on the basis of separate head, beak and neck movements all of which have been exercised *in ovo* still seems unexceptionable. In addition Kovach has pointed to continuity in righting movements across hatching and Corner *et al.* (1993a) have observed a continuation of embryonic stereotyped and co-ordinated movements in the younger sleeping chick.

Suggestions could be made of further continuities but the difficulty is to *demonstrate* their existence. New and refined techniques could be needed here. The problem of techniques applies not only to continuities across hatching.

Although qualitative accounts have been given, a detailed analysis of embryonic movements has long been thought desirable and it could be valuable to consider continuities (and also possible discontinuities) from the beginning of active movements until well after hatching.

This could show when particular behaviour patterns appear and disappear, it could take into account the direction and amplitude of movements, the elements included in the random, jerky motility and the question whether any of these are included in the stereotyped and co-ordinated movement patterns, as well as providing a basis for assessing new features in the days after emergence from the shell.

Together with the gradual development of sensory systems, of responsiveness and the effects of greatly increased sensory input after hatching, understanding of motor development is needed before we can assess the continuity between embryo and chick or say how much of chick behaviour is new.

10

HORMONAL CONTROL

If instead of reviewing the published endocrine studies *on* amphibian metamorphosis we ask questions concerning the endocrine control of the process from the viewpoint of developm-ental physiology, we become impressed more with the lacunae *in* our knowledge than with the strength of the separate threads.

In any event the general endocrine interrelations *in* this field have been reviewed in some detail Here we propose to examine metamorphosis from the viewpoint of the physiology of animal development rather than to inquire into the role of the separate endocrine glands. Before we can do that, however, we must summarize the main outlines of the *endocrine* relations among vertebrates generally in order to provide the framework within which we can ask our questions concerning control of metamorphic change by the hormones.

Endocrine Relationships

No detailed citations to the literature will be given for this discussion *of* endocrine interrelationships, but such references can be found in recent summaries of the field such as the *recent* volumes on *Neuroendocrinology* and *The Pituitary Gland* and in standard textbooks of endocrinology.

The thyroid hormones (TH) are iodine-containing derivatives of tyrosine. The principal hormone is thyroxine. The *thyroid* gland has an iodine-trapping mechanism whose efficiency *as* determined with radioiodine (1^{131}) is one measure of thyroid activity. The histological picture presented by the thyroid is also a clear indicator of its physiological state. Abundant cytoplasm and release of colloid from the follicles indicate active glands. However, where iodine uptake is prevented by goitrogens, a histological picture of activity is associated with the failure to release potent hormone.

The thyroid gland can function autonomously at a low level, but higher levels of activity are dependent upon activation by a hormone, thyrotropin, or thyroid-stimulating hormone (TSH), produced by the anterior lobe of the pituitary. The level of activity of the pituitary-thyroid (PT) axis is under control through two known mechanisms.

One of these is by way of feedback of thyroid hormones that inhibit the activity of the TSH cells, thus stabilizing the PT axis at a particular level. There is good evidence that this negative feedback of TH upon TSH production operates primarily at the pituitary level, directly inhibiting the TSH cells.

The.second mode of regulating TSH activity is by way of a thyrotropin releasing factor (TRF) produced in the hypothalamus. TRF is presumed to be a neurosecretory material reaching the anterior pituitary by way of the pituitary portal veins that drain the median eminence of the hypothalamus.

At present the chemical nature of TRF has not been clarified nor have the neural cells from which it arises been identified. It is believed, however, to differ from the peptides that constitute the classical neurosecretory substances produced in hypothalamic nuclei (preoptic nucleus in amphibians) and stored in the neural lobe of the pituitary.

In stimulating TSH production, TRF may be thought to act by partially desensitizing the pituitary to the negative feedback action of TH and thereby permitting the level of the TP axis, the so-called thyrostat, to rise.

A word of clarification is needed with regard to the concept of neurosecretion. This has undergone marked development in recent years Originally neurosecretion was identified by characteristic, although not specific, staining reactions in light microscopy, i.e., neuronal granules positive to aldehyde fuchsin or chrome alum haematoxylin.

Physiological, histochemical, and electronmicr-oscopical evidence has compelled the recognition that such material is related primarily to the pars nervosa although also occurring in the median eminence (Oota and. This latter region, however, has been shown, to be rich in monoamines and in peptides differing from those of the nervoasa. In the present discussion the neurosecretory system will be considered to include in addition to the classical neurosecretory pathway to the posterior lobe of the pituitary:

1. Neurons that produce other specific chemicals released into the bloodstream in the median eminence.
2. The median eminence with its primary capillary bed intimately related to neurosecretory fibers.
3. The portal veins that drain blood from the primary capillary bed of the median eminence into the secondary capillary bed of the pars anterior of the pituitary

These endocrine and neuroendocrine interrelations are applicable to amphibian metamorphosis. As early as 1912, Gudernatch repo-rted that the feeding of thyroid-gland preparations to young tadpoles precipitated metamorphic changes. It was soon demonstrated—particularly by M. Allen, Hoskins, and Hoskins and P. E. Smith—that the pituitary gland of the tadpole controls thyroid activity and that the latter is the immediate agent of metamorphic change.

The neuroendocrine relations governing metamorphosis were worked out more recently. The evidence is convincing here, too, that the hypothalamus influences pituitary TSH activity by way of a TRF passed into the capillaries of the median eminence.

Another pituitary hormone that recently assumed a role in our understanding of metamorphosis is prolactin, which like TSH is a protein product of the anterior lobe of the pituitary. It is known to be active in stimulating body growth in several vertebrates and has recently been found to have antithyroid activity in tadpoles. In mammals and probably in amphibians prolactin appears to be under inhibitory rather than stimulatory control by the hypothalamus.

The Metamosis in a Typical Anuran

However, the endocrine analysis with which we are concerned here has been carried out largely on a few species of anurans. To follow this analysis t will be most useful to have a more exact and quantitative delineation of the developmental pattern relative to metamorphosis in a typical Rapid, Rana *pipiens*.

The principal points to be noted here are as follows. The first post-embryonic period characterized by much growth with very little change in form. This is here designated the growth or premetamorphic period and lasts about seven weeks at conventional room temperatures (22 to 25°C). This is followed by the period designated prometamorphosis and characterized by differential growth of the hind legs.

During this period of about three weeks, the legs grow from about 2 mm to about 20 mm, whereas the body growth increases only from about 55 mm total length (body length, 18 mm; tail length, 37 mm) to 65 mm body length, 21 mm; tail length, 44 mm). It is easy to follow the progress of the animal through prometamorphosis by nothing the ratio of hind-leg length to body length (HL/BL).

The animal has definitely entered prometamorphosis when the HLJBL ratio exceeds 0.2 and it reaches the end of that period when this ratio approaches 1.0. During the latter part of prometamorphosis certain minor morphological changes may be observed in the intact animal. For example, the resorption of the anal canal piece of ACP (the basal lobe of the ventral fin that carries the anal opening out into the left side of the tail) occurs when the HL/BL ratio is about 0.6.

In a day or two thereafter at a ratio of about 0.8 the first signs of the degeneration of the opercular skin over the gill chambers (the skin window for the forelegs, SWFL) can be seen on the right side. The corresponding change on the left is the reduction of the tubular wall of the spiracular opening from the gill chamber.

In one or two days thereafter the forelegs emerge, usually within a few hours of each other, and the changes from the narrow tadpole mouth to the wide frog mouth begin with the shedding of horny teeth and beaks. The emergence of the forelegs and loss of tadpole mouthparts constitute the last phase of prometamorphosis and the beginning of metamorphic climax.

As noted in described in terms of the numbered stages of Taylor and Kollors (1946), Nieuwkoop and Faber (1956), and Gosner (1960). However, for the experimental analysis it is important to recognize the exact quantitative and temporal relations of metamorphic changes as depicted.

By use of the information in this figure it is possible to derive the concept of slowly accelerating metamorphic activity during prometamorphosis with climax recognizable as a burst of morphological change. In the design of experiments it is possible by use of these tables to select animals at particular stages and to predict the course of events to be expected in normal animals.

It should be noted that the widely used stages of Taylor and Kollros describe the entire range of tadpole development and are not directed at the analysis of metamorphosis per se. They do not recognize the differential acceleration of hind-leg growth as the initiation of metamorphosis but regard this process as beginning in what we designate

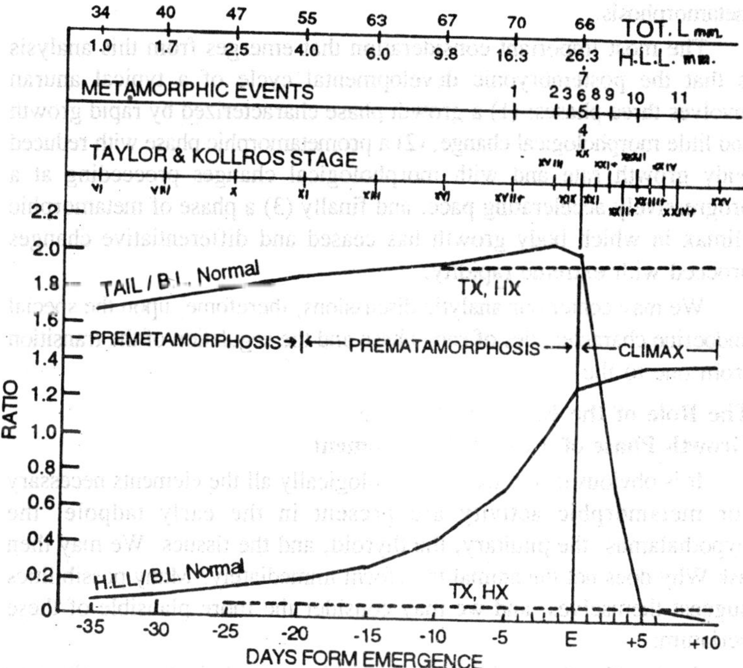

Figure 10.1 : Pattern of metamorphosis in Rana pipiens. Data from one batch of normal animals raised at 23°C = 1 shown by solid line. Comparable data for thyroidectomized animals (TX) and hypophysectomized animals (HX) are shown by broken lines. The metamorphic events indicated in the figure by number are as follows: 1. Anal canal piece-first definite reduction. 2. Anal canal piece reduction completed. 3. Skin window for the forelegs clearly apparent. 4. Loss of 2d (both) beaks. In experimental animals this is the most satisfactory criterion for the beginning in climax. 5. Emergence of first foreleg to appear. 6. Emergence of second foreleg (E). 7. Mouth widened to level of nostril. Events 4, 5, 6 and 7 usually occur within a period of 24 hours. 8. Mouth widened to level between nostril and eye. 9. Mouth widened to level of anterior edge of eye. 10. Mouth widened past level of middle of eye. 11. Tympanum definitely recognizable. By use of the data of this table it is possible to classify individual animals in terms of days before or after the beginning of climax (E).

here as late prometamorphosis. They also do not differentiate a climax phase. Descriptively, the procedure of Taylor and Kollros is highly useful, but it obscures the experimental analysis. As will be seen later, early leg growth (stage XII to XVIII) is dependent upon activation of the myroid gland and is thus considered here to be part of metamorphosis.

Furthermore, the endocrine balance undergoes a profound change at the beginning of climax so that the precise delineation of this phase is essential for an adequate understanding of the hormonal control of metamorphosis.

The most important consideration that emerges from this analysis is that the postembryonic developmental cycle of a typical anuran involves three phases: (1) a growth phase characterized by rapid growth and little morphological change, (2) a prometamorphic phase with reduced body growth rate and with morphological changes proceeding at a progressively accelerating pace, and finally (3) a phase of metamorphic climax in which body growth has ceased and differentiative changes proceed with extreme rapidity.

We may center our analytic discussions, therefome, upon the special endocrine characteristics of each phase and the regulation of the transition from one to the next.

The Role of the Endocrines in the Growth Phase of Tadpole Development

It is obvious that at least morphologically all the elements necessary for metamorphic activity are present in the early tadpole, the hypothalamus, the pituitary, the thyroid, and the tissues. We may then ask Why does not the animal transform immediately? Many possibilities suggest themselves, and we may consider the more plausible of these seriatum:

1. Are the tissues initially insensitive or relatively insensitive to thyroid hormone? Such a change in sensitivity might determine the time of initiation of metamorphic activity. The evidence, however, seems to indicate clearly that this is not the case.

It is indeed true that the tissues of the embryo seem to be insensitive to thyroxin as seen in the often repeated observation that embryos develop normally in thyroxin solution and do not show metamorphic change (e.g., leg growth or tail resorption) until reaching the definitive tadpole stage (stage I)-for earlier literature see Etkin, 1955.

On the other hand, it is also clear, as in Gudernatch's original experiments, that the postembryonic tadpole's tissue is capable of giving metamorphic responses to thyroid hormone. In a brief study reported only in abstract form, Etkin attempted to determine more precisely just when the tissues acquire thyroxin sensitivity.

He did this by exposing embryos of R. pipiens to immersion in strong thyroxin solutions (or implantation of thyroxine crystals), beginning at successively later stages. He observed that embryos exposed before embryonic stage, S-23 did not develop metam-orphic changes any earlier than did those whose exposure began at stage S-23. On the other hand, those exposed at later stages showed the first evidence of response at successively later periods. He concluded, therefore, that the tissues first became sensitive at about stage S-23 and are not influenced by exposure at earlier stages.

There is a latent period of about three days in *the* appearance of metamorphic change, even when late tadpoles are exposed to strong thyroxine-1 part per million (ppm). The first changes do not appear in the above experiment until the animals are in stage S-25. Recent experiments by Prahlad and De Lanney (1965) with embryos of the axolotl likewise indicate early acquisition of thyroxine sensitivity.

Furthermore, the pattern of change in the embryos of R. pipiens exposed at stage S-23 is the same as that in animals exposed, later. This suggests that the general tissues, as observed. externally; acquire sensitivity at the same time and approximately to the final degree.

There have been many attempts to show that sensitivity of tissues to thyroxin increases during tadpole development. However, the present author does not regard the evidence available on this point as satisfactory for two reasons: (1) When, as in the experi-ments that claim to demonstrate changes in sensitivity, normal tadpoles of different ages are compared in their reactivity, the older animals start from a more advanced baseline since their tissues have been exposed for a longer time to conditioning by low levels of thyroid hormone present in the tadpole. (2) The differences in sensitivity reported are small, less than a factor of two times and therefore as compared to the changes in thyroid level can play only a minor role in determining normal metamorphic change.

In some cases differences in time of first appearance of metamorphic change in animals treated as embryos have been interpreted as indicating differences in time of acquisition of sensitivity. However, such differences may represent merely differences in latent period of response. In summary, then, except for certain nervous structures, the tadpole's tissues appear to be ready to respond to the metamorphic stimulus throughout the growth phase of development. Changes in sensitivity play a minor, if any, role in determining metamorphic pattern.

2. A second possibility to be considered is that the thyroid gland may not be capable of forming the necessary hormone during the growth phase. However, the earliest studies have shown that the thyroid gland differentiates normally in the absence of the absence of the pituitary.

In the normal animal, moreover, the gland forms detectable levels of T_4 in the early tadpole stage (16 mm total length in R. pipiens), as shown by Flickinger (1964). The tadpole's thyroid also responds to pituitary TSH at all stage of tadpole development and even in the embryo.

Therefore, there is no reason to suspect that the thyroid of the tadpole is incapable of responding appropriately to a metamorphic stimulus during any part of the growth phase. We must look higher than the thyroid in the hypothalamuspituitary-thyroid axis for the explanation of the failure of the tadpole to metamorphose during this period.

If we ask next whether the pituitary is capable of effective levels of TSH production during this period, we must again answer that it is. The most direct and satisfactory evidence for this would be the demonstration of the responsiveness of the tadpole pituitary to a hypothalamic TRF factor.

However, as is discussed below, evidence from this source is not yet available. But other evidence does indicate clearly that the tadpole's pituitary is capable of producing an effective TSH during tadpole life. One such bit of evidence is the demonstration that when the pituitary (adenohyp-ophyseal) primordium an thyroid primordium are placed close together in the embryo, the thyroid is precociously activated.

This shows that even at early stages the pituitary is capable of producing an effective TSH. The evidence derived from the effects of goitr-ogens in tadpoles further demonstrates that this capacity for producing TSH continues through tadpole life. The administration of such goitrogens leads to hypertrophy of the thyroid.

This action, like the similar action in mammals, works by preventing the iodination of the molecule in the gland and, therefore; renders its hormone production ineffective. The lack of thyroid hormone in the goitrogen-treated tadpole removes the feedback inhibition normally exerted by thyroid hormone upon TSH production.

The excess TSH produced by the pituitary of the goitrogen-treated tadpole. stimulates the thyroid to increased growth and to the abundant secretion of its physiologically ineffective product. We have recently found this goitrogen effect to obtain even in the early tadpole (unpublished results). Thus, we are led to regard the pituitary of the early tadpole as capable of producing the TSH necessary for the induction of metamorphosis but restrained from doing so because of the inhibition of its TSH-secreting cells by negative feedback from thyroxine produced by the thyroid.

The premetamorphic period may be regarded as one in which the pituitary-thyroid axis is maintained in a steady state at an extremely low level of activity because of the great sensitivity of the TSH cells of the pituitary to negative thyroid feedback. The level of this steady state is so low that the amount of thyroid hormone produced is not sufficient to make the normal animal deviate perceptibly from the thyroid-ectomized one.

As the growth curve of the animal suggests, and as has been recognized and emphasized by Gudernatch and other early workers in the field, there is a reciprocal relation between the rate of overall body growth and the rate of metamorphic change. Therefore, we may inquire as to the significance of the endocrine factors in maintaining the high growth rate of the premetamorphic period.

It was recognized early that thyroid treatment inhibits growth even as it promotes metamorphosis. Steinmetz studied this phenomenon quantitatively by measuring the growth rate of tadpoles immersed in various concentrations of thyroxine and treated with the goitrogen, propothiouracil (PTU).

He showed that distinct inhibition of the growth rate occurs at the lowest concentration of thyroxine measurably affecting leg growth and that the goitrogen stimulates the tadpole's growth. From this he' inferred that the level of thyroid hormone in the remetamorphic animal is maintained at an extremely low level, less than that equivalent to one part thyroxine per billion parts of water.

Our own unpublished work with thyroxine concentration applied to normal tadpoles fully confirms Steinmetz's results. Thus, it is clear that the low steady state of the PT axis is an essential factor in maintaining the characteristic high growth rate of the premetamorphic period.

In recent years it has come to be realized that prolactin (or a prolactinlike molecule) plays a significant role in amphibian physiology. Berman et al. (1964) and Nicoll *et* al. (1965) reported a stimulation of growth in normal tadpoles by prolactin, and Etkin (1964) and Etkin and Gona (1967) reported such stimulation in hypophy-sectomized animals.

These studies indicate an antagonism between thyroid hormone and prolactin. The antithyroid activity of prolactin has been found at low doses to act peripherally and to act as a goitrogen at high dosage level. In any case it appears that the retardation of growth to be seen in the hypophysectomized tadpole is to be ascribed at least in part to the absence of the prolactinlike factor.

Further support of this concept of derived from the report of Etkin and Lehrer (1960). They found that the transplanted pituitary reduces growth in hypophysectomized hosts in excess of that of the normal. This

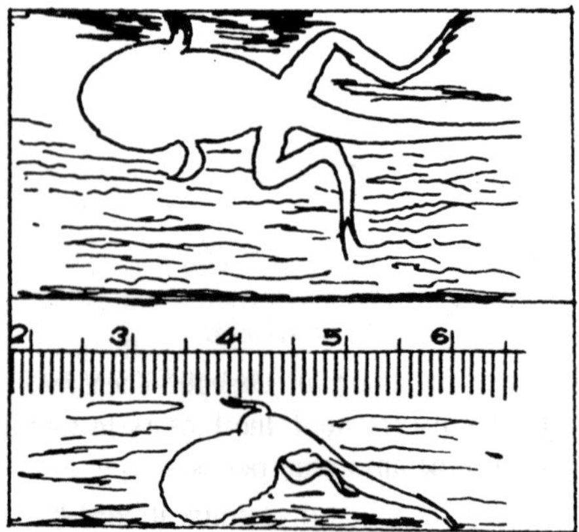

Figure 10.2 : Normal animal (above) compared to experimental animal raised in 3 to 9 ppb thyroxin.

suggests that the production of a prolactinlike hormone in the amphibian, like its production in the mammal, is under inhibitory control by the hypothalamus.

The concept of the neuroendocrine relationships responsible for the high growth rate and the low metamorphic tendency in the premetamorphic tadpole that emerges from this evidence may be summarized as follows.

The hypothalamus is only minimally active in producing a TRF or a prolactin-inhibiting factor. By this activity it only partially retards prolactin production and does not appreciably stimulate T5H production.

Consequently, the negative feedback of thyroxine upon the activity of the TSH cells prevails and keeps the PT axis at a very low level. The high level of prolactin and low level of thyroid hormone favour a high growth rate and no appreciable metamorphic change during this period.

The Regulation of the Pattern of Tissue Response

The concept of the low-level steady state of the metamorphic mechanism during the growth phase of the tadpole's life raises most directly the question of how this system is disrupted to permit metamorphosis to take place.

Betore we can examine this question, however, it is necessary to arrive at some understanding of endocrine activity during metamorphosis so that we can appreciate what the nature of the transition from one condition to the other might be. The clue to this was given by the analysis of the pattern of thyroid activity in relation to the pattern of

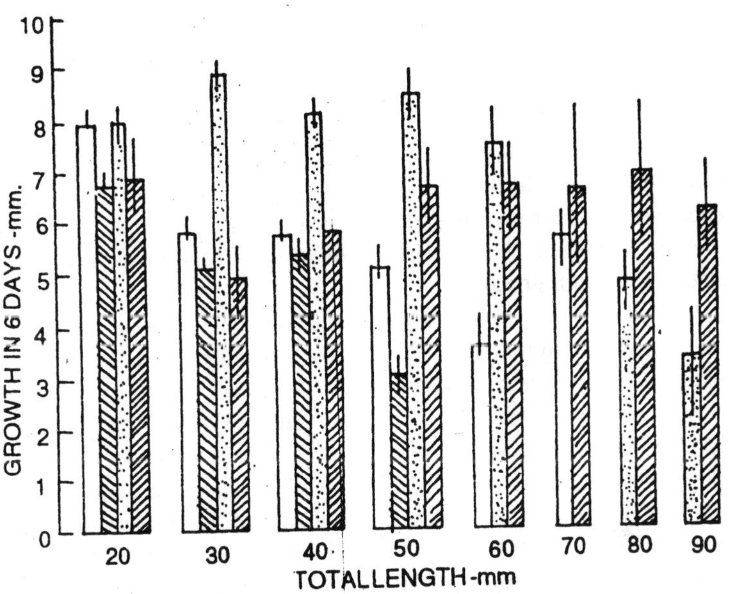

Figure 10.3 : Rate of growth of tadpoles of Rana pipiens. All animals of ,same batch. Individual growth curves kept for each animal and amount of growth in a six-day period as derived from these growth curves is plotted at different stages of development. Controls entered metamorphic climax at 65 to 70 mm. Hypophysectomized animals generally failed to grow above 6C mm.

metamorphic change. It will be recalled from a preceding section that the prometam-orphic period is characterized by rapid growth of the hind legs and other less conspicuous changes that are spread over about three weeks in R. pipiens.

This is followed by about a week of climax, during which profound changes in morphology occur. The relation of developmental and histological changes in the thyroid gland to. these events was early subjected to extensive studies.

The results in different anuran species indicate that during the period of, prometamorphosis the relative growth rate of the thyroid is high and the gland reaches a maximum size at the end of this period.

The histological picture of moderate increase in cell height accompanied by enlargement of follicles indicates cell activity and storage of hormone.

In early climax, however, the epithelium reaches a maximum height and the follicles show a reduction of colloid which, particularly in smaller species,brings on partial collapse of the follicles. This picture was early interpreted as indicating an increasing release of thyroid hormone during prometamorphosis and a very high level of activity at climax.

At the end of climax the histological picture shows a return to a flat epithelium and to follicles distended with colloid, thus indicating a deactivation of the thyroid. Modern studies have attempted to apply recently developed techniques to the analysis of thyroid activity in relation to metamo-rphosis.

These have included chemical studies with and without radioiodine electrical conductivity measurements and enzyme studies. In a general qualitative way these studies have supported the morphological investigations mentioned above since most authors interpret their findings in terms of increasing activity with the approach of metamorphic climax.

However, the quantitative refinement that one could wish for from such studies was not forthcoming. For example, Saxen et al. found no clear increase in protein-bound iodine in prometamorphosis and only a twofold increase at climax. Whereas these authors and Kaye reported a large increase in I^{131} uptake, this uptake peaked in mid-or late prometamorphosis rather than at climax.

Similarly, the conductivity measurements of Gorbman and Ueda indicated the greatest change in midprometamorphosis and a return to the premetam-orphosis level at climax. Flickinger was able to detect thyroxine in peripheral tissues only at climax. Yamamoto followed t_4 deiodinase activity through preisely defined stages of metamorphosis and reported a great increase during prometamorphosis peaking at climax, falling abruptly thereafter.

Although the physiological significance of this deiodination is not dear, the parallelism to metamorphic activity is suggestive of changes of hormone level. Unfortunately, the results of Dowling and Razevska are not in agreement with the findings of Yamamoto's study. The reason for this difference remains unexplained. Needless to say, each of these techniques has its complexities of interpretation, the analysis of which would take us far afield at this point.

Therefore, without going into detail, the present author may summarize his interpre-tation that these studies reinforce the concept that prometam-orphosis is marked by an increase in thyroid activity, much of which goes into storage of hormone in follicles. However, they fail to clarify in any quantitative way the changes of level of effective hormone. Particularly, they fail to define the effective hormone level responsible for climax changes. It is to be hoped that this admittedly personal evaluation of the results of these methods constitutes a challenge to a critical application of biochemical and biophysical methods to.this problem.

There is as much to be gained in !heislarification of the methods as well as in the understanding of thyroid action in metamorphosis by such a study. A most promising application of modem biochemical methodology to this problem is the attempt of measure T_4 and T_3 levels in the blood directly by chromatographic separation.

Preliminary results by this method support the concept of increasing concentration of thyroid hormone during prometamorphosis to a maximal at climax with a decrease thereafter. But a more precise qualification of hormone level is not yet available. An indirect method for evaluating thyroid function quantitatively is that of studying metamorphic change in young tadpoles,. prekrably thyroidectomized animals, during immerskn in thyroxine solutions of various concentrations.

An early study demonstrated that low concentrations (3 to 10 parts thyroxine per billion parts of water) yielded good leg growth, but later metamorphic change was protracted at this hormone level. Higher concentrations (100 to 1,000 ppb) on the other hand yielded climax events at near normal rates but did not permit sufficient time for leg growth to attain normal dimensions before tail resorption was induced.

The general inference drawn from these studies was that a normal pattern of metamorphic change could be induced only by an extended period of treatment with a low concentration of T_4 followed by immersion in high concentration. This inference has been repeatedly confirmed in many subsequent studies in this laboratory.

More recently we have attempted to give this concept more precise quantification by studying the relation of level of immersion to rate of a quantifiable event of prometamorphosis (leg growth) and of climax (tail resorption).

The-graph shows that the maximal rate of leg growth comparable to that in normal animals may be attained in concentrations of 10 to 20 ppb, whereas maximal rates of tail resorption require well over 200 ppb for this induction. The general conclusion that may be drawn from these studies is that normal prometamorphosis may result from a thyroid activity level equivalent to 3 to 20 ppb, whereas as climax is approached the level of activity of the gland must increase by a factor of 10 or more. This increase must occur within a day or two of the beginning of climax (emergence of the forelegs).

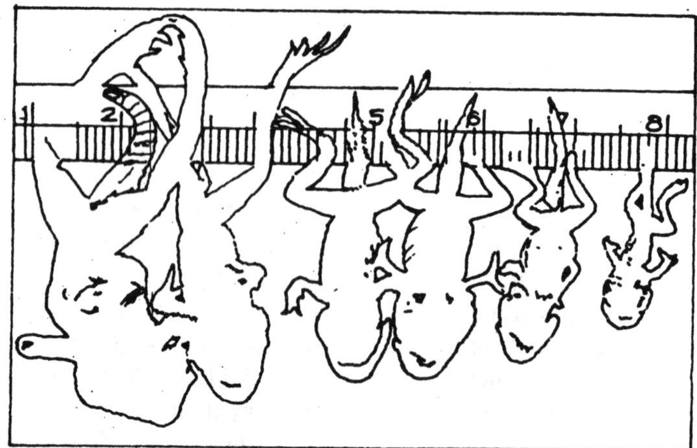

Figure 10.4 : Tadpoles of R. pipiens. Second from right is an untreated control fixed three days after emergence of forelegs when tail resorption is half completed. The maximal size attained by this animal was 68 mm. The animal at right was a thyroidectomized animal induced to metamorphose by a patterned application of increasing concentrations of thyroxine after reaching 90 mm. To the left are a series of normal animals similarly treated beginning at various early stages and fixed at middimax. Note the normal morphology induced by the pattern of increasing concentration of hormone in contrast to the abnormalities induced by single concentrations.

It should be noted parenthetically that Frieden advocates the application of thyroxine by injection rather than immersion because he finds a low level of absorption of the hormone from the medium. We cannot agree with this view-point for two reasons:

(1) the equilibrium level of absorption is irrelevant as long as it varies with the external level and

(2) after injection the internal level does not remain constant since loss by excretion and degradation are rapid, as shown by his own studies.

Of course, immersion studies do not tell us directly what the equilibrium level of hormone is within the animal's body, but Kollros (1963) has shown that the hormone's concentration is maintained at a highly constant level by the immersion method.

This type of evidence then leads to the concept that the level of thyroid hormone action rises progressively during prometam-orphosis and undergoes an explosive increase by a factor of at least 10 times at climax.

The most immediate question raised by this concept is that of the significance of this changing thyroid hormone level for the induction of

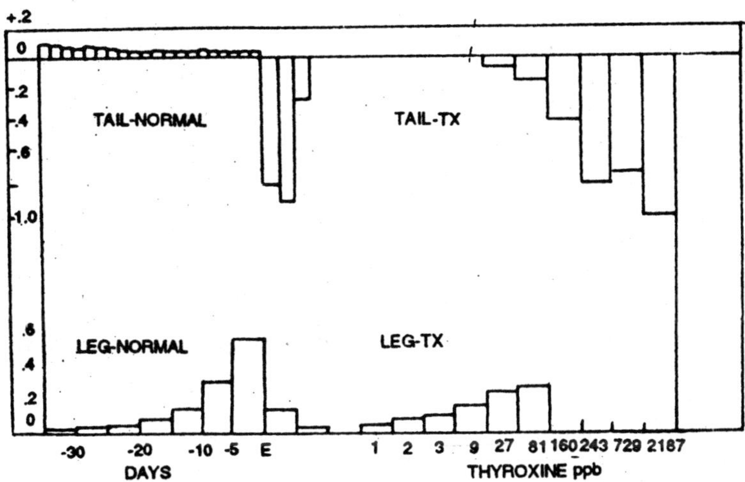

Figure 10.5 : Results of an experiment (right) to test the effect of different concentrations of thyroxine (Tx) on hind-leg growth and tail resorption in half-grown normal tadpoles. For comparison the rates of change of leg and tail in normal metamorphosis are shown at the left. The rates of change are expressed as proportions of body (snout to anus) length make the data of smaller experimental animals comparable controls.

the normal sequence and spacing of metamorphic events. We continued here with the endocrine analysis by asking how this pattern of thyroid activity is regulated.

The Control of Level in the Pituitary-Thyroid Anix

Since thyroid activity above the minimal level as found in the hypophysectomized tadpole is dependent upon TSH production, it could be assumed that the activation of the thyroid in metamorphosis is induced by an increase of TSH release from the pituitary.

Exploration of the morphology of the pituitary therefore might be expected to yield as clear a -picture of activation as did the studies of the thyroid discussed above. There have been repeated studies of this problem, but although a general qualitative agreement with this concept has been reported, the clarification of the quantitative interrelations is not impressive.

Recent studies by Kerr (1966) in *Xenopus* and by Kiremidjian and Ortman (1966) in *R. pipiens* have indicated an increase in the supposed TSH cells (identified as Aldehyde fuchsin positive basophils). Yet the amount of the increase, even admitting that the level of activity of a cell is not readily inferred from its staining reaction, does not seem entirely consistent with the inference from the thyroid pattern. Kerr emphasizes that there is no evidence of hormonal discharge in the pituitary at the time of metamorphic climax. The identification of the TSH cell as a basophil is itself uncertain. Several authors have presented evidence pointing to an acidophil as the source of THS.

The study of this problem at the electron microscope level by Dent and Gupta (1967) again indicates some increase in the supposed TSH cells but not as great a change at climax as was to be anticipated. The recent finding of the antithyroid activity of prolactin suggests that instead of a simple increase in TSH there may be a shift in proportion of TSH to prolactin production that accounts for the activation of the thyroid gland.

Indeed, the criteria for identifying the TSH cell (response to goitrogens, thyroxine, etc.) will have to be reconsidered in the light of a possible role for prolactin in controlling thyroid activity. In any case, it is apparent that further quantitative work will be required before we can specify the. nature of the change in the 'pituitary activity that is responsible for the patterning of thyroid activity· during metamor-phosis.

But whatever may be the precise nature of the change in the pituitary, it is clear that the signal initiating and controlling the pattern of metamorphosis must come from the pituitary. Therefore, we may ask whether the pituitary is autonomous, i.e., whether it determines its own rate of activity by some sort of built-in' clock mechanism or whether it is subject to some external control.

The remarkable fact that throughout the vertebrates, however diverse their brain-pituitary morphology may be, the connection between the two structures is always maintained suggests that the pituitary is under control by the CNS. Furthermore, the clarification of the morphological relationship by the demonstration of the neurosecretory system and the pituitary portal system of blood vessels strongly reinforces this viewpoint.

Before the nature of the neurosecretory system was understood, this problem had been attacked in relation to anuran metamorphosis by isolating the gland from the brain. In 1938, Etkin showed that if a tadpole's own pituitary were transplanted in the embryo the animal would usually grow very well but would show at best a retarded and protracted metamorphosis.

Nonetheless, the capacity of at least the most successful cases for inducing prometamorphosis was taken to indicate the essential independence of the pituitary from its connection to the brain, a connection then thought of in terms of conventional innervation. In 1940, Uyematsu reported the converse experiment in a toad, removing the region of the embryonic neural plate that gives rise to the infundibular lobes of the hypothalamus and thereby isolating the epithelial pituitary. He also reported his animals to show prometamorphosis, but he found them unable to complete climax changes.

This failure he ascribed to a degeneration of the isolated gland at climax. But despite this evidence, which speaks for the concept of an autonomous clock mechanism in the pituitary as the controlling factor in metamor-phosis, when the nature of the neurosecretory and the pituitary portal system and hypothalamic influence on TSH production was elucidated in mammals, the need to reexamine this question was recognized.

We were able to confirm out earlier work with embry-onic grafts and extended it to grafts of differentiated glands placed into hypophysectomized hosts. In these studies the animals were maintained into climax, and it was found that the hosts to successful grafts, though often completing prometamorphic development, invariably entered a metamorphic stasis at climax.

Clearly, then, graft activity was 'capable of activating the thyroid to the low levels responsible for prometamorphic change but not to the high level necessary for climax events. It was apparent that the failure at climax was not due to a degeneration of the gland at that time since the gland was found to be maintained morphologically and also was seen to continue the production of its growth and pigmentation-regulating factors in spite of the metamorphic stasis.

We were also able to confirm Uyematsu's findings of similar cap-acities in pituitaries isolated by removing part of the embryonic hypothalamic region. There have been many reports in the last 10 years confirming the necessity of the presence of the hypothalamus for the completion of metamorphosis.

However, Guardabassi reported completion of metamorphosis after removal of the hypothalamus. It should be understood that failure of metamo-rphosis in short-term experiments or in experiments in which the animals fail to grow vigorously does not necessarily imply hypothalamic control of TSH function since injury or starvation by themselves can inhibit metamorphosis.

However, in our experiments and those of Remy and Bounhiol there can be no doubt of the vigour of the animals since many achieved gigantic size without completing metamorphosis. At least in these experiments, then, it is clear that isolation of the pituitary from the hypothalamus specifically inhibits it from attaining a high level of TSH activity. It is possible that the discordant results reported by Guardabassi are the consequence of inmmpl removals accompanied by regenerative repair, which is possible in the tadpole's nervous system.

In the opinion of this author it appears to be clearly established that the high level of thyroid activity necessary for metamorphic climax can be maintained by a pituitary gland only if it has its normal connection with the brain. In urodeles also it' is clear that the metamorphic process that corresponds to climax in anurans depends upon normal hypothalamic relations with the pituitary.

A barrier inserted between these two structures (see frontispiece D) effectively prevents metamorphosis, provided it completely separates them. A low prometamorphic level, however, can be maintained by the isolated pituitary in anurans.

On the other hand, some evidence indicates a considerable measure of independence of the pituitary TSH production from hypothalamic control in amphibians. Dent (1966) reported that molting in the newt, in which this process is known to be dependent upon the thyroid, is maintained by pituitaries isolated from the brain. Iodine metabolism, though reduced in-toads with pituitary grafts, is higher than in hypophysectomized animals.

We regard such evidence as consistent with that derived from metamorphic studies where the isolated pituitary is found to sustain prometamorphic change. It is possible that the role of thyroid in the molting of these amphibians is merely permissive. Though necessary at a minimal level, it does not determine the rhythm of molting by its variation.

One is tempted to infer from the evidence given above that the pituitary TSH system is autonomous for low levels of activity and requires hypothalamic cooperation only for the highly activated state.

However, if we think of the hypothalamicpituitary interrelationship in terms of neurosecretion rather than conventional innervation it is apparent that inference is not justified. Presumably, the significance of the pituitary portal venous system lies in the greater efficiency it provides for the transfer of chemicals produced in hypothalamic neurons to the cells of the anterior pituitary.

Thus, it appears likely that though this system of direct transfer of neurosecretion may be essential for maximal stimulation of the pituitary,. a lower level of TSH activity may be maintained by the transfer of neurosecretion through the systemic circulation. Such a low level of stimulation could reach a graft wherever, it was located. The evidence on brain removal mentioned above either does not include the areas presumed to contain the relevant pericarya or is too fragmentary and incompletely analyzed (Voitkevich and Guardabassi) to permit judgment. Further studies should clarify this critical point.

At present it appears most logical to accept the possible transfer of hypothalamic substances in diluted form through the systemic circulation as the explanation of the low and moderate levels of TSH'function reported in isolated pituitaries.

Of course, a most direct approach to this question could be derived from experiments with extracts of the hypothalamus that have thyrotropin-releasing properties (TRF). Such TRF prepar-ations have been developed for mammals in several laboratories.

However, Bowers and Schally, working with their TRF preparatidn, failed to find evidence of TSH release in the tadpole (personal communication). In our laboratory, work with a TRF prepared by Dr. Roger Guillemin that was potent in tests on rodents likewise failed to elicit activation of metamorphosis, even when large amounts were injected directly into the heart.

Presumably the chemistry of the amphibian TRF must differ from that in mammals, but no adequate studies of an amphibian TRF are yet available. The possibility should also kept in mind that a shift in the prolactin-TSH balance rather than a simple release of TSH may be necessary for metamorphic activity.

Our consideration of the hypothalamic-pituitary-thyroidtissue axis in relation to metamorphosis thus leads us to the concept that at the beginning of prometamorphosis the hypothalamus initiates the changes, in some way stimulating the pituitary-thyroid axis.

This activation process builds up during prometamorphosis to a subtotal activation of the axis by the beginning of climax. At the end of metamorphosis the hypothalamus apparently stops its stimulation of the PT axis since the latter becomes inactive, although we have no direct evidence on this point.

Such a concept raises the question of the. control. of the hypothalaints.Does it, rather than the pittritary, contain the "clock" mechanism that determines the timing and pattern of metamorphosis?

The Activation of the Metamorphic Mechanism

Sine the evidence indicates that the control of the metamorphic process depends upon the hypothalamic neurosecretory apparatus, it was desirable to investigate the development of this mechanism. Wilson et al. (1957) reported on the morphology of this system in the western tree frog *(Hyla regilla)* and showed that the classical neurosecretory material of the posterior lobe (see above) appears early in development. However, they did not follow the differen-tiation of the portal vessels and medium eminence in detail.

When this was studied in *R. pipiens.* in correlation with metamo-rphic change, it was found that in the premetamorphic tadpole these parts of the neurosecretory apparatus are poorly differentiated. The anterior lobe of the pituitary is broadly adherent to the thin floor of the hypothalamic lobes with a diffuse network of capillaries lying between the two.

The capillaries of this network derive their blood from hypothalamic arteries and feed into the capillary system of the anterior pituitary. During prometamorphosis this system differe-ntiates in the neural floor, becoming shorter and thicker as the capillaries sink into the neural substance. The anterior lobe separates from the neural tissue except at the anterior tip, where it remains attached by the few large venous channels that drain the primary capillary bed.

These constitute the pituitary portal veins draining the now-differentiated medium eminence as found in the adult; when the animal enters climax, this medium eminence system is well advanced although not completely differentiated.

It is evident that this morphological development is entirely consistent with the concept that some hypothalamic substance is fed from the neurosecretory fibers in the hypothalamic stalk into the capillary bed of the anterior lobe. In the premetamorphic animal this system, being only diffusely organized, presumably carries a minimum of hypothalamic substance.

As the system matures during prometamorphosis and the medium eminence differentiates, the amount of material transported increases to reach a level sufficient to achieve the high-level activation of the PT axis at metamorphic climax. If this hypothesis is valid, then the question of hypothalamic control could be studied by examining the differentiation of the median eminence region in the thyroidectomized larva.

If the hypothalamus is autonomous in its differentiated, the median eminence should mature in spite of the failure of the thyroidless larva to metamorphose. The examination of this question showed that this was not so. In the thyroidectomized larva the median eminence region retains its larval character (see frontispiece B).

However, if such a larva is artificially induced to metamorphose in a normal pattern by the application of graded concentrations of thyroxine, the median eminence is seen to differentiate as in the normally transforming animal. Thus, it is apparent that the median eminence, part of the hypothalamic mechanism that is presumed to control the PT axis, is itself controlled by thyroid hormone. Evidently, then, there is a circular mechanism involving a positive feedback of thyroid hormone to the hypothalamus.

This suggests that the differentiation of the hypothalamic mechanism is gradually brought about by the action of thyroid hormone. This feedback process starts in the early larva at an extremely slow rate because of the low level of thyroid activity in the premetamorphic period.

Since it contains a positive feedback element, however, this circular system accelerates slowly at first then more rapidly and finally ends in a burst of activity as any circular system with positive feedback must. This intriguing possibility could be tested in a number of *ways*. One such test is to expose early larvae to low concentrations of thyroxine (1 to 9 ppb) for sometime.

Such exogenous thyroxine in this hypothesis, should, accelerate the maturation of the endogenous hypothalamic mechanism responsible for activating the animal's own PT axis. Animals thus treated should metamorphose earlier than controls and should show a normal pattern of climax events. Repeated attempts in which different concentrations of exogenous hormone were applied in various combinations however, failed, to produce an acceleration of metamorphosis.

Of course, the thyroxin-treated animals were smaller at metamorphosis than controls because the hormone inhibited their growth rate, as explained above. This result raised the possibility that the hypothalamic system, as certain other neural elements, does not acquire its sensitivity to thyroxine until late in the premetamorphic period. An experimental analysis of this point showed that it was indeed so.

Animals *(R. pipiens)* treated with graded concentrations of thyroxine from early larval stages and thus brought into mid-climax much earlier than normal failed to show any metamorphological advance of the median eminence region over the larval condition. On the other ha , animals that were so treated as to be only slightly precoci us did show such development.

Thus, it is clear that until a short time before the normal animals enter prometamorphosis, the median eminence region, and presumably the rest of the hypothalamic neurosecretory apparatus, is sensitive to positive thyroid feedback. The feedback cycle thus begins about a week before a commencement of metamorphosis in normal development and

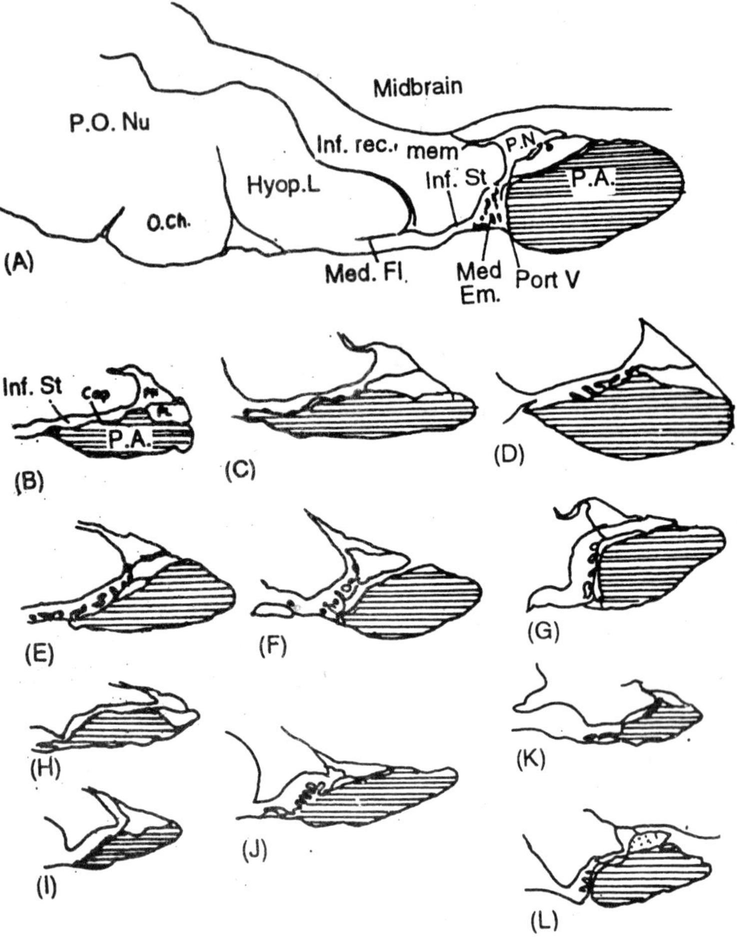

Figure 10.6 : Camera lucida drawings of sagittal sections of the pituitary region. A: Adult frog, B: premetamorphosis; C: early prometamorphosis; D: late prometamorphosis; E: beginning of climax; F: midclimax. G: post-climax. A to G show Rana pipiens. H: Early prometamorphosis; I: late prometamorphosis; J: one day before E. H to J show Rana sylvatica. K: midprometamorphosis; L: early climax. K to L show Bufo americanum.

builds up quickly through prometamorphosis to climax. Thus, we are led to consider that the timing of the beginning of metamorphosis is regulated by a clock mechanism in the hypotha-lamus that times the acquisition of hormone sensitivity in that organ.

On the other hand, the pattern of thyroid hormone build up that, as we have seen previously, determines the pattern of metamorphic change is determined by the positive feedback system of thyroxine to the hypothalamus. In order to form a more comprehensive picture of the mechanism regulating development in larval life of the frog we must now add the newly acquired knowledge of the role of prolactin in promoting body growth and inhibiting metamorphosis in tadpoles as discussed above.

Since prolactin is under inhibitory control by the hypothalamus, the positive feedback of T_4 to the neurosecretory system presumably inhibits prolactin activity while at the same time stimulating the PT axis. In Fig. 10.9 an attempt is made to depict the interaction of these elements in the life history of the anuran larva.

METAMORPHOSIS AS A DEVELOPMENT PHEOMENON

I propose in this section to speculate more broadly than, in the stress of scientific investigation and writing, scientists commonly permit themselves to do. Since the ideas expressed are very general, no detailed documentation will be offered.

Evolution and Developmental Mechanisms in Amphibians

Although a few fossil larvae are known, they contribute little to an understanding of the evolution of metamorphosis beyond the fact that it was present is paleozoid labyrinthodonts. General biological considerations, however, permit some reasonable speculations. Metamorphosis in urodeles is a developmental adaptation permitting the change from water to land environment.

The basic mode of feeding by predation and the associated locomotor mechanisms, however, remain unchanged. Presumably this was the situation is primitive amphibians. In anurans, on the other hand, the larva generally feeds as a free-swimming scavenger, herbivore, or filter feeder. In metamorphosis it changes to a predator which depends upon powerful leg action for its saltatory locomotion.

The significance of the prometamorphic phase in anuran transformation appears to lie chiefly in providing time for the legs, whose development had been repressed during the larval phase, to grow before the transition to land is made at metamorphic climax. It urodeles the limbs develop in the way usual among vertebrates, during embryonic stages, and are not involve in metamorphosis.

This evolutionary process must, of course, have operated through modification of the mechanics of development. From this point of view, the changes of prometamorphosis, principally limb growth, have been brought into the metamorphic picture by the evolution of a developmental mechanism separating limb differentiation from the general embryonic processes and suppressing it during the premetamorphic period.

The action of thyroid hormone, then, appears as the factor which removes this initial inhibition of limb development. We see here an example of a phenomenon appearing frequently in the mechanics of develop-ment. This is that control of development is effected by a balance between inhibiting and disinhibiting factors operating on a protoplasmic mechanism with very broad developmental capacities.

We see this type of development control in gene action at the molecular level, in fertilization phenomena, and in field effects in organ determination. At the level of endocrine action we again see that develop-mental control is effected by a balance between inhibition and disinhibition rather than by simple stimulation.

The PT axis is kept at a low level of activity in the growth phase of the tadpole's develo-pment by negative feedback, and at metamorphosis its activation is brought about by disinhibition of this feedback by hypothalamic action. It is possible that the hypothalamic controlling mechanism of the pituitary has been inhibited for the duration of the larval period by some evolutionary change in the genome.

In that case what we have described above as the positive feedback of thyroxine to the hypothalamus would be another instance of the disinhibition of development by thyroxine.

The complexity of the push-pull type of interaction in governing metamorphosis is further emphasized by the discovery of the role of prolactin as a thyroid antagonist in amphibian development. Whether this substance acts at the peripheral level or as a goitrogen or in both ways its role again emphasizes that development is controlled by a dynamic balance of plus and minus factors.

A fascinating aspect of this particular interaction is the manner in which it appears to have been seized upon as the mechanism for the evolution of the "second metamorphosis" in the common newt.' After a first metamorphosis that produces the land form, this animal returns to the water for breeding in a second period of transformation.

This second metamorphosis has been shown to be under the influence of prolactin. We must now view this change as a shift in the balance between the pituitary factors, prolactin and TSH, rather than as an activation of one of them. First metamorphosis is induced by a shift in favour of TSH, second metamorphosis by a return to the predominance of prolactin. What brings on this second shift is yet to be explored.

Insect and Amphibian Metamorphosis

The analogy between insect metamorphosis and that of amphibians has often been commented upon. In both, the brain acting through a neuro-secretary mechanism controls the endocrine apparatus which governs tissue response through hormone release.

Until recently one feature of the insect system seemed to be lacking in vertebrates. In insects the hormone of the corpus allatum acts as a modulator inhibiting the action of ecdysone. We now see, however, that there is an analogous situation in amphibians, with prolactin playing the role of modular to thyroid activity.

Perhaps the analogy will suggest ways in which factors that are found to regulate balance in one system can be applied to the other.

Metamorphosis and Puberty

A final area of speculation deserving comment is the analogy between amphibian metamorphosis and puberty in mammals. There is striking similarities in the mechanism of sexual regulation and metamorphosis.

In each, the hypothalamic neurosecretory system appears dominant, controlling the pituitary (gonadotropin or TSH) which controls the basic endocrine gland (gonal or thyroid). This is turn regulates the response of the peripheral tissues. Within this system, likewise, there is a negative feedback relation which makes this, too, a balanced system.

In the prepuberal mammal in the amphibian larva the lower levels

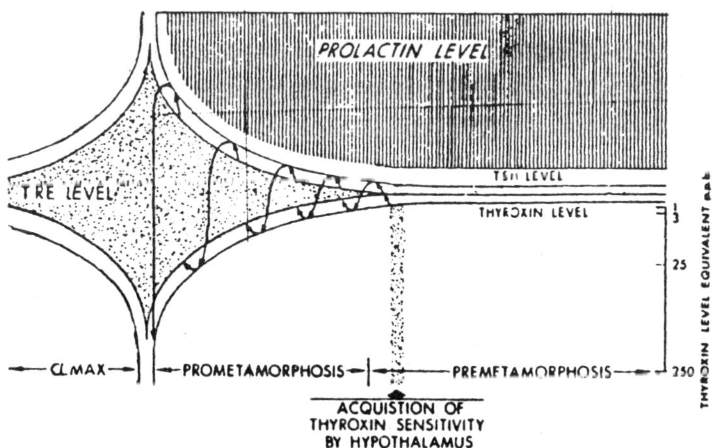

Figure 10.7 : Diagram illustrating the interaction of endocrine factors in determining the time and pattern of anuran metamorphosis.

of the control sequence are capable of response much before they are actually called upon to act. Thus, the sex accessories and secondary sex characters can respond to gonadal stimulation long before puberty in most mammals.

Likewise the gonads and the pituitary show early responsiveness. The critical factor determining the time of puberty then lies in the activation of the hypothalamus just as it does in metamorphosis. Beyond this point, however, the analogy appears to break down. Property placed lesions in the hypothalamus have been shown to accelerate puberty so that the hypothalamic gonadotropic releasing mechanism appears to be under inhibitory control of other neural centers.

This has never been shown to apply to amphibian metamorphosis. On the other hand no positive feedback· system of hormones to the hypothalamus has been described for the mammal as it has for the amphibian larva. Could this be because the possibilities suggested by the analogy have never been explored?

ENDOCRINE REGULATION

The normal human menstrual cycle is dependent upon a complex system containing several components, including higher brain centers, the hypothalamus, the pituitary gland (hypophysis), the ovaries, and the uterus. The cohesive function of these components is integrated by positive and negative feedback signals, and comp-rises the endocrine system in the female.

In the male the same system operates, only the testes take the place of the ovaries as the major steroid-hormone-secreting organs. Briefly, the reproductive endocrine system functions as an integrated unit in which a releasing hormone (known as gonadotropin releasing hormone [GnRH], or luteinizing hormone releasing hormone or factor [IRF or LHRH] is produced and released by the hypothalamus under the stimulatory influence (positive feedback) of steroid hormones produced by the target organs (ovaries and testes).

LRF then stimulates the anterior pituitary gland to release the gonadotropic hormones-follicle-stimulating hormone (FSH) and/or luteinizing hormone (M-which stimulate production of steroidal hormones from the target organs. In the case where steroid hormones inhibit release of hypothalamic releasing hormone, this is known as negative feedback (see the section on Integration).

Recent technological advances which have facilitated the measurement of hormones in biologic fluids, coupled with advances in the study of the interaction between hormones and target organs, have opened the way to a better understanding of the regulatory systems. We now enjoy a reasonable complete understanding of the endocrinology of the menstrual cycle.

This chapter will first discuss the individual components of the system, then their integrated function, and finally the effects of the hormones on their target organs.

THE COMPONENTS

The Central Nervous System

Although there are conflicting data and hypotheses, the majority of evidence suggests that transmitter substances such as dopamine and norepinephrine, in the brain (particularly in the hypothalamus) affect the specialized nerve cells which contain the hormone-releasing and inhibiting factors.

Current data support the existence of a dual system: dopamine inhibits, and norepinephrine stimulates, gonadotropin releasing factor secretion. The cell bodies of the neurons containing norepinephrine are largely located in the medulla and pons, whereas those containing dopamine are found in highest concentration in the arcuate and periventricular nucleii.

This type of neuronal activity is largely regulated directly by changes in the circulating levels of the steroid hormones—estrogen and progesterone. The hypothalamus secretes a releasing factor for each of the tropic hormones produced by the anterior pituitary gland except the gonadotropins, which are probably regulated by a single releasing hormone.

The structures of two of these hypothalamic releasing hormones, thyrotropin releasing factor or hormone (TRF or TRH) and gonadotropin releasing factor or hormone (LRF, GnRh, or LHRH), have been elucidated.

Thyrotropin releasing factor is a tripeptide that causes release of thyroid-stimulating hormone (TSH), as well as prolactin (PRL)—another trophic hormone which has a major action on the mammary gland. However, it may not be the primary prolactin releasing factor. Gonadotropin releasing factor is a decapeptide.

Although it causes synthesis and secretion of both follicle-stimulating hormone (FSH) and luteinizing hormone (LH), it is a much more potent stimulator of the latter.

In addition to its inhibitory action on release of gonadotropin releasing factor, dopamine has an inhibitory effect on the synthesis and release

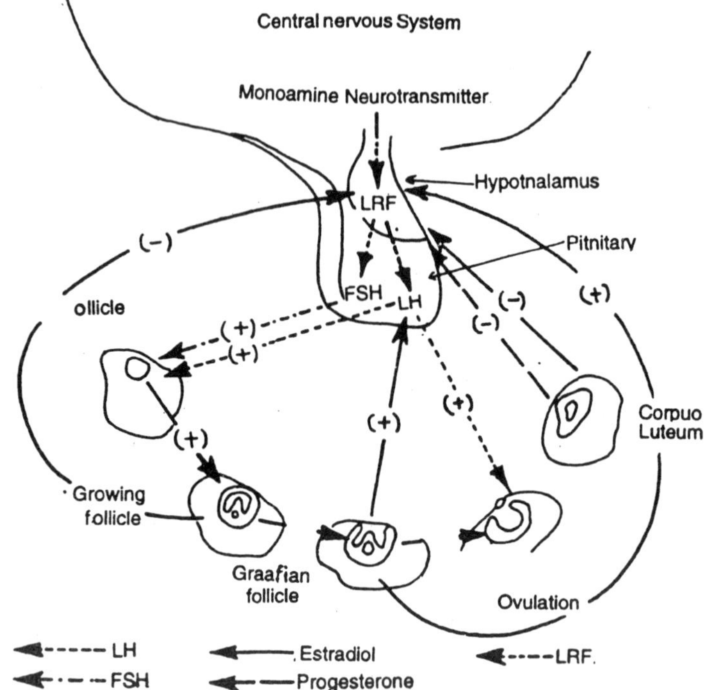

Figure 10.8 : Positive (+) and negative (–) feedback (long-loop) pathways in the central nervous system and the ovary. Arrows point in direction of feedback. See text for explanation of the pathways.

of prolactin, probably through synthesis of an as yet uncharacterized prolactin inhibiting factor (PIF).

The LRF neuronal system has been described in the monkey and in the human. The cell bodies of LRF-containing neurons are found in the anterior hypothalamus and in the tuberal hypothalamus. LRF is synthesized in these neurosecretory cells of the hypothal-amus, and then released into the hypothalamic-hypohyseal portal system, in which they are transported to the anterior pituitary. At this site, they stimulate synthesis and release of LH and FSH.

The Pituitary

Three gonadotropic hormones of the anterior pituitary have been identified: follicle-stimulating hormone (FSH), luteinizing hormone (LH), and prolactin (PRL).

FSH and LH are glycoprotein hormones containing alpha (a) and beta (a) subunits. They are closely related, structurally and chemically.

The a subunits of these hormones are essentially identical. The immunologic specificity of FSH and LH is invested in the B subunit. FSH and LH are synthesized in basophilic cells, called *gonadotropes,* which are scattered throughout the anterior pituitary.

There has been debate about the source of FSH and LH. The critical question was whether both hormones are made in the same gonadotropes, or whether there are two separate populations of gonadotropes, one for LH synthesis and one for FSH synthesis.

This question remains incompletely settled, but recent evidence strongly supports the contention that FSH and LH are located in the same gonadotropes in the human pituitary LRF seems to control both FSH and LH synthesis and release.

The third hormone produced by the pituitary, prolactin, is a single-chain polypeptide composed of 198 amino acid residues. It is synthesized and released by acidophilic cells of the anterior pituitary. The role of PRL in the normal human menstrual cycle has been less precisely defined than the roles of FSH and LH.

Recently Vekemans and his coworkers reported a midcycle PRL peak, with continued elevated levels during the luteal phase. However, other investigators have found no cyclic variation in PRL. Current opinion holds that prolactin is necessary for ovarian and testicular steroidogenesis.

FSH and LH exert two types of influence on the reproductive cycle. Their primary effect is stimulation of target cells in the gonad. Although there may be some overlap of activity, because of the structural similarities of the two hormones, each gonadotropin has a specific role in the regulation of gonadal function.

FSH stimulates follicular growth and maturation in the female, and initiates spermatogenesis in the male. In addition, FSH catalyses the aromatiz-ation of testosterone to estradiol in both sexes. LH stimulates the conversion of cholesterol to pregnenolone in the gonads of both sexes.

LH stimulation is indispensable to the complete maturation of germ cells in male and female gonads. However, steroidogenesis proceeds to functional levels under the stimulation of relatively low levels of LH, FSH alone produces neither complete germ cell maturation nor complete steroidogenesis.

Thus, it appears that LH, even if FSH is present only at very low levels, can induce steroido-genesis, but that both trophic hormones must be present in adequate amounts, in order for reproduction to be normal.

Thus, the gonadotropins are ultimately responsible for the production of the ovarian hormones, which, in turn, control LRF release through the "long-loop" positive or negative feedback system.

In addition to their primary effects on the gonads, the pituitary gonadotropins exert a negative feedback effect on the hypothal-amus, through a retrograde "short-loop" vascular system connecting the pituitary and the hypothalamus.

Recent animal experiments have demonstrated a neurohypophyseal capillary network common to the median eminence. and the infundibular stalk and neural lobe of the pituitary. Three vascular routes have been demonstrated: fenestrated portal vessels to the anterior lobe of the pituitary; capillary connections to the medial basal hypothalamus, with orientation of the capillary loop toward the arcuate nucleus; and an internal plexus to the ependyma of the median eminence.

Additional data have been reported demonstrating retrograde transport to the hypothalamus of LH and prolactin, in the rat. Thus, there is strong evidence for short-loop feedback of the *gonadotropins* to the hypothalamus.

The Ovary

The human menstrual cycle is under ovarian control, in the sense that the ovarian steroids profoundly influence the entire complex. For clarity of exposition, it is convenient to begin with a description of the hormonal events of the menstrual cycle.

The average length of the menstrual cycle is 28 days. By convention, the day of initiation of bleeding is designated day 1. The new follicle for a given cycle starts to develop during the luteal phase of the preceding cycle. Just before and during menses, the cells of the developing follicles proliferate (the follicular phase), and the follicles increase in size, soon to exceed 1 mm in diameter.

Many do not attain this size, but become atretic. The follicle that is most sensitive to FSH stimulation gets a lead that it never relinquishes. The rising FSH stimulates the follicle to grow, by inducing mitotic proliferation of the granulosa cells and formation of the theca. The early follicular phase ends when a definite rise in plasma estrogen is noted.

The locally produced estrogen also stimulates the follicle to grow, so that the mature follicle, i.e., the Graafian follicle, reaches a diameter of 6 mm. The next largest follicles become atretic at a size of 1 to 3 mm. The concentrations of estradiol (the major biologically active estrogen) in the general circulation rise from 50 pg./ml. on day 1 to about 75 pg./ml. on day 6 of the cycle. Estradiol concentrations increase more sharply to reach levels of about 150 pg./ml. on day 9. A sharp rise, usually called a "surge," then occurs, to reach a peak of about 3.50 pg./ml. on day 11. Estradiol levels decline rapidly, to values of about 250 pg./ml. on day 14. The estradiol concentration then gradually rises again.

The central nervous system-hypothalamic-pituitary axis becomes increasingly sensitive to the positive feedback action of estradiol as the concentration of estradiol in the bloodstream increases during follicular development. An acute surge of LH and FSH at mid-cycle is induced by estradiol.

This positive feedback requires a circulating level of estradiol of 100 to 200 pg./mL, which is sustained for 36 to 42 hours (8, 9, 23, 26, 29). Current evidence suggests that estradiol is required for the initiation, but not for the maintainence, of the LH surge. In primates, including man, the interval between the estrogen peak and the LH peak is 14 to 27 hours. Ovulation follows the LH peak within 11 to 24 hours.

Shortly before ovulation, plasma levels of 17 hydroxypro-gesterone, produced by the ovary, increase. Just after the onset of the LH surge, levels of progesterone, coming from the ovary, also begin to rise.

Around the time of ovulation, the granulosa cells luteinize. Luteinization is initiated by LH. For the first three postovulatory days (cycle days 15 to 17) the granulosa cells proliferate, and the corpus luteum reaches a size of 1 cm. During postovulatory days 4 to 9 (cycle days 18 to 23) capillaries, previously found only in the thecal cells, grow to reach the granulosa, and peak vascular-ization begins. After day 23, regression begins. The highest levels of plasma progesterone are reached on cycle days 18 to 23.

To summarize, the following endocrinologic events are hallmarks of the normal menstrual cycle. The estrogen peak precedes the LH peak- 2) estrogen secretion attains appropriate levels; 3) the LH peak occurs at least 13 days prior to the onset of menses; 4) initial rises in plasma 17-hydroxy-progesterone and progesterone levels occur, coincident with the LH rise;-5) a second rise in plasma 17-hydroxyprogesterone occurs later, coincident with the marked rise in plasma progesterone; 6) plasma progesterone begins to rise with the LH surge, and reaches a maximum 6 to 8 days after the LH peak.

The Endometrium

The cyclic changes in the ovary elaborating first estrogen, and then estrogen and progesterone, induce, in the endometrium, first, proliferation, and then differentiation. These changes are reflected in the endometrial morphology, and the functional state of the ovaries can be inferred with reasonable accuracy from the histologic appearance of the endometrium.

During the first half of the cycle, endometrial glandular and stromal cells undergo replication. The endometrium thickens, and the glands begin to elongate. The epithelial cells become taller, and mitotic activity is seen in glands and stroma. The firstmorphologic change after ovulation is the formation of subnuclear vacuoles in the glandular epithelium. This is first seen on the 15th to 16th day of a 28 day cycle.

The vacuoles are caused by the accumulation of glycogen, which on the 19th or 20th day is deposited in the glandular lumen. From the 19th to the 23rd day, no mitoses are seen in either the glands or the stroma, suggesting that progesterone causes cessation of hyperplastic growth. The earliest stromal change seen under progesterone influence is progressive loosening of the stroma, with accumulation of fluid between the cells.

This is due to a change in the ground substance, presumably related to preparation for implantation. At about the 23rd day, evidence that progesterone causes hypertrophic growth and differentiation can be inferred from the appearance of changes in the stromal cells surrounding the blood vessels which will allow for implantation.

These decidual changes progress over the next few days, until the stroma has undergone generalized decidual transformation. If pregnancy does not occur, the stroma becomes infiltrated with inflammatory cells.

The premenstrual endometrium consists of thousands of tiny units composed of large numbers of stromal cells and exhausted glands surrounding markedly coiled arterioles. The first event in menstruation is vasoconstriction of these arterioles.

The resulting ischemia results in dissolution of the architecture of the endometrium. Foci of swelling and distortion appear, the stromal cells begin to clump, hemorrhage occurs within the stroma, and the endometrium is shed as the menstrual discharge or flow.

The raw, denuded surface of the endometrium is rapidly covered by epithelium, so that within three days healing is complete. Thus, the changes caused by progesterone in a non-fertile cycle are peak secretion, followed by peak edema in the endometrium, followed by decidual formation: If conception occurs, instead of each of these endometrial changes peaking and waning in sequence, all continue and peak together.

This lends a characteristic appearance to the endometrium, which has been described as "gestational hyperplasia."

INTEGRATION

Integration of the components of the endocrine system is controlled by the hormones described in the first section of this chapter. The same hormones are synthesized by both men and women, but the patterns of secretion are different in the two sexes.

The gonadotropins are secreted in pulses, at intervals of about 90 minutes. Pulse amplitude is greater for LH than for FSH. In males, the pulses are maintained in a tonic pattern, whereas in females the cyclic gonadotropic peak that occurs before ovulation is superimposed upon the tonic pulsatile pattern.

Tonic secretion of LH and FSH is regulated by a feedback loop involving central nervous system components (i.e., dopamine, norepinephrine, and LRF), ovarian steroids, and the gonadotropins. Estradiol is the most potent gonadotropic inhibitor.

A quantitative relationship between the negative feedback action of ovarian steroids and gonadotropin release can be demonstrated by interrupting the negative feedback loop by withdrawing estradiol through ovariectomy; this results in significant increases in circulating LH and FSH concentration.

The rise in gonadotropins continues until a plateau is attained at approximately ten times preoperative levels about 3 weeks after the operation. The increase is greater for FSH than for LH. This differential effect may reflect preferential inhibition of FSH by estradiol (25, 27, 28) and possibly, by follicular inhibin (6).

There is some evidence that inhibin, a non-steroidal gonadal factor, is involved in the feedback control of gonadotropic secretion. A protein which inhibits FSH secretion has been identified in the testes and in testicular secretions (1,7,11).

Because the increase in FSH levels in menopausal women is disproportionate to the increase in LH levels, an inhibin factor has also been postulated to be present in women. Some evidence for the presence of inhibin in ovarian follicles has been obtained in animals (6, 12, 22).

Although incomplete, the inhibin story may prove to be an exciting breakthrough, because the presence of a specific inhibitor of FSH secretion and release would explain the differential levels of FSH and LH observed under certain circumstances.

Other observations provide further evidence of the feedback effects of estradiol on gonadotropin output: 1) during the first week after ovariectomy, a greater rise in LH and FSH is seen in women operated upon during the follicular phase of the cycle than in those operated upon during the luteal phase; 2) withdrawal of estradiol results in elevated secretion of hypothalamic LRF and thus of pituitary gonadotropins (5, 10); 3) there is also a rapid decline of circulating gonadotropin levels in response to estradiol administration to ovariectomized and postmenopausal women.

Moderate levels of estradiol, as seen in the early follicular phase of the menstrual cycle, maximally inhibit gonadotropin output. In the normal menstrual cycle, a change in estradiol levels in either direction reduces inhibition and stimulates the hypothalamic-hypophyseal axis. As mentioned above when the hypothalamic-pituitary axis is maximally stimulated through ovariectomy or the menopause, then the action of estradiol will be solely inhibitory.

The negative and positive feedback actions are not interrupted phenomena, but rather a continuum. The positive feedback effect of estrogen on gonadotropin release is always preceded by a phase of negative feedback.

To summarize, in response to a nadir in circulating levels of estrogen, gonadotropin releasing hormone is secreted by the hypothalamus. The pituitary, in turn, releases FSH, which stimulates follicular growth. LH, secreted by the pituitary in basal amounts, stimulates synthesis of estradiol by the theca interna of the developing follicles. Estradiol synthesized by the Graafian follicle has a local trophic effect on the follicle and an inhibitory effect on pituitary release of FSH.

AsTSH declines, the remaining follicles regress and become atretic. Estradiol synthesis in the ovulatory follicle increases dramatically. Peak levels of estradiol are reached approximately 2 days prior to ovulation.

In response to the sustained estradiol peak, a surge of stored LH is released from the anterior pituitary, presumably due to increased secretion of gonadotropin releasing factor by the hypothalamus. A smaller increase in FSH concentration is noted at the same time. Thus estradiol inhibits the hypothalamus in the follicular phase, but stimulates the hypothalamus at mid-cycle.

After ovulation, the granulosa cells become vascularized as the corpus luteum forms. Basal levels of LH stimulate secretion of estradiol and progesterone by the corpus luteum. In combination, these steroids are potent inhibitors of the synthesis of gonadotropin releasing factors. Progesterone may also, inhibit further follicular maturation. In the absence of a luteotrophic factor (hCG), the corpus luteum ceases to function after 12 to 14 days.

The cells of the corpus luteum show evidence of degeneration. Although the yellow colour is maintained for months, the structure becomes smaller and is eventually replaced by hyaline tissue. The resultant structure is called a corpus albicans.

In response to falling levels of estradiol and progesterone, FSH secretion increases, and a new crop of follicles begins to grow and synthesize estradiol, thus perpetuating the cyclic pattern. The nadir of steroid synthesis occurs approximately 2 days before the onset of menses. The rise in FSH begins shortly before this nadir is reached.

Variations in cycle length are generally considered to take place at the expense 4 the follicular phase. Thus, in normally ovulating women, the length of the postovulatory phase was thought to remain constant from cycle to cycle, whereas the duration of the preovulatory phase was thought to be variable.

However, recent data suggest that the luteal phase may also vary in duration. These variations may be a reflection of very fine adjustments between PSH, secretion and follicular response which take place until one follicle becomes large enough to assume major estradiol synthesis. Once this critical size' has been achieved, only modest amounts of FSH and LH are, required for continued growth of the follicle.

THE TARGET ORGANS

The Testis

The fetal testis differentiates earlier than the fetal ovary and secretes testosterone and dihydrotestosterone by the end of the first trimester of pregnancy, at which time differentiation of the external genitalia also occurs. Fetal gonadal steroidogenesis is thought to occur in response to hCG secreted by the placenta.

This mechanism is substantiated by the identification of Leydig cells in the fetal testis from the end of the first trimester. They revert to mesenchymal cells shortly after birth.

From birth until shortly before puberty, the testis is inactive. The germinal epithelium, although present, displays no active spermatogenesis. Levels of FSH and LH rise perceptibly as puberty approaches and induce increased testosterone synthesis and spermatogenesis.

As adult males reach middle age,, dihydrotestosterone levels may increase, testosterone levels decrease, androgenbinding globulin levels increase, and spermatogenesis becomes less efficient. However, there is no sudden cessation of reproductive function in the male such as occurs in the female.

Medullary elements dominate in the adult testis. Each testis is composed of specialized cell lines, the Sertoli cells, which nourish and support spermatogenesis, and the germinal epithelium, from which the spermatozoa develop.

Scattered about the stroma surrounding the seminiferous tubules are interstitial (Leydig) cells which synthesize and release testosterone. The seminiferous tubules converge into a secon collecting tubular system called the rete testis, and from there merge into a single long convoluted tubule, the epididymis, in which spermatozoa complete maturation and are stored.

Testosterone secreted by the Leydig cells diffuses directly to the Sertoli cells. It does not become absorbed into the bloodstream until it reaches the region of the rete testis. An effective blood testis barrier appears to exist up until this point.

FSH is necessary for the initiation of spermatogenesis, and testosterone is required for the later steps in spermatogenesis. Both FSH and testosterone stimulate Sertoli cells to elaborate androgen-binding protein (ABP). The exact role of this protein in spermatogenesis is uncertain.

Sertoli cells also convert testosterone to 5a-dih'ydrotestosterone and estradiol. Testosterone, rather than 5a-dihydrotestosterone, appears to be important in spermatogenesis. Estradiol may be the primary mediator of feedback inhibition from the testis to the hypothalamus. However, as noted with respect to the ovary, inhibin may, either alone or in concert with estradiol, be the mediator of testicular negative feedback.

The spermatogenic cycle in man takes 74 ± 6 days. All stages of spermatogenesis can be observed simultaneously in the same seminiferous tubule. Spermatogenesis involves production of haploid germ cells (spermatids) from diploid stem cells (spermat-ogonia). Spermiogenesis is the process by which the spermatids mature into cells which are capable of independent motility and transport (spermatozoa).

The spermatogonia are divided into two groups: those which constantly divide to maintain the stem cell population and those which undergo growth preparatory to meiotic division, producing two secondary spermatocytes. Each secondary spermatacyte, in turn, undergoes reduction division to produce two spermatids. Development of head, midpiece, and tail then occurs, resulting in mature spermatozoa.

Spermatozoa are released into the lumen of the seminiferous tubuless and transported to the epididymis, where they undergo further maturation and storage. At ejaculation, they are propelled as a bolus through the vas deferens. Seminal plasma secreted by the seminal vesicles and prostate is added to the bolus of sperm as it traverses the prostatic urethra, thus completing the composition of the ejaculate.

As noted above, the testis produces testosterone and other androgens, which are secreted in pulsatile fashion. The testis also produces estrogen, which may be involved in the feedback 'mechanisms at the hypothalamic and pituitary levels.

Thus FSH is important for spermatogenesis and LH for production of testosterone. The regulation of the components of the hypothalamic-pituitary testis system is similar to that in the female, with testosterone and testicular inhibin playing major roles in the long-loop feedback system.

The short-loop system of pituitary tropic hormones (FSH and LH) feeding back on the hypothalamus and its releasing hormone (LRF) is also probably operative. Since the production of LH in the male is only tonic, and not both tonic and cyclic as in the female, the regulatory system is simpler and less susceptible to disruption.

Testosterone has two major roles in the male. It stimulates development of male secondary sex characteristics, such as enlargement of the genitalia, increase in laryngeal size with concomitant deepening of the voice, increase in muscle mass, and growth of body hair. It also is indispensable to normal spermatogenesis. The effects of androgens on other tissues and organs are discussed later in this chapter.

The Ovary

The fetal ovary does not demonstrate the hormonal activity seen in the fetal testis. Ovarian steroid synthesis is not necessary for differentiation of female external or internal genitalia. However, the fetal ovary actively produces germ cells. The full complement of ova is achieved by 20 weeks of intrauterine life, and the process of follicular maturation and regression is evident by the third trimester of pregnancy.

The ovary is composed of numerous follicles, in varying stages of maturation, surrounded by stromal cells. Follicular units consist of an ovum surrounded by a layer of granulosa cells, and internal and external layers of thecal cells.

The maximum number of follicles (millions) is reached at approximately 20 weeks of intrauterine life. From this time, follicles become involved in an inexorable cycle of partial maturation followed by atresia. Most follicles become atretic; only about 500 develop to the stage of ovulation in a lifetime.

Gonadotropin levels in females are low in infancy and rise gradually as puberty approaches. Levels of FSH are slightly higher than those of LH, but the ratio is nearly one to one. In contrast,, during reproductive life LH is always higher than FSH, except early in the follicular phase of the cycle, when growth of a new crop of follicles is stimulated.

The rising gonadotropin levels of late childhood stimulate estrogen synthesis in the ovarian stroma. The secreted estrogen causes maturation of the external genitalia and stimulates growth of the breasts. Eventually, when there has been sufficient estrogen stimulation of the endometrium, uterine bleeding occurs.

The first episode of bleeding is called the menarche. The initial endometrial shedding probably occurs in response to a drop in circulating levels of estrogen. The first several "menstrual" episodes may occur at irregular intervals because ovulation has not yet been initiated. Once an ovulatory pattern has been established, it recurs with consistent periodicity in an individual woman.

Toward the end of the reproductive age, ovulation and menstrual cycles again become irregular. At menopause, estradiol synthesis decreases drastically, and gonadotropin levels become markedly elevated.

The pattern of ovarian steroid secretion fluctuates throughout the menstrual cycle. The ovulatory follicle and corpus luteum are the major sources of estradiol synthesis and secretion.

Average production rates range from 0.07 mg./day in the early follicular phase to 0.8 mg. /day at the time of the preovulatory peak. Estrone is also secreted by the ovary, but most of the circulating estrone is derived from peripheral conversion ofandrostenedione.

About half of circulating androstenedionee of ovarian origin and half is of adrenal origin; it is secreted by the ovary in a cyclic pattern which, parallels that of estradiol and estrone. Mean testosterone levels in the normal woman are about 40 ng./ml. of serum.

Approximately half is secreted by the ovary and the remainder by peripheral conversion of androstenedione and of dehydroepian-drosterone, an androgen mostly produced (in normal circumstances) in the adrenal gland.

About 50 per cent of the estrone and estradiol secreted daily undergoes 16-hydroxylation to form estriol. Estriol is then conjugated with sulfate or glucuronide to facilitate excretion.

Progesterone and 17-hydroxyprogesterone are predominantly secreted by the corpus luteum during the luteal phase of the cycle. Low levels of progesterone found in the follicular phase of the cycle are presumably of adrenal origin. As noted earlier, ovarian progesterone and 17-hydroxyprogesterone secretion become detectable in the periovulatory period concomitant with the LH surge.

Peak secretion of progesterone and 17-hydroxyprogesterone in the mid-luteal phase is about 24 mg./24 hours and 4 mg./24 hours, respectively. Progesterone gradually rises, beginning in the periovulatory period, whereas 17-hydroxyprogesterone rises sharply before ovulation, and again during the midluteal phase.

Mechanisms of Hormone Action

Gonadotropins and steroids select their target cells because of the presence of specific receptors on, or in, these cells. The mechanism of hormone receptor interaction differs for each class of hormones.

The protein hormones, FSH and LH, attach to receptors located on the plasma membrane. The resulting membrane receptor-hormone complex, in turn, activates a membranebound enzyme, adenylcyclase. In the presence of intracellular magnesium, adenyl-cyclase acts on ATP to produce cyclic AMP.

Cyclic AMP then activates protein kinases, which mediate the biologic responses of the cell. In the case of LH, steroidogenesis is initiated, and in the case of FSH, specific proteins are elaborated which affect germ cell growth and development.

In contrast, steroid hormones combine with specific receptors in the cytoplasm of target cells and are carried by these receptors to the nucleus of the cells, where they interact with specific sequences of DNA to effect messenger RNA synthesis and prot-ein synthesis.

Steroids secreted by gonads are transported to target tissues in the bloodstream in free and bound forms. Testosterone and estradiol bind to a specific binding globulin known as TEBG (testosterone-estradiol binding globulin) or sex steroid binding globulin.

Progesterone has no specific binding globulin of its own but appears to bind preferentially to corticosteroid binding globulin (CBG). Only free steroids are biologically active in target organs and in the feedback mechanism.

In addition to feedback relationships with the hypothalamus and pituitary, gonadal steroids affect a variety of target organs including the skin and its appendages, subcutaneous fat, muscle, breast, larynx; uterus, cervix, vaginal mucosa, external genitalia, and the gonad itself.

Androgens exert their primary tropic effect on skin, muscle, laryngeal tissue, and male external genitalia. Androgens are responsible for the laryngeal enlargement which produces the deep male voice and for the increased muscle mass characteristic of the male. Androgens stimulate penile and testicular growth and are responsible for rogations of the scrotum.

They stimulate growth of body hair and secretory activity of sebaceous glands and, paradoxically, are also responsible for the temporal recession of the hair line seen in males. At the peripheral level, the aetive androgen is 5a-dihydrotestosterone, which is usually derived by peripheral conversion of testosterone through an irreversible reaction mediated by 5ȧ-reductase.

Estrogens are responsible for growth of the breast and develo-pment of the ductile system, and for the deposition of fat *in* the breasts, abdomen, hips, and things which produces the characteristic female body habitus. Estrogens also induce hyperplastic growth of the endometrium, increased production of cervical mucus, and thickening and rugation of the vaginal mucosa.

Estrogen induces the growth of estrogen and progestrone receptors in target tissues, whereas progesterone eliminates these receptors. Progesterone is a complement, as well as an antagonist, to estrogen. In the breast progesterone induces growth of the lobule-alveolar complexes after estrogen-induced growth of the ductile system has taken place.

Estrogen causes increased water content of cervical mucus, resulting in a copious, clear mucoid secretion, and progesterone induces increased viscosity and production of an opaque, gummy mucoid material. The endometrium is probably the single most important tissue that the responsive to the effects of both estrogen and progesterone.

Estrogen induces mitotic growth of the endometrial glands and stroma, and increased endometrial blood flow. Progesterone halts proliferation of endometrial tissue and induces hypertrophic growth and differentiation of the existing tissues. The interaction of these two steroids is finally responsible for the cyclic bleeding referred to as "menstrual periods."

Gonadal steroids also affect target tissues that. are not related to sexual or reproductive function. The anabolic action of testosterone results in positive nitrogen balance and retention of potassium, phosphorus and calcium. Testosterone also stimulates hemoglobin synthesis. Both testosterone and estradiol are vasoactive and abrupt depletion, as experienced in loss of gonadal function, produces vasomotor instability, with accompanying "hot flashes."

Estradiol increases serum levels of trigly-cerides and hormone binding globulins, including those for sex steroids, cortisol, progesterone and thyroxine.

Estrogen increases insulin response to carbohydrate and causes sodium retention, through interaction with the renin-angiotensin system; it also affects the ability of the liver to secrete substances such as bromosulfophthalein. Proges-terone induces natriurisis, probably by competing with estradiol.

All three major sex steroids may cause changes in mood. Estradiol and testosterone are alleged to enhance libido and contribute to a feeling of optimism and well being. Progesterone is said to contribute to depression. Evidence for these effects is largely inferential, and requires further documentation.

INDEX

A

E

F

G

N

O

P

R

S

T

U

V

W

X

Y